The Terrestrial Biosphere and Global Change
Implications for natural and managed ecosystems

This new synthesis summarizes the international research effort in the Global Change and Terrestrial Ecosystems (GCTE) Core Project of the International Geosphere Biosphere Programme (IGBP). Five major thematic areas are covered: ecosystem physiology; ecosystem structure and composition; terrestrial production systems; global biogeochemistry; ecological complexity (biodiversity). A summary of the integrated and interactive effects of global change on the terrestrial biosphere for four key regions of the world is presented, as well as a projection of future trends in the terrestrial component of the global carbon cycle. The book also includes a section on tools developed or modified for global change research.

BRIAN WALKER is Chief of the CSIRO Division of Wildlife and Ecology in Canberra, Australia and former Chair of the GCTE Core Project.
WILL STEFFEN is IGBP Executive Director at the Royal Swedish Academy of Sciences in Stockholm.
JOSEP CANADELL is GCTE Executive Officer at CSIRO Wildlife and Ecology in Canberra, Australia.
JOHN INGRAM is GCTE Focus 3 Scientific Officer at the NERC Centre for Ecology and Hydrology at Wallingford, UK.

The International Geosphere–Biosphere Programme was established in 1986 by the International Council of Scientific Unions, with the stated aim

To describe and understand the interactive physical, chemical and biological processes that regulate the total Earth system, the unique environment that it provides for life, the changes that are occurring in this system, and the manner in which they are influenced by human activities.

A wide-ranging and multi-disciplinary project of this kind is unlikely to be effective unless it identifies priorities and goals, and the IGBP defined six key questions that it seeks to answer. These are:

- How is the chemistry of the global atmosphere regulated, and what is the role of biological processes in producing and consuming trace gases?
- How will global changes affect terrestrial ecosystems?
- How does vegetation interact with physical processes of the hydrological cycle?
- How will changes in land-use, sea level and climate alter coastal ecosystems, and what are the wider consequences?
- How do ocean biogeochemical processes influence and respond to climate change?
- What significant climatic and environmental changes occurred in the past, and what were their causes?

The International Geosphere–Biosphere Programme Book Series brings new work on topics within these themes to the attention of the wider scientific audience.

Titles in the series

1 Plant Functional Types
T. M. Smith, H. H. Shugart & F. I. Woodward (Editors)
0521 48231 3 (hardback)
0521 56643 6 (paperback)

2 Global Change and Terrestrial Ecosystems
B. H. Walker & W. L. Steffen (Editors)
0521 57094 8 (hardback)
0521 57810 8 (paperback)

3 Asian Change in the Context of Global Climate Change
J. N. Galloway & J. M. Melillo (Editors)
0521 62343 X (hardback)
0521 63888 7 (paperback)

4 The Terrestrial Biosphere and Global Change
B. H. Walker, W. L. Steffen, J. Canadell and J. S. I. Ingram (Editors)
0521 62429 0 (hardback)
0521 63888 7 (paperback)

INTERNATIONAL GEOSPHERE – BIOSPHERE PROGRAMME BOOK SERIES

The Terrestrial Biosphere and Global Change

Implications for natural and managed ecosystems

Edited by

Brian Walker, Will Steffen,
Josep Canadell and John Ingram

CAMBRIDGE
UNIVERSITY PRESS

PUBLISHED BY THE PRESS SYNDICATE OF THE UNIVERSITY OF CAMBRIDGE
The Pitt Building, Trumpington Street, Cambridge, United Kingdom

CAMBRIDGE UNIVERSITY PRESS
The Edinburgh Building, Cambridge CB2 2RU, UK   http://www.cup.cam.ac.uk
40 West 20th Street, New York, NY 10011-4211, USA   http://www.cup.org
10 Stamford Road, Oakleigh, Melbourne 3166, Australia

Printed in the United Kingdom at the University Press, Cambridge

Typeset in Ehrhardt 11/14pt [VN]

A catalogue record for this book is available from the British Library

ISBN 0 521 62429 0 hardback
ISBN 0 521 62480 0 paperback

Contents

J.-C. Menaut
ORSTOM, 213 rue La Fayette, F-75480 Paris Cedex 10, France

H. A. Mooney
Department of Biological Sciences, Stanford University, Stanford, CA 94305-5020, USA

D. Murdiyarso
Southeast Asian Impacts Centre (IC-SEA), SEAMEO BIOTROP, PO Box 116 Bogor, Indonesia

I. R. Noble
Research School of Biological Sciences, Australian National University, Canberra, ACT 0200, Australia

W. J. Parton
Natural Resource Ecology Laboratory, Colorado State University, Fort Collins, CO 80523-1499, USA

L. F. Pitelka
Appalachian Laboratory, University of Maryland Centre for Environmental Science, Gunter Hall, Frostburg, MD 21532, USA

P. S. Ramakrishnan
School of Environmental Sciences, Jawaharlal Nehru University, New Delhi-110067, India

O. E. Sala
Department of Ecology, Faculty of Agronomy, University of Buenos Aires, Av. San Martin 4453, Buenos Aires 1417, Argentina

R. J. Scholes
Division of Water, Environment and Forestry Technology, CSIR, PO Box 395, Pretoria, 0001, South Africa

E.-D. Schulze
Max Planck Institute für Biogeochemie, Sophienstr. 10, 07743 Jena, Germany

H. H. Shugart
Department of Environmental Sciences, 356 Clark Hall, University of Virginia, Charlottesville, VA 22903, USA

M. Stafford Smith
CSIRO National Rangelands Program, PO Box 2111, Alice Springs, NT 0871, Australia

W. L. Steffen
GCTE, CSIRO Wildlife & Ecology, PO Box 84, Lyneham Act 2602, Australia

R. W. Sutherst
CSIRO Entomology, c/o CRC Tropical Pest Management, 5th Level Gehrmann Labs, University of Queensland 4072, Australia

C. Valentin
ORSTOM, BP 11416, Niamey, Niger

B. H. Walker
CSIRO Wildlife & Ecology, PO Box 84, Lynneham Act 2602, Australia

F. I. Woodward
University of Sheffield, School of Biological Sciences, PO Box 601, Sheffield, S10 2UQ, UK

X. Zhang
Institute of Botany, Chinese Academy of Sciences, No. 141 Xizhimenwai Avenue, Beijing 100044, China

1 The nature of global change

B.H. Walker and W.L. Steffen

1.1   Introduction

Despite the growing prominence of global change as an environmental issue, the term 'global change' and the phenomena it encompasses are still widely misunderstood. Global change is often referred to as 'global warming' or 'climate change', but there is much more to it than a change in climate and, in fact, climatic change is now and will continue to be for the next few decades one of the less important components of global change in terms of effects on the structure and functioning of terrestrial ecosystems. Other more immediate and more certain components are the direct human conversion and modification of terrestrial ecosystems, especially the accelerating loss of natural biological communities, and the alteration of the chemical composition of the atmosphere.

This book synthesizes the current understanding of global change interactions with terrestrial ecosystems, those systems that humanity depends upon for the bulk of its food and fibre, for the provision of clean air and water, for the absorption of much waste material, and for recreational and inspirational amenities. In addition to this anthropocentric perspective, terrestrial ecosystems are comprised of millions of other species with which *Homo sapiens* shares the planet and which, according to the ethical beliefs of many human cultures, also have a right to exist.

The chapter begins with an overview of the nature of global change itself, focusing on the fundamental driving forces. The three well known components – (i) land-use/cover change; (ii) changes in atmospheric composition; and (iii) climate change – adopted by the International Geosphere–Biosphere Programme (IGBP) as an operational definition – are then described. To these a fourth factor is added – changes in biological diversity – which, from the perspective of terrestrial ecosystems, should be considered as a component as well as a consequence of global change. Following the description of these global change components individually, their certainties, time scales and interactions are discussed.

Next the chapter examines the science of global change interactions with terrestrial ecosystems, the emerging discipline of global ecology, and the research approach of the GCTE (Global Change and Terrestrial Ecosystems) Core Project of the IGBP. The chapter closes with a rationale for the structure of this synthesis of GCTE and related research, and a 'roadmap' of what follows.

1.2   The driving forces and components of global change

For nearly all of its existence *Homo sapiens* has been a hunter–gatherer, with population size, movements and lifestyles controlled by the same ecological constraints working on other large mammals. The human population was no more than a few million at the beginnings of agriculture about 10 000 years ago. Since then the population has risen exponentially to nearly 6 billion. Although the growth rates have tended to decrease over the past few decades, the population itself will continue to rise rapidly for some time into the future, given the inertia effect built into the current age structure. Projections of the human population by the end of the twenty-first century range from about 6 billion to about 19 billion, depending on the assumed fertility. Probabilistic projections (Lutz, 1996) indicate that there is a 95% probability that the population will lie between 10.0 and 12.0 billion in 2050, and between 15.7 and 17.3 billion in 2100.

Over the next 30 years, the divergence in estimates is much less in both percentage growth and absolute numbers. The population will increase by about one billion per decade for the next two or three decades, at least.

These global aggregates are misleading, however, as there is a great heterogeneity of growth rates around the world (Table 1.1). Nearly all of the increase will occur in the developing regions of the world, where the rates are higher than the global mean. There will be significant regional differences, with large percentage increases in Africa and West Asia; Central Asia will decline in share of population (but not absolute numbers) and South and Southeast Asia will remain almost constant as a proportion of the total.

The impact that humans have on ecosystems is determined by the so-called 'PLOT' equation (CSIRO, 1995); a function of **P**opulation size, **L**ifestyle, **O**rganization and **T**echnology. Clearly lifestyle plays a key role in determining the size of the impact, with developed countries having impacts per capita at least an order of magnitude greater than those of developing countries. Social and political organisation are also important. For example, Western European countries consume considerably less energy per capita than the United States or Australia because of more efficient and widely used public transport systems and a greater emphasis on energy conservation. Technology itself has a strong influence on the level to which human activities affect the Earth system (Lutz, 1994).

Table 1.1 *Population projections for selected regions of the world (based on FAO, 1993a).*

	Population (billion)	
	1989	2010
World	5.21	7.21
All developing	3.96	5.84
Developed	1.24	1.37
Africa (S. Sahara)	0.47	0.92
East Asia	1.56	2.00
South Asia	1.14	1.73

With the globalization of economies, the 'ecological footprints' of particular groups of people and of nations now extend to distant places. With the rapid increase in human populations and in their activities, these footprints are also increasing in intensity, overlapping in area and amplifying in effect. Taken together, they are so pervasive that they are measurably affecting major compartments of the Earth system, such as the atmosphere, soils, and terrestrial and coastal biospheres, and the processes that move material and energy between these compartments. This is the essence of global change.

1.2.1 Land-use and land-cover change

Human-driven changes in land use and cover are by far the most dominant component of global change in terms of impacts on terrestrial ecosystems (Turner II *et al.*, 1990; Houghton, 1994). Land cover refers to the structure of the soil/vegetation system (e.g. forest, tundra, cropland), while land use refers to the ways in which human societies use a land cover type (e.g. forests can be used for timber production or conservation; croplands may be regularly tilled to control weeds, or zero-tillage may be practised).

Over the past several thousand years more than half of the Earth's terrestrial surface that is potentially suitable for agriculture, and virtually all of the most fertile land, has been converted to that cover type. The rate of this conversion process has increased dramatically in recent times; over half of the world's croplands have been created during the last century (Houghton, 1994).

The land-cover change process is often not a simple, single act of conversion of natural vegetation to cropland, but rather follows a more complex trajectory through stages of clearing, active agriculture, abandonment, secondary growth, and re-clearing, as shown in Fig. 1.1, to produce a dynamic matrix of cover types. On a broader scale, land-cover change in the tropics is concentrated on the conversion of forests to more intensively managed systems, while in the

From 1986 to 1988

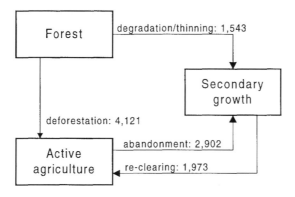

From 1988 to 1989

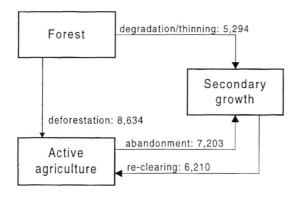

Figure 1.1   Transition rates between three land-cover classes in the Brazilian Amazon Basin. Values are hectares per year (from Skole *et al.*, 1994).

temperate regions abandonment of agricultural lands and their return to native vegetation is an increasing phenomenon.

One of the most prominent types of land-use change is the intensification of agriculture on lands which have already been converted for agricultural use. This intensification process is set to increase sharply in the future, given that the most fertile lands have already been converted. Thus, most of the increases in food production required for the expanding human population will have to be obtained from increased productivity of existing agricultural lands, with significant impacts on local and global biogeochemical cycles, as discussed below.

Although land-use change can occur independently from land-cover change, the two often occur together and are commonly referred to as the same process in the global change literature. The process of land-use/cover change almost always has profound effects on the structure and functioning of terrestrial ecosystems. Such effects include dramatic changes in the complexity of ecological systems through the elimination of existing species or whole groups of species and the introduction of exotic species (Vitousek *et al.*, 1997*a*; Vitousek, 1994, and references therein); the loss of soil carbon and nutrients (e.g. Burke *et al.*, 1991); changes in forest biomass (e.g. Brown *et al.*, 1991); increased release of radiatively active trace gases to the atmosphere (e.g. Crutzen *et al.*, 1986); and direct effects on regional climate (e.g. Pielke & Avissar, 1990).

Although the effects of land-cover change on vegetation composition and structure (the aboveground component of terrestrial ecosystems) are usually clearly visible, the often unseen impacts of land-use/cover change on soils (the belowground component) are just as important. Significant changes have occurred to the physical, chemical and biological structure and functioning of soils as the result of land-use/cover change, most notably the conversion of natural ecosystems to agricultural uses.

Changes to the physical structure of soils range from total loss for another use (e.g. urbanization) through partial loss due to erosion, to damage from mechanical agriculture and overgrazing. Erosion of soil from cultivation is ongoing, and it has been estimated that about one-third of the world's croplands suffer soil losses beyond the maximum level at which the soil's fertility can be maintained (R.W. Arnold *et al.*, 1990). There is some evidence, however, that, in the absence of soil erosion, soil loss associated with agriculture is not an indefinite process; a new equilibrium level of soil organic matter (SOM), a key component of soils, is eventually reached (e.g. Donigian *et al.*, 1995). Management practices, such as conservation tillage, are undoubtedly important in reducing SOM losses, but there is not yet enough long-term data on them to give reliable information on their effects on soil losses (Gregory & Ingram, 1996).

One of the most prominent features of global change is the very rapid rate at which it is occurring. For soils, rates of change are especially important as many key soil processes occur at slow rates, especially soil formation from parent material. Thus, on human lifetime scales, the loss of soil through erosion, for example, can be considered to be an irreversible process.

1.2.2 Changes in atmospheric composition

The second major component of global change is the alteration of the chemical composition of the atmosphere by human activities. The best-known change is the build-up of carbon dioxide, primarily due to fossil fuel burning. Increasing atmospheric CO_2 concentration is of particular importance to terrestrial ecosystems, as carbon obtained from the atmosphere through photosynthesis is the basic building block of Earth's vegetation. However, it is useful to consider changes in atmospheric composition in the broader context of changes to biogeochemical cycles.

The movement of chemical substances between the various compartments of the Earth system (e.g. atmosphere, oceans, terrestrial biosphere), and the transformations they undergo during these movements, are often referred to as 'biogeochemical cycles' (see Chapter 10). The 'bio' part of the term recognizes the importance of biological processes in storing, transforming and transporting chemical substances in these cycles. The cycling of these substances is a natural phenomenon which has been occurring throughout Earth's history; however, human activities have recently intensified enough to have discernible effects on biogeochemical cycles.

Much of the interest and research effort in biogeochemical cycles has focused on the well-documented and very rapid (in geological time scales) accumulation of carbon in the atmosphere due to human activities, as shown in Fig. 1.2 (Webb & Bartlein, 1992; Raynaud *et al.*, 1993; Vitousek, 1994). Table 1.2 gives an overview of the perturbation to the carbon cycle by human activities as an annual average for the 1980s (Schimel, 1995). Terrestrial ecosystems are involved as both a source and a sink for atmospheric carbon. Land-use change, primarily in the tropics, releases carbon to the atmosphere, while a number of processes are thought to enhance the sequestration of atmospheric carbon by the terrestrial biosphere. It should be emphasized that there is still considerable uncertainty regarding the existence and magnitude of these terrestrial sinks. Much of the evidence for them has been deduced by difference from carbon budget calculations or is derived from inverse calculations based on atmospheric transport modelling (e.g. Tans *et al.*, 1990; Enting & Mansbridge, 1991, Keeling *et al.*, 1996*b*). Conclusive evidence from ecological process studies has proved difficult to obtain.

Another way to view the human perturbation to the terrestrial carbon cycle is to consider changes in its pathways. Humans depend upon the net assimilation of carbon by the terrestrial biosphere (annual net primary production, NPP) for most of their food and fibre. Vitousek *et al.* (1986) estimated that about 40% of the total potential NPP of the terrestrial biosphere is now used or dominated by human activity or is foregone as a result of land-use change. They estimated that about 4% of terrestrial NPP is consumed directly by humans and their domestic animals, about 27% occurs in systems dominated by humans, and about 12% is 'lost' in terms of the difference between the NPP of croplands in comparison with the natural ecosystems that they replaced. This analysis implies that, in general, all of the other species that depend on terrestrial vegetation for sustenance have only 60% of NPP available for their needs. Furthermore, with the human population expected to double by the middle of next century and *per*

Table 1.2 *Average annual budget of CO_2 perturbations for 1980–1989.*

CO_2 sources	
Emissions from fossil fuel combustion and cement production	5.5 ± 0.5
Net emissions from changes in tropical land use	1.6 ± 1.0
Total anthropogenic emissions	7.1 ± 1.1
CO_2 sinks	
Storage in the atmosphere	3.2 ± 0.2
Oceanic uptake	2.0 ± 0.8
Uptake by Northern Hemisphere forest regrowth	0.5 ± 0.5
CO_2 fertilization	1.0 ± 0.5
N deposition	0.6 ± 0.3
Residual (source)	(0.2 ± 2.0)

Fluxes and reservoir changes of carbon expressed in Gt C y^{-1}. Numbers are IPCC (1994) plus Schimel's estimates for terrestrial sink terms. Errors are accumulated by quadrature. From Schimel (1995).

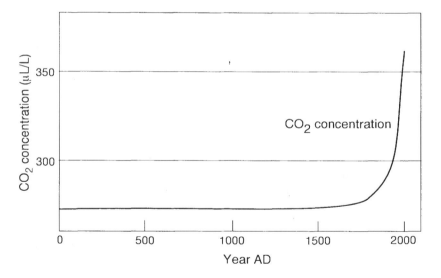

Figure 1.2 Concentration of CO_2 in the atmosphere over the past 2000 years (from Vitousek, 1994).

capita usage of natural resources expected to rise for much of that population, the pressure on other species will increase dramatically. Thus, the increasing co-option of terrestrial NPP by humans will likely be the dominant perturbation of the terrestrial carbon cycle for the next several decades at least.

The impacts of human activity on the global nitrogen cycle have been even

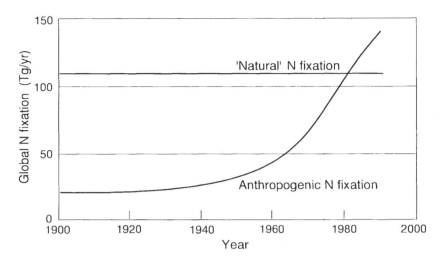

Figure 1.3   Human alteration of the global biogeochemical cycle of nitrogen. The 'natural N fixation' line represents biological N fixation in natural terrestrial ecosystems plus fixation by lightning. The 'anthropogenic N fixation' line represents the sum of industrial N fixation for fertilizers, fixation during fossil fuel combustion, and fixation by leguminous crops (from Vitousek, 1994).

more pronounced, as shown in Fig. 1.3 (Vitousek, 1994, Vitousek *et al.*, 1997*b*). At present human activity is responsible for more nitrogen fixation than are natural processes, and this ratio is increasing sharply. This human-fixed nitrogen is often transported large distances around the world, first as fertilizer and then as agricultural products. Again, given the projected growth in human populations and economic activity over the next few decades, this alteration of the global nitrogen cycle will undoubtedly become even more severe. An important perturbation to the nitrogen cycle caused by human activities is the increased emission of volatile compounds (e.g. N_2O, NO, NH_3) to the atmosphere. These gases, like CO_2, are radiatively active and are contributing to climatic change (see below).

The biogeochemical cycles of phosphorus and sulphur are also important. The N, P and S cycles, and the human modifications to them, are discussed in more detail in Chapter 10.

Another global atmospheric phenomenon with potentially severe consequences for all species is the depletion of stratospheric ozone. This is caused by the release of human-made chlorofluorocarbons (CFCs), and the consequent increase in UV-B radiation. Despite a reduction in the rate of increase in CFCs since the signing of the Montreal Protocol, the actual concentration of the gases in the stratosphere will continue to increase until sometime near the middle of the next century, and the effects of UV-B will therefore get worse before they get better. The main effects of UV-B radiation are on genetic mutations, photosynthesis (especially of algae) and skin and eye diseases of humans and animals.

1.2.3 Climate change

Climate change is the most well publicized component of global change. Changes in the atmospheric compartments of global biogeochemical cycles are predicted to affect the Earth's climate because many of the C- and N-based gases that are increasing in atmospheric concentration (e.g. CO_2, CH_4, N_2O) absorb long-wave radiation emitted from the Earth's surface and thus change the heat balance at the surface. This enhanced 'greenhouse effect' is the basis for predictions of increases in global mean temperatures.

Although there is now general consensus within the scientific community about the scientific basis for climate change, there is uncertainty about attributing the recent observed increase in global mean temperature to human factors (Fig. 1.4). In its most recent assessment of climate change, the Intergovernmental Panel on Climate Change concluded that '. . .the balance of evidence suggests that there is a discernible human influence on global climate' (IPCC, 1996b). The IPCC considered a wide body of evidence in coming to this conclusion, but one of the key pieces is the ability of General Circulation Models (GCMs) to simulate the stall in the temperature increase during the 1940 to 1980 period (see Fig. 1.4) when the appropriate effects of atmospheric aerosols are included in the models. Aerosols have a net cooling effect. Another strong piece of evidence is derived from pattern-based studies, in which the modelled climate response is compared with geographical, seasonal and vertical patterns of atmospheric temperature change. The studies show good agreement between model output and observations, and show also that this correspondence is increasing with time, as expected for a strengthening anthropogenic signal (IPCC, 1996b).

A major problem in pinpointing human activities as the cause for increasing global mean temperature and other climate anomalies is that the climate is subject to natural variability, and it has proved difficult to separate the signal of human-induced changes from the background noise of these natural climatic fluctuations. One of the most significant features of natural climate variability is the El Niño-Southern Oscillation (ENSO) phenomenon, which affects the climate over much of the globe but has a particularly strong impact on the moisture regimes in the subtropical zones. Since 1977 there have been frequent ENSO warm phase episodes, relative to the 120-year instrumental record, although this cannot be definitely linked to anthropogenic climate change. It is not yet possible to determine or predict with any confidence whether the climate has or will become more variable, with more storms and extreme events such as droughts and floods. It is often an extreme event, or a sequence of such effects,

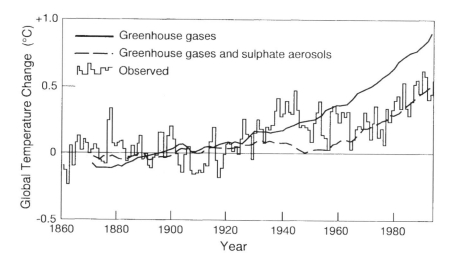

Figure 1.4   Simulated global annual mean warming from 1860 to 1990 allowing for increases in greenhouse gases only (solid curve) and greenhouse gases and sulphate aerosols (dashed curve), compared with observed changes over the same period (from IPCC, 1996*b*).

that causes significant changes to the structure and functioning of ecosystems, rather than a gradual change in a long-term mean climatic value.

One of the most difficult features to predict is the onset of rapid, large changes in climate, as apparently have occurred in the past. There is some evidence from Greenland ice cores that suggests the existence of climate 'flickers', in which radical excursions (e.g. 6 or 7 °C) from long-term mean temperatures, over just a few decades, have occurred in the Northern Hemisphere (e.g. Grootes *et al.*, 1993). The causes of such rapid changes are still unknown, but, if confirmed, they highlight the surprises that can be expected when nonlinear systems are subject to strong, rapid forcing.

1.2.4   Changes in biological diversity

The alteration to the composition and spatial arrangement of the Earth's biota is an additional, and very significant, component of global change. It is nevertheless often neglected as a 'global' process or a component of global change. There is a developing concensus that the planet is currently in the midst of the sixth major extinction event in the history of life (e.g. Leakey & Lewin, 1995). Although the causes of earlier extinction events (e.g. that of the dinosaurs) are uncertain, they are probably due to sudden changes in the physical environment caused by factors such as meteor impacts, pulses of volcanism, etc. The current extinction event is being driven by the activities of humans. The loss of species as a result of extinctions is unique among major global and biotic changes

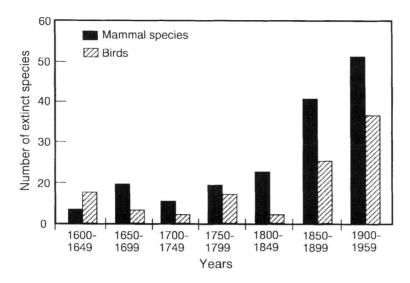

Figure 1.5   Trend in extinctions of bird and mammal species (from Reid & Miller, 1989).

because it is irreversible. For this reason, it is critical to understand the causes and consequences of species loss.

   Although there is now little doubt that the rate of species extinctions is high, it is difficult to quantify precisely. Acknowledging the uncertainty in the estimates, current extinction rates appear to be 100–1000 times higher than pre-human levels, and the expected extinction of currently threatened species could increase this rate by a factor of 10 (Pimm *et al.*, 1995). There is good evidence that the rate, whatever it might be, is accelerating (Fig. 1.5); extinctions of both bird and mammal species have increased sharply in the last 100 years over their previous rates (Reid, 1992), and it is estimated that up to half of all terrestrial plant and animal species could become extinct over the next 50 years (May, 1992). In addition to complete extinctions, many species are being driven locally extinct or occur in only a small fraction of their previous ranges.

   Apart from the intrinsic loss of species itself (which should not be down-played, given the strong worldwide interest in biodiversity conservation), the consequences of changes in biological diversity are primarily their effects on ecosystem functioning. The few studies conducted to date indicate that changes in functioning occur with changes in species diversity, although the nature and magnitude varies with ecosystem type, functional response meas-ured, and experimental conditions (e.g. Fig. 11.5). There have been too few studies to draw any general conclusions. All studies so far have been done on relatively simple systems, where a change in diversity is most likely to have an effect on functioning. Although much attention on the conservation of biodiversity has focused on species-rich ecosystems, the results from the simpler systems suggest that the functional consequences of species gain or loss may be expressed most quickly in species-poor ecosystems. Thus, from the perspective of ecosystem functioning, global 'hot spots' for biodiversity loss will probably not correspond to the current areas of interest in biodiversity conservation.

1.3   Certainties, time and space scales, and interactions

Although there has been considerable debate on the uncertainties associated with anthropogenic influence on climate, there is no doubt that all of the other components of global change are real, are occurring now, and are caused directly by human actions. Land-use and cover change is pervasive on all continents except Antarctica, is clearly driven by direct human actions, and will probably be the dominant component of global change for the coming decades, especially in the tropics (Turner II *et al.*, 1990). Alterations to the global carbon and nitrogen cycles have, through a variety of analyses, been unequivocally attributed to human activities (Vitousek, 1994). Changes to soil structure and chemistry are also beyond dispute, and are clearly caused by humans, primarily through agricultural activities. (It is important to note, however, that not all changes represent degradation. There are many examples of well-managed agricultural lands where soils, though somewhat different from their native state, are still fertile and 'healthy'.) Although precise rates of the loss of biodiversity are difficult to measure, there is a strong consensus within the scientific community that they are widespread, are accelerating, and are caused by human actions (Heywood & Watson, 1995).

The time scales associated with the individual components of global change vary widely. Although land-use change is dominant now, it is likely, if the scenarios generated by GCMs are realistic, that climate change will become increasingly important into the twenty-first century as its signal emerges more strongly from the present-day natural variability. Changes in atmospheric composition, especially CO_2 concentration, are possibly more important than climate now, but will be less so as climate change gathers pace.

Two aspects of the spatial character of global change deserve mention. First, the global averages that are often used as a measure of a changing global environment mask large geographical variations and do not show the relative importance of global change components at a regional level. For example, mean temperature increases are predicted to be least in the tropics and greatest in the high latitudes, while land-use change is currently most important in the tropics.

Thus, although land-use change will probably be dominant for much of the Earth for the foreseeable future, increasing temperature is important now in the high latitudes and its importance is likely to increase. For example, there is already evidence of sharp changes in the disturbance regimes of high latitude terrestrial ecosystems (see Chapter 12), possibly as a response to increasing temperature, which have significant consequences both for the sustainable management of those systems and for the terrestrial carbon budget (Kurz *et al.*, 1995).

The second aspect of the spatial character of global change is that it is caused by a myriad of individual, site-specific actions occurring daily around the planet. Fossil fuel emissions arise from a number of intermittent point sources operating around the globe. Through the mixing power of the atmosphere, they lead to systemic changes in the Earth's environment. Land-use change is also highly distributed around the world, and is usually driven by a complex interactive set of demographic, socio-economic and political factors that are site- or region-specific (although with the globalization of economies, the drivers of land-use change are assuming an increasingly global character).

Within the range of scales encompassing global change, resource management issues are often addressed at the landscape scale (tens of kilometres), and at this scale global change tends to be considered as just one more perturbation in a complex suite of factors affecting the management regime – often less imminent than the others. While some global change components can be dealt with as global issues (e.g. atmospheric CO_2 concentration), most cannot. Global scale scenarios of land-use change, for example, while important for Earth system modelling, are less relevant for analysing impacts. More relevant are local and regional scale land-use changes, where, for example, changes in one part of a catchment can have significant impacts downslope. Also, finer scale regional climate scenarios are far more relevant for impact studies than global means.

It is important to match the time scale of the nature of the resource management issue with the appropriate global change component. While an agricultural manager concerned with food production over the next few years may not be interested in a predicted change in mean global temperature for 2040, he or she will be interested in the changing frequency of regional droughts in the coming decade, and in local temperature changes. However, nature conservation managers designing a reserve network are interested in long-term scenarios of climate and of land-use/cover change leading to various landscape fragmentation patterns.

The drivers of global change have been presented individually, as separate effects, but they do not act in this way. The interactions among them will strongly determine how terrestrial ecosystems respond to an environment of

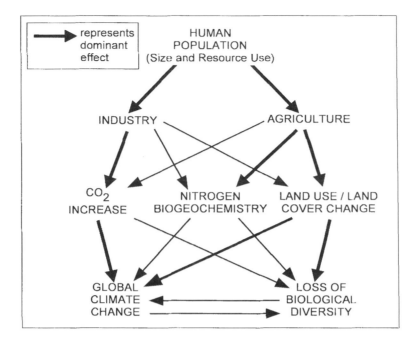

Figure 1.6   Interactions among global change 'drivers', showing relationships among human population and activity and the components of global change discussed in this chapter (from Vitousek, 1994).

accelerating change. This is a re-statement of a fundamental principle of ecology; the outcome of several environmental factors acting on an ecological system is determined by their interactive effects, and cannot be deduced without considering the nature and magnitude of these interactions. This principle applies pre-eminently in global change ecology. Figure 1.6 (Vitousek, 1994) gives a schematic overview of some of the important interactions.

Some examples help to illustrate these interactions:

■ Land-use/cover change leads to direct impacts on soils (e.g. erosion), which influence further land-use change, and also affect soil biodiversity and biogeochemistry, which in turn can affect the emission of trace gases from soils, which influences atmospheric composition and climate.
■ The effects of changes in atmospheric CO_2 concentration on ecosystems is strongly linked to the nitrogen and hydrological cycles, as shown in Fig. 7.1. Changes in climate affect plant available moisture, which influences system response to CO_2. Increased CO_2 leads to higher water use efficiency, and therefore NPP. However, increased temperatures associated with higher

CO_2 can also lead to a higher potential evapotranspiration/precipitation ratio, and therefore a lower NPP. The magnitude of the CO_2 response of vegetation depends on its nitrogen status, and so changes in the global nitrogen cycle, especially through fertilizer use and atmospheric deposition, will affect CO_2 response. Changes in CO_2 uptake by the terrestrial biosphere can influence atmospheric CO_2 concentration, with a feedback effect on climate.

■ Land-use types are broadly constrained by climate. Changes in land use lead to emissions of trace gases that affect climate; changing climate then influences the distribution of future land uses, which again affect climate, and so on.

The components of global change are often both drivers and consequences. They are not related in linear cause–effect sequences, but are interlinked in numerous feedback loops in a complex, interactive system.

1.4   The science of global change interactions with terrestrial ecosystems

Given the complex nature of global change, the key to understanding its interaction with terrestrial ecosystems lies in a systems–oriented approach (as distinct from a reductionist 'cause–effect' approach) focused on the structure and functioning of the terrestrial ecosystems themselves. GCTE's research programme was developed on this basis (Steffen et al., 1992). Its objectives are:

■ to predict the effects of changes in climate, atmospheric composition, and land use on terrestrial ecosystems, including (i) agriculture, forestry, soils, and (ii) ecological complexity;
■ to determine how these effects lead to feedbacks to the atmosphere and the physical climate system.

The research effort is organized around four major thematic areas: Foci 1 and 2, ecosystem physiology and global change, and change in ecosystem structure, respectively, study the fundamental processes of the interaction between ecosystem composition, structure and functioning with the components of global change. Foci 3 and 4, global change impact on agriculture, forestry and soils, and global change and ecological complexity, respectively, address the impacts of global change on two issues of importance for human well-being – the production of food and fibre from terrestrial ecosystems and changes in the composition and spatial arrangement of biological diversity, and their implications for ecosystem functioning.

More details on GCTE research are given in Box 1.1.

Box 1.1   Global Change and Terrestrial Ecosystems (GCTE)

GCTE is a component of a larger international global change research effort, the International Geosphere–Biosphere Programme (IGBP). The goal of IGBP is:

■ *to describe and understand the interactive physical, chemical and biological processes that regulate the total Earth system, the unique environment that it provides for life, the changes that are occurring in this system, and the manner in which they are influenced by human actions.*

Within IGBP, in addition to GCTE, other components address global change-related questions in atmospheric chemistry, biospheric aspects of the hydrological cycle, the coastal zone, land-use/cover change, oceanic carbon fluxes, marine ecosystems and palaeo–environmental sciences.

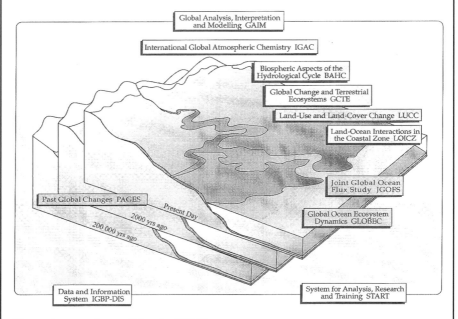

Component projects of the the IGBP.

GCTE's research is built around two major themes – global change impacts and feedbacks – and includes the entire range of terrestrial ecosystems, from pristine natural systems to intensively managed agricultural systems. GCTE's structure is shown in the figure opposite.

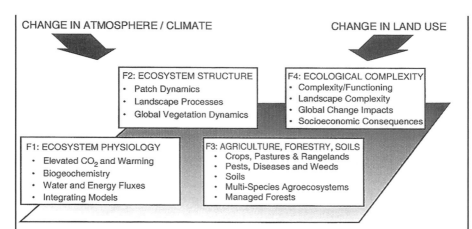

Structure of GCTE, showing the four Foci and the Activities within each Focus.

The strategy for GCTE's research agenda is built around several key elements:

- *GCTE Operational Plan.* This report (Steffen *et al.*, 1992) describes in detail the scientific framework (Foci, Activities, Tasks) and implementation plan for executing the research agenda. It is the document against which relevance to GCTE Core Research is evaluated.
- *Task Leaders.* One or two scientists are invited to lead each of the Tasks within the GCTE Core Research Programme. Their role is to organize the individual contributing projects to their Task into a coherent programme through mechanisms such as common experimental protocols, standardized methodologies, model comparisons, and synthesis workshops.
- *Acceptance of existing research.* Existing research projects constitute the bulk of GCTE Core Research. They are normally submitted by individual scientists on behalf of the project, evaluated and accepted (or otherwise) by the GCTE Scientific Steering Committee, and join others to form networks designed to address particular Tasks. Increasingly, the Task Leaders form initial networks, and then submit the network as a package for approval by the SSC.
- *Initiation of new research.* Where there are obvious gaps in the programme, GCTE, in partnership with national and regional agencies, attempts to initiate appropriate new research projects.

GCTE has been operational for six years. As of July 1997, its Core Research Programme consists of 55 contributing projects involving over 1000 scientists and technicians from 44 countries around the world. This research, with an aggregate funding of about $US 44.2 million annually, receives support from a large number of national and regional agencies. GCTE's role is to add value to these individual efforts by coordinating them into international networks and consortia, thereby building up a global synthesis from individual projects and accelerating our understanding of global change interactions with terrestrial ecosystems.

Further information on GCTE can be found on the homepage at:
http://Jasper.Stanford.EDU/GCTE

1.5   The structure of this book: a synthesis of GCTE and related research

This book is a synthesis of research undertaken within GCTE since the beginning of its implementation phase in 1990, and of related research. It focuses on those areas where GCTE has made a contribution or where it is establishing coordinating structures and is involved in project development. It also draws on related research, where appropriate, and places the GCTE work in the context of the broader research effort on global environmental change and the terrestrial biosphere.

The first section of the book, following this introductory chapter, deals with the tools that have been developed to tackle the challenges of research on global change and terrestrial ecosystems. It includes chapters that discuss: (i) approaches to experimental and observational studies at the ecosystem level (Chapter 2); (ii) the use of networks and consortia (Chapter 3); (iii) the IGBP Terrestrial Transects (Chapter 4); (iv) global databases (Chapter 5); and (v) the development of quantitative models for global change research (Chapter 6).

The next five chapters synthesize research results and highlight major new insights and understanding. The first three (Chapters 7, 8 and 9) are based on the work of GCTE's Foci 1, 2 and 3, respectively. Chapter 10 addresses a key cross-cutting issue to which the three preceding issues all contribute: the role of terrestrial ecosystems in global biogeochemical cycles. This chapter is not so much a synthesis of GCTE and related research, but rather provides the global, long-term context into which much of this research can be placed. Chapter 11 describes the major issues, and initial research, in the new area of global change and ecological complexity (GCTE Focus 4) and outlines the next steps in this challenging research field. The final chapter (12) synthesizes the major findings of the previous chapters and then uses a selected set of the IGBP Terrestrial Transects as an organizing tool to examine the implications of the findings for particular regions of the world. It then discusses the global implications of the synthesis and concludes with an identification of key future research issues.

2 The study of ecosystems in the context of global change

E.-D. Schulze, R.J. Scholes, J.R. Ehleringer, L.A. Hunt,
J. Canadell, F.S. Chapin III and W.L. Steffen

2.1 Introduction

Global change research requires not only knowledge of how individual species
(e.g. pests) respond to climate and land-use change, but also an understanding
of the responses of whole systems to their multiple and interacting drivers. An
upscaling from single systems to landscapes and continents is an additional
essential component of global change research. In contrast to the well-established research methodology in the ecophysiology of species (see Pearcy *et al.*,
1989), the approaches to studying ecosystems as a whole, and the theory
required to identify key parameters that drive the multiple interactions at the
ecosystem level, are less developed (see Schulze, 1995*a*). The available tools
become even more limited for the study of responses at geographically broader
scales. One of IGBP's (International Geosphere – Biosphere Programme)
products has been the contribution to methodology and theory for extrapolations from ecosystems to continental scales.

Experimental investigations of ecosystems started between 1960 and 1970
with the International Biological Programme (IBP), and the ecosystem-level
global change studies of GCTE (Global Change and Terrestrial Ecosystems)
have built strongly on that heritage. Global change research has reiterated that
ecosystems are difficult to study for several reasons:

Large-scale processes: Important ecosystem processes, such as biogeochemical
  cycles, disturbance regimes and demographics, take place at scales much
  larger than the typical experimental plot of a few hectares. These large-
  scale processes are only partially manifested at the smaller scale.

Slow and often highly buffered processes: Many ecological processes, including
  succession and changes in the soil, require decades to centuries for
  completion and may even be in a transitional dynamic state. Nevertheless,
  there may be no perceptible change for many years following the application of a treatment, but rapid change may occur once the buffering

capacity of the system is exhausted. Ecosystem changes may not be fully revealed even in the lifetime of the researcher, which is a very different scale than the average time of support for typical research projects.

Complexity: Ecosystems contain unknown numbers of organisms. This results in large numbers of variables that affect ecosystem fluxes, and these variables are connected to one another typically in nonlinear relationships. This makes it difficult to isolate single factors to be systematically varied in an experiment without simultaneously disrupting many others. The number of interactions swiftly renders factorial experiments unwieldly, especially since the nonlinearity of the responses requires several treatment levels for each factor, rather than just a 'high or low' or 'present or absent' treatment.

Fluxes are difficult to measure: Ecosystem processes often cannot be directly observed but must be inferred from their consequences. For material flows, this usually involves observing changes over time in pools. Where the pool is large relative to the flow, the small result is easily lost in the noise created by sampling and analytical variation. In addition, the net flux in and out of systems does not reveal the internal circulation of substances.

True replication is difficult: Classical statistics assume multiple identical units, randomly assigned to treatments. The high spatial heterogeneity characteristic of ecosystems means that every patch is in some respect unique. Furthermore, the experimental treatment is often applied by nature, rather than by the researcher, and is thus not randomly assigned. Finally, there are ethical constraints in applying large-scale manipulative experiments to what are frequently already-threatened ecosystems. Truly global experiments (such as the inadvertent experiment we know as 'global change') have no control and no replicates, but contain only a time series in a naturally highly variable environment. Conducting such an experiment may threaten not only our own well-being, but also that of millions of other species.

Based on these intrinsic problems, ecosystem research started in the IBP, including some of the most important long-term observations of ecosystem pools and processes, is still used by IGBP. Examples are the long-term observations of soil solution chemistry and acidification (Ulrich, 1989) or the monitoring of the atmospheric composition of CO_2 (Conway *et al.*, 1994). While IBP was focused on understanding mainly productivity, the study of ecosystems emerged in the following decade of acid rain and air pollution research. In this period it was demonstrated for the first time that phenomena such as forest decline cannot be explained by single processes and linear regressions. Rather, to interpret the decline in forest health adequately, a systems approach was needed that included direct and indirect effects, feedbacks and interactions (Last & Watling, 1991). Air pollution research demonstrated also the necessity of ecosystem-scale experiments in addition to the field observations (Mooney *et al.*, 1991). Thus, the following earlier approaches are now used for the study of ecosystems in the context of global change: (i) investigations of natural extreme events (e.g. volcanic eruptions, El Niño); (ii) long-term observations at the plot level; and (iii) ecosystem manipulation (Mooney *et al.*, 1991). The global scale of the questions addressed by IGBP required an additional tool. It is impossible to study all ecosystems globally with the same level of detail. This led to the idea of establishing continental-scale transects as a basis for replicated observations (see Chapter 4).

This chapter gives an overview of the research approaches and techniques used and developed by GCTE, from the plot-level to the global scale. Databases and model development will be discussed in Chapters 5 and 6, respectively.

2.2   Approaches to the study of ecosystems

Figure 2.1 presents a schematic summary of approaches to the study of ecosystems in the context of global change. In the following sections each type of approach is described with a number of examples of its use and a discussion of its strengths and weaknesses.

2.2.1   Sampling problems in global change research

Every study plot is a unique sample from a highly variable environment. For example, when a forest transect through Europe was studied (NIPHYS/ CANIF project – GCTE, 1996), plots were selected so as to maintain soil conditions as similar as possible in order to detect effects of climate and nitrogen deposition. Following the selection, it emerged that the local variability in nitrogen concentrations of needles and leaves was larger than the variability along the transect as a whole (Bauer *et al.*, 1997). The same problem occurs again, although at a different scale, with respect to the heterogeneity within a plot.

Oren *et al.* (1989) made use of this variability at the plot level in order to extract general regression relationships that describe the variation in a functional manner. It became obvious that a series of smaller plots, representing local variability, is more useful than the study of a single large plot. Another approach using regional variability for a mechanistic understanding of processes was that of Durka *et al.* (1994). Based on the regional variability of 360 forest surface springs, six plots were identified that represented the regional range of conditions, and for these plots those processes that caused the observed

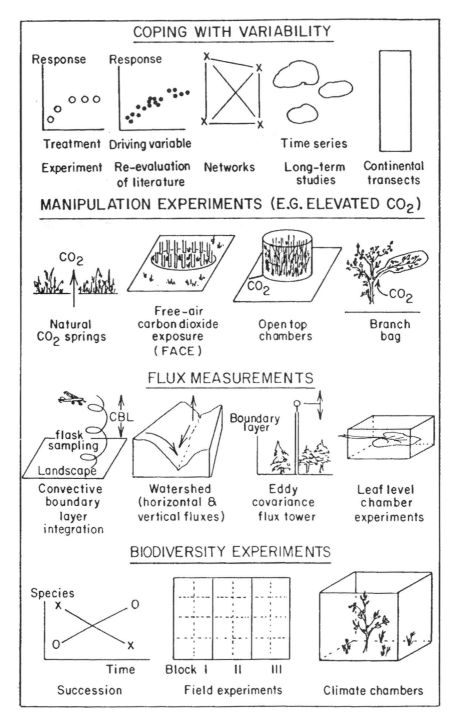

Figure 2.1   A summary of approaches to the study of ecosystems in the context of global change. Each of the approaches is described in more detail in the text.

variation in outputs were identified. Thus, local versus regional or continental variability can help in identifying key processes. However, there are limits to this approach. Hornung (1992) studied some 40 watersheds throughout Europe, but it became difficult to identify single processes that explained the variability, because the number of parameters causing change in a network of sites (not a transect) was greater than the number of plots being studied.

The statistical problems of how to deal with variability at different scales are a serious issue for global change research, and have not been resolved. The Long-Term Ecological Research Program (LTER) of the United States, the Chinese Ecological Research Network (CERN) and the biosphere reserves of Russia are networks of sites representative of biomes. They can serve as ground truth for global modelling, but a comparison of these sites cannot produce response curves for variables that drive the change from one biome type into another because their selection was biome- and not gradient-oriented. This was one of the reasons that IGBP adopted the transect approach which allows an *a priori* definition of a driving variable against which the responses (often nonlinear) of ecosystem parameters can be examined (see Chapter 4).

In view of the immense problem of local and regional heterogeneity, an approach has emerged in GCTE for ecosystem studies at global scale. Processes have been studied, independently of plots, by focusing on the ecology of species. For example, the re-examination of data on CO_2 and water exchange of plants and their assessment in relation to other driving variables, such as nutrition, has led to new information about the general relations between parameters that can then be tested at a plot or transect scale (Körner, 1994; Woodward & Smith, 1994*a*; Schulze, 1994*c*). This approach has the advantage that it contains, by definition, species- and habitat-related variability. This gives more power for the use of these functions in global models than functions based on observations on a single plot. The disadvantage of this approach is also obvious; the variables are not measured at the same time on the same object. It is necessary to integrate this information, for example, according to plant functional types (see Section 2.2.2 and Chapter 8). The combination of (i) development of general and global relations; (ii) their test in specific situations; and (iii) their use in modelling on landscape or regional scales has become a new approach in the context of global change research.

With respect to ecosystem experiments, a different set of statistical constraints and considerations emerge (Jeffers, 1993). Classical statistics, such as the Analysis of Variance (ANOVA), were developed for the analysis of agricultural trials, where true replications, factor separation and discrete treatment levels are possible. This design proved to be successful in applied research. In fact, a treatment effect can readily be produced if the treatment is strong enough in relation to the control. For many ecological applications where these conditions are not met, the more general approach offered by regression analysis (including nonlinear and multivariate forms) is more appropriate. In general, the focus in ecological studies is less on proving that two treatments are statistically different from one another, but on establishing the form of the relationship between dependent and independent variables, and identifying critical thresholds. Where the nature of the treatment is not known *a priori* – as is frequently the case in gradient studies – there are several indirect gradient techniques to reveal the relationship between environmental drivers and ecosystem responses. Thus, for ecosystem experiments in the context of global change, it is more appropriate to quantify a response surface than a particular treatment effect.

Numerical simulation modelling is indispensable for exploring the interactions in even moderately complex systems, especially where nonlinearities and feedbacks render them intractable to analytical mathematical techniques. Furthermore, models act as hypotheses which test whether current understanding of the processes at work is sufficient to recreate the observed phenomena. This set of techniques has become so important for global change research on terrestrial ecosystems that a full chapter – Chapter 6 – is devoted to a treatment of the topic.

2.2.2   Ecosystem comparative studies

Biome comparisons and networks

The comparison of widely separated ecosystems which are acknowledged to be different in their details provides a test of the robustness of our understanding of systems in relation to soils and climate. The approach is especially powerful when one treatment is repeated in several sites, across a range of environments and ecosystems. This obviously requires standardization of treatments and measurements. An example is the experiment designed by the GCTE Rice Network for implementation at four tropical and three temperate sites (Ingram, 1995). The experiment involves determining the effects of higher temperatures and nitrogen application levels on the grain yield of rice. The association of related sites into networks has become an important component of international collaborative science efforts (see also Chapter 3).

The difficulty of replicating an ecosystem experiment spatially may sometimes be avoided by replicating in time instead if all other conditions are constant. The main problem with this approach is in ensuring that the time-separated treatments are independent of each other, and that factors such as climate differences are accounted for. The comparison of observations taken over long periods of time within one ecosystem is a powerful technique for detecting change, provided that the statistical techniques appropriate to time series analysis are used and that methodological consistency in observation techniques is maintained. The time series of soil solution chemistry in the Solling study of Ulrich is a classical example of this approach (Ulrich, 1989). The LTER sites in the USA and the CERN sites in China have been set up for this purpose.

Very long-term processes can be explored using paleoecological data. For example, the rate of species migrations has largely been inferred from palynological records of the changing distribution of species in response to past climate (Firbas, 1949). In an analogous way, the growth rate of trees over the past few centuries can be inferred from tree rings (e.g. Spieker *et al.*, 1996).

Functional types

It is practically impossible to study all species involved in an ecological process on an individual basis. One approach to tackling this problem is to classify species according to their functions rather than using the traditional phylogenetic approach. While the study of functional types goes back to the beginning of ecology in the last century (see Schulze, 1982 for a review), GCTE has taken up the challenge to develop a scheme of functional types for process-related studies (Woodward & Cramer, 1996*b*; Smith *et al.*, 1997). It emerges that structural characteristics have an overriding importance for classifications based on physiology. In fact, the classification by Monsi (1960) may encompass most requirements (annual/perennial; herbaceous/woody; evergreen/decidu-ous), except for a few physiological traits such as C_4/C_3 metabolism and nitrogen fixation, which add a new physiological quality.

Although the use of functional types is proving useful in global modelling (see Chapter 8), studies of community ecology have demonstrated that individual species also matter in a global change context (Mooney *et al.*, 1995). A good example is rooting depth. In grasslands, 90% of the root biomass occurs in the top 20 cm of the soil profile (Jackson *et al.*, 1996*b*), while the average maximum rooting depth reaches 3 m (Canadell *et al.*, 1996*a*). From a closer inspection of the grassland data from the root atlas by Kutschera (1960), it emerges that 187 species reached 70 cm rooting depth on average, but 10 of the 187 species reached 223 cm. In case of increasing drought, those 10 species have a much larger chance of survival because of this species-related trait.

Certainly a new functional type could be defined for deep-rooted herbaceous species. However, if functional groups are defined for all parameters that are important at the community level (e.g. fecundity, leaf orientation, phenology, etc.) a species-level classification may result, because the many different combinations of the functional attributes will eventually describe a specific species. Thus, the earlier expectation that species-specific studies could be avoided is dampened with the knowledge that combinations of species-specific traits are often important in understanding the response of terrestrial ecosystems to global change (see also Chapter 11). The problem can be resolved to some extent by defining two kinds of functional types – 'control' and 'response' types (Walker, 1997). The importance of species has been specifically a problem with pests and parasites, which almost always act by a species–species interaction. Nevertheless, in many cases for testing focused hypothesis at the ecosystem level, functional types remain an important research tool.

Gradient studies

Observing ecological responses along gradients of environmental drivers has been one of the basic tools of ecology since the earliest years of the discipline. Many ecological responses in nature take the form of a continuum, rather than discrete steps, making a gradient a natural way to examine them. Gradient analysis is also a way of using spatial variability for testing hypotheses. A long-established natural gradient allows for the long adaptation times necessary for many important processes to adjust to the gradient, which would not be possible to recreate in a newly established treatment. In many cases the gradients are complex (involving more than one driving variable, varying either in parallel or independently). Sometimes the gradients are continuous in space, as, for example, along the catena sequence of soils that form down a slope (Giesler et al., 1998). However, more often gradients are not continuous in space. A special case of a gradient study is the chronological sequence of sites of different age since disturbance (e.g. Schulze et al., 1995).

Carbon dioxide emitted by natural springs diffuses away from the source, providing a gradient of elevated atmospheric CO_2 which has proved unique for examining the long-term effect of elevated CO_2 on vegetation (Koch, 1993; Miglietta et al., 1993). Variations in wind speed and direction disrupt the diffusing plume, causing large fluctuations of the CO_2 concentration at any given point, but averaged over the day or the growing period of a plant, there are clear patterns of exposure. Gradients of partial pressure of CO_2 can also be obtained with increasing altitude in mountains, but in this case the gradient is confounded with increasing UV-B and decreasing temperatures, and often with changes in precipitation. Altitudinal gradient studies play a key role in a major IGBP research effort on global change in mountainous regions (Becker & Bugmann, 1997).

GCTE adapted the well-established ecological technique of gradient analysis to the global scale by developing the idea of 'megatransects', a concept subsequently adopted by the IGBP as an inter-core project research platform (Koch et al., 1995a,b). The rationale for the IGBP transects and their use in global change research is described more fully in Chapter 4. An example of the use of the transects for analysing the interactive effects of global change on terrestrial ecosystems is presented in Chapter 12.

2.3 Manipulative experiments

Building on knowledge from forestry and agriculture, ecosystem-level manipulative experiments were started in the period of IBP (Tamm, 1991), where one major question was a global quantification of productivity. However, the reservations of ecologists about such experiments have often been cited: (i) There may be indirect treatment effects which make it impossible to manipulate just a single factor. For instance, the effect of fertilization on yield may be overlaid by its effects on humus decomposition, soil acidification and pollution of ground water, which, at the ecosystem level, override the immediate effect of fertilizer on growth. (ii) The history of processes on a plot may influence the result (e.g. fire history; a dead animal may have been decomposing by the tree selected for study). Results may thus be idiosyncratic and of a correlative nature rather than reflecting mechanisms.

Nevertheless, manipulative experiments at the ecosystem level are becoming an increasingly important tool in global change research. This section describes such experiments in five key research areas – effects of elevated atmospheric CO_2, impacts of warming, nutrient and water limitations, effects of ultraviolet radiation, and biodiversity interactions with global change.

2.3.1   Elevated CO_2

The fact that atmospheric carbon dioxide will continue to rise for the foreseeable future is one of the certainties in global change. Thus, a large number of studies have used experimental approaches ranging from cellular to ecosystem-level in order to predict plant and ecosystem responses to elevated CO_2. Almost all studies have focused on the effects of an instantaneous doubling of CO_2. The bias and the problems using this approach are quite obvious. A review of the published literature shows the uneven distribution of information along a response surface (McGuire *et al.*, 1995*a*). It emerges from the large body of elevated CO_2 research (see Chapter 7 and Section 12.2.2) that the hypothesis of Gifford (1994*a*) still holds, namely, that an effect at the protein level will continually shrink when investigating the same response at the leaf level, the whole plant or the ecosystem level. Thus, the approaches for studying elevated CO_2 in the context of global change must emphasize natural (low nutritional) ecosystems, with the aim of understanding the response surface of CO_2, nutrient, and water interactions at the system level.

With respect to crop species, additional problems emerge. Much recent CO_2 work has been undertaken using major crop species as the experimental target, and in particular, wheat. All such experimentation has been organized in accordance with the basic principles of research design, with randomization and replication of experimental treatments, and has endeavoured to accomodate the widely recognized impact of different base environments by continuing over at least two growing seasons. Such work has demonstrated effects of CO_2, but has not provided response curves because only high and low levels were tested in a block design, which is inappropriate for generalizations. Because of limited funds, the studies have also been conducted with a limited array of cultivars, and generally with only one cultivar in a particular study. The experiments thus do not allow for examination of genotype–environment interactions, which are generally quite significant. To help overcome this limitation, individual experiments at specific sites are being complemented by coordinated experiments in which a common genotype with seed from a standard source is examined at different sites (e.g. ESPACE, the European Stress Physiology and Climate Experiment, H.J. Jüger, personal communication). Such work is currently concentrated on one genotype; it will need to be extended to an array of genotypes if the full significance of differential genotype response to CO_2 and other environmental variables is to be fully understood.

In the absence of coordinated multi-site and multi-genotype experiments, some indication of the magnitude of the interaction terms can sometimes be obtained by analysing jointly the results from different experiments. Unfortunately, this task is often made difficult by the different modes in which data have been recorded and subsequently collated by individual researchers and research groups. To help overcome the problem, attempts have been made recently to develop some standard file structures that can be used for storing experimental details and data, for input of data into various analysis programs and simulation models, and in particular, as transfer files for the interchange of data between experimental groups (see Chapter 3 for more details).

Beyond the problem of genotype–environment interactions and of data storage and use, interpretation of the results of field experiments has often been limited by inadequate information on parts of the total system. In an attempt to overcome the problem, efforts have been made by a number of individuals concerned with agricultural experimentation to define a 'Minimum Data Set' that should be recorded in all experimental work (see Hunt *et al.*, 1994). In such a minimum data set, which perhaps could better be termed a 'balanced data set', an effort is made to encompass information on the soil, on conditions at the start of the experiment or study (e.g. soil surface conditions, soil water contents, soil inorganic N contents), on management interventions, and on various aspects of the growth and development of the plant cover. The concept is applicable to all ecosystem studies, whether managed or natural, and should be followed widely to be able to interpret results at an ecosystem level.

The need to investigate ecosystem-level responses to elevated CO_2 has required the development of new experimental technologies (Schulze & Mooney, 1993). Most early research on effects of elevated CO_2 utilized plants grown in pots in greenhouses or growth chambers and involved exposing plants to elevated CO_2 for periods ranging from minutes up to a single growing season (Strain & Cure, 1994). While these early experiments provided important insights into short-term responses of individual plants, they revealed little about responses of entire ecosystems over long time periods. Technologies that have been developed or employed over the past decade to study ecosystem responses include ecocosms, closed chambers (CC), open-top chambers (OTC), and free-air carbon dioxide enrichment (FACE). Natural CO_2 springs can also provide a valuable understanding of long-term and evolutionary responses to elevated CO_2. Here the technologies and their respective advantages and disadvantages are briefly described in relation to sample/plot size, expense, degree of environmental control, and 'naturalness' of the system.

Ecocosms are essentially elaborate growth chambers in which investigators attempt to recreate conditions found in natural ecosystems (e.g. Körner & Arnone, 1992). This can include attempts to establish natural soil conditions and a representative assemblage of species, including microbes. A key advantage of ecocosms is that the investigator has a high degree of control over all aspects of the experiment including species composition, soil conditions, and environmental variables. The major disadvantages include the constraints on plot size and maximum plant size; the expense of constructing numerous large chambers, controlling all important environmental parameters, and operating the experiment for long time periods; and the difficulties in creating truly natural conditions.

The first attempt to evaluate *in situ* ecosystem responses to elevated CO_2 employed closed-top chambers (Oechel *et al.*, 1994). The first CCs were small and were used only in low stature ecosystems such as tundra or grassland. More recently larger chambers have been constructed for use in chaparral or even with trees. CCs are expensive to construct because of the need to control all environmental variables; operating costs may be relatively low for CO_2 but are high for the electricity needed to control the chamber environment. Chamber effects in CCs inevitably result in alteration of the radiation and wind regimes and pollinator access.

Open-top chambers were first developed for use in air pollution studies (Heagle *et al.*, 1974) and more recently have been utilized in CO_2 research. OTCs can be up to several meters in diameter and height, and thus can be used with plants as large as tree saplings. They typically are constructed of inexpen-

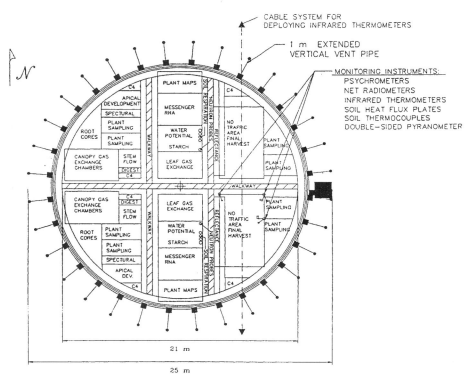

Figure 2.2   Overview of a FACE ring for the 1992–93 FACE wheat experiment in Maricopa, Arizona. Location of the assigned biological activity areas in both irrigation subplots of the ring. (From Wall & Kimball, 1993.)

sive materials (e.g. metal tubing and polyethylene sheeting) making it possible to use large numbers in factorial experiments. The concentration of CO_2 (or other gases) in the chambers is increased by adding controlled amounts of the gas to ambient air, which then is introduced to each chamber through a plenum around its base. Other environmental variables generally are not actively controlled, and natural precipitation can enter through the open top. Because of their low cost and relatively large sizes, OTCs have been used in many elevated CO_2 experiments (see Chapter 7). As with CCs, wind regimes, radiation, and pollinator access are altered within the chamber. One variant on the OTC is a tunnel system in which gradients of CO_2 are maintained, allowing plants to be exposed to a full range of CO_2 levels (Mayeux *et al.*, 1993; Hadley *et al.*, 1995). The gradient is established by blowing CO_2-enriched air through the tunnel with dilution caused by plant uptake and leakage.

The newest and best technology for investigating the responses of natural ecosystems to elevated CO_2 is FACE (Lewin *et al.*, 1992) (Fig. 2.2). In FACE systems, CO_2 concentration within the experimental plots is elevated by intro-ducing gas to the plots from a series of vertical pipes arranged around each plot. Wind direction, wind speed, and CO_2 sensors within the plots are used to control which pipes emit CO_2. There are two key advantages of FACE technol-ogy. First, plot size can be large (e.g. 30 m in diameter) so that a reasonable number of large plants, including relatively mature trees, can be studied in natural ecosystems. Second, since no chamber or barrier is constructed around the experimental plots, virtually all environmental factors other than CO_2 concentration are unaffected. However, during still nights, air blowers are needed to ensure a proper distribution of CO_2 over the plot, and this has been reported to have an effect on air temperature (called the 'blower effect'). The crucial disadvantages of FACE are the cost of the experimental facilities and the very high, ongoing cost of CO_2. As a consequence FACE experiments initiated to date involve few treatments and little replication. In any case, FACE is the best technology available for the most realistic field experiments calling for enrichment of intact ecosystem units. A variation on FACE is to construct a screen around the plots to reduce wind speed through the plot and thus reduce the amount of CO_2 used.

The use of natural CO_2 springs is also an important approach to study system responses to elevated CO_2. Their clear advantage is that plants native to these sites should have adapted to these conditions over long periods of time. Special caution, however, needs to be paid to the possibility that these natural springs may carry other gasses in addition to CO_2 which, even at low concentrations, may have an important effect over long periods of time. Highly variable CO_2 concentrations over time may be also a problem (Koch, 1993).

2.3.2 Warming

Over the last few years a range of techniques has been used to study the impact of global warming on ecosystem functioning. The most common techniques are buried heating cables, air-heated open-top and closed chambers, infrared heaters, and simple passive greenhouses (Fig. 2.3). A brief description of each technique is given below.

Buried heating cables have been used to increase soil temperature between 5 °C and 10 °C above ambient in various ecosystems (van Cleve *et al.*, 1990; Peterjohn *et al.*, 1993; Lükewille & Wright, 1997). This technique is commonly used in forests because the warming of an entire system would be costly and technically unfeasible in most of the cases. Because ecosystem responses to warming are expected to be largely modulated by altered soil processes, this technique can provide significant insights to better understanding of warming responses. However, the extent of the decoupling effect between the below-ground (temperature treated) and the aboveground components (non-tempera-

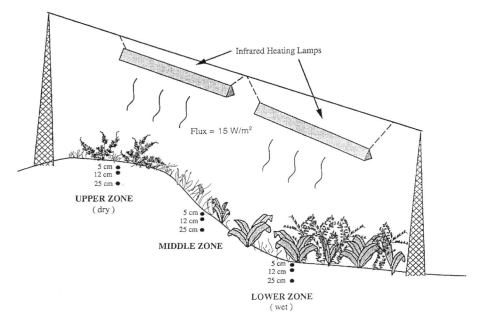

Figure 2.3   Typical heated-plot experimental profile of a montane meadow, Colorado, USA (Harte *et al.*, 1995).

ture treated) remains as one of the major concerns in this type of experiment (NSF, 1992).

Passive greenhouses have been extensively used to increase air and soil temperature (Shaver *et al.*, 1986; Debevec & MacLean, 1993; Kennedy, 1994; Wookey *et al.*, 1995; Chapin & Shaver, 1996). It is an inexpensive technique which allows extensive replication and multi-site experiments (e.g. ITEX – Welker *et al.*, 1995). However, passive greenhouses alter the micro-environment inside the enclosure, an effect which needs to be carefully monitored (Kennedy, 1995). Another problem associated with this technique is the restricted access for pollinators, dispersal agents, grazers, and pathogens.

Air-heated open-top and closed chambers have been recently used in a variety of experiments (J. Melillo & C. Neill, personal communication; S. Linder, personal communication). They provide very realistic warming, but they also suffer from the problems associated with enclosures. A variant of the open-top chamber is the gradient temperature tunnel (Hadley *et al.*, 1995). A large version of a closed chamber is the catchment enclosure used in the CLIMEX project, which studies ecosystem-level processes including soil and catchment hydrology (Jenkins & Wright, 1995; Lükewille & Wright, 1997). Because of the costly setup required for this kind of experiment, replication is usually insufficient or is lacking.

Infrared heaters suspended over relatively short vegetation are also a very useful technique (Harte *et al.*, 1995). Infrared radiation warms the surface of plants and soils directly but does not warm the surrounding air, which has some drawbacks for reproducing the most realistic warming of a convective nature. This technique, however, allows the study of soil processes in addition to plant community dynamics (Harte & Shaw, 1995).

Another technique that has been successfully adopted is the use of a pipe grid with inside running water that controls the temperature of the soil and the immediate air above the soil surface (Hillier *et al.*, 1994).

2.3.3 Nutrient and water interactions

The potentially large impact of global change on the hydrological and bio-geochemical cycles has prompted many manipulative experiments examining the effect of irrigation and fertilization on ecosystems. Such experiments have long been used to design management regimes for achieving optimal crop yield, and have more recently been applied to non-agricultural systems, including forests (Tamm, 1991). These experiments normally have a factorial design with treatments consisting of irrigation only, fertilization only, and both irrigation and fertilization, along with control plots. Both treatments are usually applied at several levels to determine the effect on the system response, with fertilizer applied either in liquid or solid forms. An innovative approach to study the interactive effects of CO_2, temperature, nutrients and water on forests is based on the addition of a soil-warming experiment and whole tree chambers with CO_2 and temperature control to irrigation and fertilization studies on Norway spruce at Flakaliden, Sweden (Linder, 1987, 1995).

2.3.4 UV-B

Concern over the effect of increasing UV-B radiation (the biologically most harmful component of the ultraviolet band) resulting from decreasing stratospheric ozone concentrations has led to numerous experiments studying the effects of enhanced UV-B on terrestrial ecosystems. Such experiments are often carried out in growth chambers or greenhouses, but more recently there has been an increasing number of field studies in more natural conditions (Caldwell *et al.*, 1995). Growth chambers have the advantage that other factors, such as temperature and humidity, can be controlled or manipulated. However, both UV-A and visible radiation are usually less than in sunlight.

UV-B radiation is controlled either by increasing it over ambient by use of fluorescent lamps or by filtering natural light to remove UV-B using either plastic filters or ozone gas. When natural light is filtered, the comparison is made between systems being exposed to ambient levels of UV-B and those exposed to reduced levels (Barnes *et al.*, 1995; Flint & Caldwell, 1996). Experiments are now being undertaken to examine the interactive effects of enhanced

UV-B with other global change factors. For example, a joint British–Swedish experiment is using lamps over OTCs to study the interactive effects of elevated CO_2 and UV-B on mountain heath vegetation (Caldwell *et al.*, 1995).

In any experiment it is important to achieve a realistic balance between UV-B (280–315 nm), UV-A (315–400 nm) and visible (400–700 nm) radiation. Visible radiation can have strong ameliorative effects and so experiments undertaken in chambers, which reduce the amount of ambient sunlight, need to ensure that supplement radiation in the 315–700 nm range is supplied. Another important consideration, especially in field experiments, is to adjust radiation for cloudiness and other atmospheric conditions; otherwise exaggerated UV-B effects may be found (Caldwell *et al.*, 1995).

2.3.5   Biodiversity

The growing recognition that ecosystems are not simply uniform surfaces acting as sources and sinks of fluxes (the 'green slime' concept), but that species diversity within ecosystems is a major factor influencing ecosystem functioning (see Chapter 11), has led to a novel type of global change experiment. Based on the earlier Rothamsted (UK) grassland experiment, in which various treatments were maintained over more than a century and the effects on diversity and species composition observed, Tilman (1993) initiated experiments in which species were exposed to a new competitive environment of neighbours. Naeem *et al.* (1994) extended this design to a statistical random selection of species from a suite within a natural community. The selected species were then grown in various combinations of density and species number with the aim of quantifying the effects of species numbers on ecosystem fluxes, a novel connection between ecosystem complexity and material fluxes (ecosystem functioning). Initial results show effects of species composition on drought tolerance (Tilman & Downing, 1994) and on ecosystem processes such as nitrate leaching (Tilman *et al.*, 1996). The problem of co-varying confounding factors in the experiments has, however, drawn criticisms of the interpretation of the results. For example, Huston (1997) re-evaluated some of the data and showed that it is not the biodiversity *per se* that drives the observed outputs, but rather stand biomass, density and other structural components of the vegetation which are correlated with the biodiversity treatment. In order to avoid such indirect responses, the recent BIODEPTH (Biodiversity and Ecological Processes in Terrestrial Herbaceous Ecosystems) experiment (J.H. Lawton, personal communication) is based on a statistical and repeated selection of species groups from a natural community. A more detailed account of the current understanding of the relationship between ecological complexity and ecosystem functioning is given in Chapter 11.

2.4   Observational studies

Despite the major advances and innovations in ecosystem-level experimental techniques described above, there are nevertheless significant remaining methodological difficulties and considerable expense associated with manipulative experiments. Thus, observational studies of ecosystems continue to play a crucial role in global change research.

Observational studies are a central activity in many areas of global change research, in GCTE and also in atmospheric chemistry (rates of trace gas emissions), hydrology (rates of evapotranspiration from ecosystems) and coastal zone studies (delivery of sediments and nutrients from upland areas). Technologies for observing ecosystem functioning, such as eddy correlation, aircraft transects and satellite observations for measuring the interaction between terrestrial ecosystems and the atmosphere, are advancing rapidly and have been described elsewhere (e.g. Chen & Cihlar, 1996; Hall *et al.*, 1996). This section emphasizes the interactions within the ecosystem itself, those that must be taken into account in designing and interpreting observational studies of ecosystem functioning, especially its interaction with the abiotic environment.

2.4.1   Pools and fluxes

Quantifying changes in the pools and fluxes of important elements is fundamental to understanding global change effects on the functioning of terrestrial ecosystems. Here some common problems associated with measuring ecosystem pools and fluxes are described, following which several recent innovations designed to overcome some of these problems are briefly presented.

Problems with a non-systems approach (components only)

At a global scale, the estimates of the size and location of the important biogeochemical pools in terrestrial ecosystems, and of their rates of change, are still inadequate, hence the continuing conjecture about the 'missing sink' (see Chapters 7, 10 and 12). Early global change studies emphasized only carbon and not the associated fluxes and pools of nutrients or water. Very often only parts of ecosystems were investigated (e.g. soil carbon, neglecting the humus layer and aboveground biomass). Other parameters important for ecosystem fluxes are still under debate (e.g. what is the Leaf Area Index of a *Sphagnum* bog?). Thus, any ecosystem study will have to start with an inventory of pools and fluxes that should be as complete as possible, i.e. it should reach beyond the focused objectives of a particular study in order to create a minimum dataset which is useful in a broader global change context (see also Section 2.3.1). Various approaches to statistical sampling have been developed to deal with these issues (e.g. Schulze *et al.*, 1995).

Initially it was thought that a 'black box approach', in which nothing needs to be known about the mechanisms within the box, would be sufficient to model the transfer function that connects input and output fluxes of a system (Jarvis, 1987). However, it has become clear that this is not the case if a predictive capability is required. While it is relatively easy to make predictions about the physico–chemical aspect of fluxes through boundaries, it becomes very difficult to extrapolate across the range of common perturbations, such as drought, if, for example, the process that determines the flux is a hormonal signal from roots to shoots (Schulze, 1994*d*). Thus, the ecological control of fluxes has remained an area of uncertainty because it involves biological regulation, acclimation, adaptation, changing composition of species including replacements and invasion, and interactions between trophic levels.

The most basic method of estimating the magnitude of a flux is to measure the rate of increase of the pool into which it flows, or the rate of decrease in the pool from which it originates. The interpretation of these data is simplified when other fluxes into or out of the pool can be blocked. For example, nitrogen mineralization is generally measured as the rate of increase of the inorganic nitrogen pools, following incubation of a soil sample in the absence of root uptake. This procedure, however, introduces new problems. By sieving soil and extracting roots, the natural soil density and structure are altered, which affects the process to be measured. Thus, it is not generally possible to block a component flux without altering the process under study in other ways.

New approaches to pool and flux observations

In terms of carbon fluxes, Net Primary Productivity (NPP, in most cases defined as seasonal aboveground growth, see also Chapter 7), with its associated flux of nutrients, is the main parameter to be measured in plot studies. Various methods have been developed to assess this quantity, most of them ignoring the important belowground component. The GCTE objective of developing general relations for processes which control fluxes has led to one solution to this problem; based on relations of root distribution with soil depth in different climates (Jackson *et al.*, 1996*b*), predictions of global fine root turnover can be made (Jackson *et al.*, 1997). This is an example of GCTE-initiated research in which global datasets combined with functional relations and consideration of the statistical variation have led to global predictions of parameters which previously were largely unknown.

So far biogeochemical models concentrate on assimilation and respiration of resources in the soil–plant–atmosphere system. They totally neglect the effect and function of herbivory and pests. At this stage there is insufficient detailed understanding of foodweb structures (Pimm, 1993) and their control features for fluxes (Zwölfer & Arnold-Rinehart, 1993) beyond plot scales. This, however, will be needed for predictions of the effects of global change on pests and herbivores and the consequences for global biogeochemistry (Burdon, 1993; McNaughton, 1993). General overriding relations, as for fine root turnover, are useful also in this case to identify nonlinearities and discontinuities of environmental or system-related parameters on the overall flux. Schulze (1995a) has suggested the use of the 'flux control' approach of biochemistry. In this approach ecosystem-level fluxes are related to some key state variables or species diversity, and the slope of the response curve will identify the sensitivity of the process under investigation to that driving variable. Several possible response curves emerge, e.g. in relation to biodiversity, depending on slow- and fast-growing species and the presence or emergence of keystone species effects (Bond, 1993a).

In general, the extrapolation of flux measurements of NPP over a season on a single plot to an estimate of the long-term carbon sequestration of the system, although sometimes attempted, is fraught with difficulties. In addition to the complications noted above, there are a number of processes, affecting both ecosystem structure and functioning, operating at larger space and longer time scales which make such extrapolations impossible to do accurately. A recent breakthrough in tackling this problem is the introduction of the concept of Net Biome Productivity (NBP, Schulze & Heimann, 1998), which incorporates flux terms due to disturbances such as harvest by man (e.g. in agriculture and forestry) and fire, and thus provides a biome-level framework for estimating changes in terrestrial pools and fluxes of carbon over long timeframes.

Results and interpretations of measurements of terrestrial carbon pools and fluxes are dealt with in detail in other parts of this book, particularly Chapters 7 and 10 and Sections 12.2.1 and 12.2.2.

2.4.2 Vertical flux measurements

Ecosystem studies require a quantification of the exchange of material across the boundaries, most importantly in a global change context the vertical fluxes between the land surface and the atmosphere. The study of the net exchange of gases between terrestrial ecosystems and the atmosphere has developed rapidly during the past decade, mainly due to the availability of suitable instrumentation (see special issue *Global Change Biology* Vol. 2, No 3, 1996). In this context, ecosystems are often treated as networks of parallel and serial resistors and capacitances (e.g. evaporation from soil surface versus through the plant). There are, however, limitations to the 'big leaf' model of the vegetation canopy (Raupach & Finnigan, 1988) and new models have been developed which avoid this significant simplification (De Pury & Farquhar, 1997). Such models become especially important for the interpretation of mass and isotopic fluxes between the terrestrial surface and the convective boundary layer (Lloyd *et al.*, 1996).

Despite the emphasis on integrating fluxes over an entire vegetation patch ('footprint' about 1 km^2) by micrometeorological techniques such as eddy covariance, leaf-level parameters such as the response of assimilation to leaf internal CO_2 concentration (Farquhar & Sharkey, 1982) are still required in order to simulate global responses of ecosystems to elevated atmospheric CO_2. The combination of canopy-level gas exchange measurements by eddy co-variance and leaf-level measurements has produced response curves of gas exchange and climatic variables which can be used in global models (Leuning *et al.*, 1995). These models can simulate leaf-level responses, such as stomatal closure by air humidity and associated decreases in transpiration, which were not formerly expected to even exist at the canopy level.

Measurements of canopy-level CO_2 exchange give a first estimate of the net balance between assimilatory and respiratory processes of ecosystems. They reveal whether a patch of vegetation is a net carbon source or sink at the time of measurement. Thus, while leaf-level measurements and harvests can quantify NPP, canopy-level eddy covariance measurements quantify Net Ecosystem Exchange (NEE) as a first estimate of Net Ecosystem Productivity (NEP, see also Chapters 7 and 10). Long-term measurements of NEE contain information on climate effects on vertical carbon fluxes, but they do not include major exports of carbon, such as by dissolution in surface or ground water, or by harvest and fire (see Section 2.4.1). Also, by selection of 'uniform stands' (necessary to satisfy the fetch requirements for eddy covariance measurements), canopy-level measurements have not yet dealt with gas-exchange consequences of patchy disturbance, such as tree fall by wind.

A new approach for scaling vertical fluxes from plots to regions has been developed during the past decade, namely the integration of processes at the landscape scale by using daily changes in the Convective Boundary Layer, CBL (Raupach *et al.*, 1992). The CBL can be thought of as a giant cuvette in which gases accumulate over the day, as they cannot escape through the inversion layer at top of the CBL (Lloyd *et al.*, 1996). If the remaining theoretical problems of calculating fluxes from CBL measurements are solved (M.R. Raupach, personal communication), then CBL measurements will allow a quantification of regional fluxes at scales of up to 10 to 50 km^2. Thus, the CBL–derived flux from terrestrial surfaces could include the effects of disturbances and thus would move closer to a measure of NBP (Hollinger *et al.*, 1995; Crawford *et al.*, 1996).

2.4.3   Soil processes and watersheds

Availability of nutrients is a key factor controlling the functioning of terrestrial ecosystems, and is already being strongly affected by global change (see Chapter

1). Although manipulative experiments can help determine the overall system response to changing availability of nutrients (see Sections 2.3.3 and 7.5), observational studies are important in elucidating the mechanisms that affect nutrient uptake and loss by ecosystems.

From earlier studies of the effects of air pollution, soil water solution chemistry emerged as an integrative parameter useful in quantifying nutrient availability. Although relationships determined from solution chemistry appear to hold for most cations, including the toxic effects of aluminium, they do not hold for nitrogen, a key nutrient. Only the nitrate form of nitrogen is mobile in soil water solution. The ammonium form is generally bound to soil or humus particles, and is replaced (mobilized) by a proton in an equilibrium reaction regulated by roots or mycorrhizae. This becomes even more complicated in nitrogen-limited soils, where mycorrhizae and roots of some plant species directly recover organic matter without microbial mineralization (Chapin *et al.*, 1993; Read, 1993). Thus, a low concentration of nitrogen-containing ions in soil solution may express either low availability or high turnover. This creates significant problems for quantifying nitrogen availability in a global change context. For example, through a change in the plant functional type composition of an ecosystem through land-use change, the same soil can be converted from a habitat of high resource availability and turnover (e.g. deciduous forest) into a habitat of low resource availability and turnover (e.g. coniferous forest) (Schulze & Chapin, 1987). Thus, global models that link canopy gas exchange and nitrogen nutrition are based on leaf traits rather than soil characteristics; Lloyd & Farquhar (1996) have made an initial, general attempt to quantify the linkages between soil properties and vegetation growth.

The export of carbon to soils and its immobilization there is a critical component of the terrestrial carbon cycle but remains a big unknown in most flux studies. It is only recently that rooting depth of plants and vegetation types has been reviewed on a global scale (Jackson *et al.*, 1996a,b; Canadell *et al.*, 1996a). Work reported in these reviews shows that roots can be extremely deep (more than 68 m in the Kalahari sands); the carbon turnover rates and contribution to the water and nutrient budgets of the plant by such deep roots remain unknown.

Much of our knowledge about ecosystem-level biogeochemistry comes from watershed studies (e.g. Bormann & Likens, 1979), including their experimental manipulation (e.g. Bayley & Schindler, 1991; Schindler, 1991). The desire to measure all components of the hydrological and biogeochemical balances has led to the use of watersheds as the bounded unit of an ecosystem for such flux studies. The investigation of watersheds may be the only possible way to integrate horizontal biogeochemical fluxes across land surfaces (although one should be aware that any watershed contains a gradient in ecosystems, including anoxic zones which may buffer and alter the chemistry of the water flow). In the context of global change, the study of watersheds gives insight into element pathways and losses at the plant/soil interface, which is complementary to the vertical flux measurements at the canopy scale (Hornung & Reynolds, 1995).

However, the interpretation of the input/output analysis of watershed biogeochemical studies faces the same problems as the canopy level gas exchange by eddy covariance (Hornung & Reynolds, 1995). The biological regulation inside the 'black box' defined by the catchment influences the output in an unpredictable manner. This is especially true for the nitrogen cycle. For example, the chemical environment in soil macropores differs from the environment in small-diameter soil pores. The watershed chemistry may be driven more by the rapid flow of water through macropores than through fine pores; however, the internal chemistry of the system (e.g. denitrification) may depend strongly on local anoxic zones in soil aggregates (Horn, 1993). In the context of air pollution studies, this made the definition of critical loads difficult where limits were set based on watershed results (Schulze *et al.*, 1989). After sulphur deposition terminated, forests recovered faster than expected. This again illustrates the need to know details of the processes in order to make appropriate predictions, despite the appeal of ecosystem-level flux measurements.

There are other considerations which must be taken into account when using the watershed approach for global change biogeochemical research. Although overall elemental output can be measured accurately, major problems remain with the quantification of inputs because of canopy processes which cannot be detected in the canopy throughfall. Most important is the uptake of N- and S-containing gases through the stomata and their metabolysis in the leaf. It is estimated that the amount of N taken up in the form of NO_x and NH_3 may provide about 25% to 30% of the demand for growth in Central Europe (Schulze, 1994c). Furthermore, the heterogeneity of soil often prevents reliable, direct measurements of the ground water flux at a plot scale. As recent techniques have considerably improved the reliability of canopy water flux and tree sapflow measurements, the ground water flux is often estimated by difference from precipitation and canopy water loss (Köstner *et al.*, 1992; Kelliher *et al.*, 1997).

Where an understanding of the impacts of land-use/cover change on horizontal transport processes at larger scales is important, the use of the 'catchment cascades' approach (Fig. 2.4) is a powerful tool for integrating larger regional effects which could either enhance or buffer disturbances. The catchment cascade is based on a gravity-driven chain of mass and energy flows, with the output from one subsystem serving as the input to the next. Measurements of both horizontal and vertical fluxes of important elements are measured at key points along the cascade to identify those points at which elements are lost to the

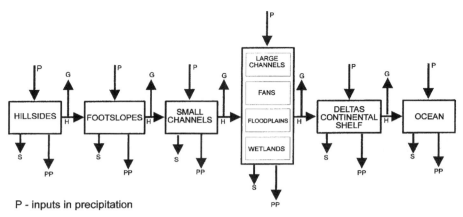

P - inputs in precipitation

S - stored in soils/sediments

PP - uptake by primary producers

G - loss as gas to atmosphere

H - horizontal flux in water

Figure 2.4 A schematic diagram of a 'catchment cascade' approach for estimating both horizontal and vertical fluxes (R.J. Wasson, personal communication).

atmosphere or stored within the catchment. The catchment cascade approach is being used in Southeast Asia to determine the impact of changes in land-use/cover in upland areas on lower lying plains, the coastal zone and the continental shelf area (Lebel Steffen, 1998).

2.4.4 Isotopes

The analysis of naturally occuring stable isotopes (e.g. ^{2}H, ^{13}C, ^{15}N and ^{18}O) is proving increasingly useful in linking ecophysiological processes, ecosystem dynamics and global ecology. The development of this technique has been driven by both the recognition that the isotopic composition of a material contains information about its source as well as the ecological processes that lead to its formation, and by technological improvements that allow samples to be measured more rapidly and accurately than before (Rundel *et al.*, 1989; Ehleringer *et al.*, 1993). Stable isotope analysis constrains uncertainties about source/sink relationships within ecosystems; it allows a separation of net mass flow and one-way fluxes (Lloyd *et al.*, 1996); and stable isotopes can be applied as tracers (e.g. Buchmann *et al.*, 1995).

Fractionation events associated with ecophysiological processes often scale to, and have significant impacts on, processes at the ecosystem level. Carbon isotope ratios and the resulting differences in discrimination by C3 and C4 canopies permit scaling of ecophysiological processes to the canopy level, which allowed Lloyd & Farquhar (1994) to estimate the global impacts of C3 and C4 ecosystems on biosphere–atmosphere gas exchange. At the ecophysiological level the carbon isotope ratio of C3 plants varies with the ratio of assimilation to leaf conductance (Farquhar *et al.*, 1989), which leads to differences in the ratio of CO_2 to water vapour exchange at the ecosystem level (Farquhar & Lloyd, 1993). More recently, CO_2 ratios have been used to separate respiration from the one-way fluxes of CO_2 into and out of canopies (Lloyd *et al.*, 1996).

The abundance of carbon and oxygen isotopes in CO_2 varies significantly on both a seasonal and a latitudinal basis, reflecting the annual dynamics of productivity and decomposition in different ecosystems (Francey & Tans, 1987). Farquhar *et al.* (1993) developed a mechanistic model showing that the biosphere should have a substantial impact in modifying the ^{18}O content of atmospheric CO_2 and that terrestrial systems differed from marine systems in their impact on the ^{18}O signal (Keeling *et al.*, 1996*b*). Ciais *et al.* (1997) combined this model with global inverse modelling approaches to confirm that terrestrial ecosystems (and not marine ecosystems) are responsible for the missing carbon sink (see Chapters 10 and 12). This has been quantified by Keeling *et al.* (1996*b*) based on isotope measurements in conjunction with long-term measurements of O_2/N_2 ratios.

Deuterium and oxygen isotopes in water are proving to be equally useful tools for understanding the water-related dynamics of ecosystems (Dawson & Ehleringer, 1991; Ehleringer & Dawson, 1992). Plants do not discriminate during water uptake, allowing patterns of water uptake and loss by different ecosystem components (Schulze *et al.*, 1996*a*) to be quantified. These approaches show that not all ecosystem components respond equally to variations in winter–summer precipitation events (Ehleringer *et al.*, 1991) and, in some cases, long-held views on the sources of water for trees have had to be modified (Dawson & Ehleringer, 1991; Thornburn *et al.*, 1992). At the ecosystem level, isotopic analysis of atmospheric water vapour allows the impact of transpirational fluxes on regional processes to be constrained (Brunnel *et al.*, 1991).

Over the past several years there has been substantial progress in isotope ratio mass spectrometer instrumentation. The capacity to analyse isotopic composition rapidly and the capability to measure small samples have reshaped sampling approaches. Carbon and nitrogen isotope ratios of organic materials are now routinely analysed by linking elemental analysers to mass spectrometers (Gebauer & Schulze, 1991). An automated approach allows 75 or more samples per day to be routinely measured. Linking continuous-flow mass spectrometry and gas chromatography permits even faster analysis of organic material and allows for high-precision analysis of smaller samples, including atmospheric and trace gases. Equally important has been the use of tracers, especially ^{15}N, to quantify ecosystem internal fluxes and identify mechanisms of turnover (Davidson *et al.*, 1992).

2.4.5 Remote sensing and geographical information systems

Remotely sensed data, particularly when manipulated in Geographic Information Systems (GIS), have given ecologists a tool to undertake spatially integrated analysis at regional scale. The regular and frequent overpass of sensors carried on satellite platforms allows consistent repeated observations of a large number of locations. At the most basic level, it has allowed the cost-effective mapping of spatial patterns over large areas. The comparison of maps for periods a decade apart provided an accurate way to determine land–cover changes (e.g. Skole & Tucker, 1993). The characteristics of the sensors has allowed them to be calibrated to important ecosystem variables, such as green leaf cover (Field *et al.*, 1993) or the detection of fires (Scholes *et al.*, 1996*b*). Chapter 5 discusses global databases in more detail.

2.5 Conclusions

Technological advances, especially in the last three decades, have provided powerful new tools for ecosystem analysis. These include:

- micrometeorological and isotopic techniques for estimating the flux of matter and energy within ecosystems, and between ecosystems and the atmosphere, geosphere and biosphere;
- techniques for altering selected environmental factors such as CO_2 and temperature at the plot scale;
- remote sensing to provide detailed spatial and temporal data over large areas;
- statistical and numerical modelling techniques capable of analysing multivariate, nonlinear problems;

The value of the information which can be obtained from studies on small plots can be greatly enhanced by:

- subjecting the whole plot to manipulations, such as elevated CO_2, fertilizer, irrigation or warming, especially if response surfaces and not only factorial combinations are obtained;
- consistent observation over long periods of time;
- comparison of similar experimental treatments in different parts of the world (transects, see Chapter 4);
- using 'natural experiments' provided by gradients of environmental drivers.

A number of parameters and processes remain uncertain and will need further investigations for their quantification. The most important of these include:

- fluxes in the soil;
- a quantification of resource availability for the vegetation cover.

3 Networks and consortia

J.S.I. Ingram, J. Canadell, T. Elliott, L.A. Hunt, S. Linder,
D. Murdiyarso, M. Stafford Smith and C. Valentin

3.1   Introduction

The successful implementation of GCTE's (Global Change and Terrestrial
Ecosystems) large and complex programme can only be accomplished by
involving many members of the global research community and by bringing
together the enormous amount of pertinent national, regional and international
research. This chapter describes how GCTE is achieving this integration by
using networking as a fundamental research strategy, and how the networks and
consortia are established and managed. Much of the science achieved in this way
is reported elsewhere in this volume, but examples are given that are relevant to
the various types of network now established.

3.2   Advancing science by networking

Research programmes in terrestrial ecosystems science are increasingly dependent on national, regional or global collaboration. A common mechanism adopted
by many groups to achieve this collaboration has been the use of networks, i.e.
groups of researchers linked or brought together to exchange ideas and to
undertake joint research. If well designed, this allows two major benefits to be
derived: First, networks allow for research that could not be accomplished
independently (e.g. experiments where many individual sites or studies are
required, enabling the generation of results greater than the sum of the parts.
Second, 'added value' of a financial nature can be achieved, as a result of sharing
data, models and facilities, all three of which can be expensive to obtain,
develop, and build and maintain.

Experience with many networks has identified a number of requirements for
success:

■ The network's goals must be clearly stated and scientifically exciting in order
   to engage good scientists.
■ The individuals in the network must benefit in addition to the group as a

whole; the initial enthusiasm and momentum of the network will only be maintained if the individual researchers continue to benefit from their active participation.

■ It must be clear what benefits (both for the individual and the network) will be gained, and also what the responsibilities are (see Box 3.1); a wide spectrum of network approaches means that interpretation of the concept may vary.

■ Success and momentum are maximized when there is a sense of 'ownership' of the network by the members. This can be achieved by jointly designing the network, and by maximizing participation in synthesis workshops, publications, etc. Although the network may be centrally administered, it cannot have 'top down' management. Rather, the network members need to decide collectively what should next be addressed.

Box 3.1   Benefits and responsibilities of participation in GCTE Core Research

To achieve its objectives, GCTE is developing a coordinated research programme built around 45 Tasks, each addressing specific scientific questions within GCTE's broad objectives. Each Task consists of a limited number of complementary research projects coordinated by a Task Leader(s), which collectively meet the Task's objectives. Within each Task a variety of mechanisms, such as common experimental protocols, standardized methodologies, model comparisons, and integrating workshops and symposia, are used to knit the contributing projects together into a coordinated effort.

Although GCTE aims for the highest scientific quality in its portfolio of contributing projects, it does not carry out full peer review evaluations. That process is undertaken in the normal way by the grant-giving body that funds the project. GCTE works with many of these granting agencies to ensure that the highest standards of scientific quality are maintained in projects accepted in the GCTE Core Research Programme, and to maximize the complementarity of the agency's scientific objectives with those of GCTE and other IGBP Core Projects.

Participation in the GCTE Core Research Programme has a number of significant benefits for component projects. The Programme:

■ facilitates the planning of research projects addressing aspects of global change and terrestrial ecosystems by providing a soundly based intellectual and organizational framework for research, with overall aims, approach and implementation developed and endorsed by the international science community;

■ adds to the scientific value of individual experimental, observational and modelling studies and assists in their interpretation by organizing networks and consortia which widen the range of observational studies and extend their temporal and spatial coverage, promote common methodologies and protocols, and provide datasets for model validation and intercomparison;

- promotes the rapid communication of scientific ideas and results through meetings and publications, and by facilitating disciplinary and interdisciplinary liaison at the international level between individuals and research groups;
- provides assistance to contributing projects in obtaining funds from national and international sources by writing letters of support and lobbying, where appropriate;
- assists in the cost-effective deployment of major capital equipment and facilitates such as FACE by providing the scientific framework for their phased deployment and assisting in their collaborative use;
- encourages the full involvement of developing countries through GCTE participation in the START regional research networks;
- promotes close working links with other relevant international programmes and studies, particularly those of other IGBP Core Projects and Framework Activities, and of IHDP, WCRP and other international research organizations;
- promotes the concepts of GCTE and IGBP science, and the results obtained from the GCTE Core Research Programme, to ensure their wider recognition among the international research and policy communities.

Participation in the GCTE Core Research Programme requires a commitment to:

- participate in the relevant GCTE Task(s) for which the project was accepted through activities such as implementation and synthesis workshops, model intercomparisons, and joint observational studies;
- carry out the project in accordance with the relevant aspects of the GCTE Operational Plan using agreed methods and protocols wherever possible;
- make data and models available to the wider GCTE community, in accordance with protocols for data and model exchange developed by GCTE networks and consortia, and with due regard to publication 'rights';
- keep the GCTE Core Project Office informed on an annual basis of (i) major changes to the project objectives, description, and participating research organization and changes in the annual budget and the major funding agencies; and (ii) changes to the number of scientific and technical staff working on the project, and provide the Core Project Office a list of publications arising from the project;
- acknowledge participation in the GCTE Core Research Programme in publications arising from the project by inserting in the Acknowledgements the sentence 'This work contributes to the Global Change and Terrestrial Ecosystems (GCTE) Core Project of the International Geosphere–Biosphere Programme (IGBP)'.

The establishment and running of successful collaborative research networks requires two essential inputs: a commitment on the part of the network members to help keep the network 'alive' through active involvement; and sufficient resources (in both funds and skilled time) to administer the network's establishment and operation.

3.3   GCTE networks

GCTE is *itself* a network, established by IGBP (International Geosphere –
Biosphere Programme), with clearly defined (if broad) goals. The design and
refinement of the GCTE agenda (Steffen *et al.*, 1992) through a series of
international planning workshops was wholly dependent on discussions among
a very large number of scientists from around the world, a process which
resulted in GCTE's internationally agreed-upon research agenda. GCTE keeps
this group informed of developments and progress by publishing a newsletter
(now mailed to over 1800 recipients worldwide), and has established a home
page on the worldwide web (http://jasper/stanford.edu/GCTE/). The number of scientists becoming actively involved in GCTE research is rising sharply,
often as a result of participation in one of the networks described below.

There are three specific scientific reasons why the network approach is
particularly valuable for GCTE: First, much of the GCTE research agenda
addresses global-scale issues (e.g. increasing CO_2 and mean global temperature),
or more site-specific issues which are so widely manifest as to be considered a
global phenomenon (e.g. soil erosion); in both cases a wide geographic spread is
central to the research, especially when responses across a range of temperature
and precipitation are needed to represent anticipated climate change. Second,
many of the research themes, by their very nature, require a multi-disciplinary
approach. Third, since GCTE aims to develop quantitative tools, it is essential
that experimentalists and 'tool developers' (e.g. modellers) are brought together, preferably also with potential users (e.g. the policy and resource management communities).

GCTE networks (Table 3.1) vary considerably in their objectives and degree
of formality. Some aim to compare and synthesize information. Others, in
addition, have more infrastructure for cataloguing and collaboratively designing
research, and require members to use agreed, standard experimental protocols
and data formats, or exchange data and model code. Those of the first type have
been termed 'consortia' rather than networks, but this mainly reflects history
rather than differences in operation.

Working within the general area identified by the early planning workshops,
inaugural network meetings refine the individual components and determine
the specific objectives for given networks. Common experimental protocols (if
appropriate) are agreed upon and a time-scale for network activities established.
A network Steering or Working Group is usually appointed, comprising
specialists from a range of relevant disciplines, and networks are typically
launched with a small number of contributing projects and then grow as more
groups worldwide become involved.

Networks initiated by Focus 3 ('Global Change Impact on Agriculture,

Forestry and Soils') often start with the collation of existing experimental datasets and models, and the collation of 'metadata' (data describing a piece of research, as distinct from the data or model code itself). This initiation serves more than just the establishment of network membership; it helps determine what GCTE-type studies are ongoing, where coordinating effort should be placed, and where new projects should be initiated. Questionnaires soliciting metadata are widely distributed, together with an invitation to apply to join the network. All applications are reviewed by an international expert panel (Network Working Group), and the conditions of acceptance are twofold: the quality of science and the pertinence to network objectives. Experiments contributing to the network must have sufficiently detailed measurements to enable the models to be evaluated and developed, and models must be mechanistic in design. The evolution of a typical GCTE network, that for Soil Erosion, is described in Box 3.2.

GCTE research networks fall into three categories: (i) networks of sites, where the research depends wholly or substantially on a geographical spread of experiments; (ii) thematic networks, where work is linked according to the topic being researched; and (iii) regional networks of researchers undertaking studies of the impacts of global change on terrestrial ecosystems. There are, however, cases where studies fall under more than one category.

3.3.1 Site networks

The GCTE Soil Organic Matter Network (SOMNET, part of Focus 3) includes a worldwide set of long-term soil organic matter management experiments. The datasets from these are an essential resource for the development of the soil organic matter (SOM) turnover models (also included in SOMNET) which aim to simulate SOM dynamics over decades to centuries. Many suitable experiments have been identified so far (mostly in the US, Europe and Australia) and other sites – particularly in the tropics – are being sought. To be included in SOMNET, the experiments, some of which are over one hundred years old, must provide datasets suitable for model evaluation.

Another network, about to be launched, is the Long-term Agroecological Experiments Network. Experiments will be included where two or more species have been grown together for at least 10 years. Data from these will be used to evaluate the role of planned diversity in sustainable production. Several of these long-term agroecological experiments may also prove valuable for SOM studies and links between the two networks will be promoted.

The role of networks of sites is further discussed in Chapter 4.

Table 3.1 *GCTE networks and consortia*

Title	Objective(s)	No. of Members	No. of Countries	No. of Models	No. of Datasets
Site Networks					
SOMNET	To use data from long–term experiments to evaluate and develop Soil Organic Matter (SOM) turnover models	40	16	22	34
Thematic networks					
Elevated CO_2 Consortium	To foster, coordinate and integrate research on ecosystem response to increasing atmospheric $[CO_2]$, with main emphasis on whole–ecosystem experiments in which other environmental factors, biotic interactions and long–term feedbacks are accounted for	27	15	n/a	n/a
DGVM consortium	To develop and improve Dynamic Global Vegetation Models (DGVMs) by performing intercomparisons using standardized input data	5	4	4	1
Wheat Network	To refine and adapt current wheat production models for use in global change studies in a wide variety of conditions; and to design and undertake experiments to provide improved mechanistic understanding of global change impact on wheat production, to aid in model development	39	17	20	53
Rice Network	To develop, improve and evaluate models of the rice crop and agroecosystem; to conduct experiments to test scientific hypotheses and to provide data that contribute to the development and evaluation of rice crop models; and to foster improved experimental protocols that support both scientific hypotheses and crop modelling needs	12	8	7	8

Network	Objectives				
Potato Network	To refine and adapt current potato production models for use in global change studies in a wide variety of conditions; and to design and undertake experiments to provide improved mechanistic understanding of global change impact on potato production, to aid in model development	10	6	5	11
Pastures and Rangelands Network (CRP1)	To predict the effects of global change on pasture and livestock production at the paddock scale	16	10	6	15
Experimental Forest Network	To design and undertake experiments, where long-term feedbacks are accounted for, to provide improved mechanistic understanding of global change impact on structure and function of managed forests; and to use results from such experiments to develop, improve and evaluate models for use in global change studies in a wide range of managed forests	8	7	4	n/a
Soil Erosion Network	To refine and adapt current soil erosion models for use in global change studies in a wide variety of conditions; and to design and undertake experiments to provide improved mechanistic understanding of the relationships between global change and soil erosion, to aid model development	24	10	9	16
Impacts Networks					
Impacts Centre, Southeast Asia	To build the capacity of Southeast Asian scientists to undertake global change studies; to conduct policy-driven studies; and to provide an advisory network with the policy community	n/a	7	n/a	n/a
Information Networks					
Pastures and Rangelands	To disseminate GCTE findings more broadly to the scientific community and to advise appropriate end users during the course of the research activities	312	48	n/a	n/a

Box 3.2 Key stages in the development of the *GCTE Soil Erosion Network* (part of GCTE Task 3.3.2 'Soil Degradation')

Date	Event	Outcomes/Product	Ongoing activities commenced
Jan 1991	GCTE Open Meeting (Brighton, UK): GCTE general planning	Draft GCTE Operational Plan	
Sept 1991	Focus 3 Open Workshop (Lunteren, NL): Focus 3 general planning/scoping exercise	Draft Focus 3 Operational Plan	
Mar 1993	Internal planning meeting (Oxford, UK)	Task 3.3.2 (Soil Degradation) Inaugural Workshop designed	Inaugural Workshop planning
Mar 1994	Inaugural Workshop (Paris, France)	Task's major objectives agreed; major erosion models and monitoring projects identified; metadatabase structure developed; Task Working Group convened	Metadatabase structures developed for models, manipulative experiments and monitoring programmes; invitation to join network widely distributed; follow–up workshop preparation
	Publication of Paris Workshop Report (GCTE Working Document 11, Ingram, 1994)		
Feb 1995	Network Planning Workshop (Corvallis, US). Task Working Group meeting	Model sensitivity analysis planned for water erosion at plot scale. 4 Core Research Projects (CRPs) planned; criteria for network membership established; inaugural membership agreed; Special Issue of *J. Soil & Water Conservation* designed	Preparation of inaugural members' metadata for submission to GCTE Scientific Steering Committee for ratification as GCTE Core Research; plot scale model sensitivity analysis workshop planning; JSWC Sp. Issue editing

Date			
Apr 1995	GCTE Scientific Steering Committee meeting (Bariloche, Argentina)	Soil Erosion Network (GCTE-SEN) formally recognised as GCTE 'Core Research'	
Sept 1995	Network Workshop: 'Global Change: Modelling Soil Erosion by Water' (with NATO; Oxford, UK). Task Working Group meeting	Results of model sensitivity analysis reported and outlines of NATO ASI book chapters produced. Model sensitivity analysis planned for water erosion at catchment scale	Editing NATO ASI book chapters; catchment scale model sensitivity analysis workshop preparation
Sep 1995	Publication of revised Focus 3 Operational Plan including revisions to Task		
Jan 1996	Network Workshop: 'SALT' Experimentation Planning (Senegal, Burkina Faso & Niger). Task Working Group meeting	Field collaboration planned along SALT transect (see Chapter 4)	
Apr 1996	Publication of Soils Activity (Activity 3.3) Implementation Plan (GCTE Report 12)		Editing metadata. Distribution of A3.3 Implementation Plan to current and prospective members of GCTE-SEN; selected projects invited to contribute to GCTE-SEN; detailed planning for Utrecht workshop
Oct 1996	Publication of 5 papers as Sp. Issue of *Journal of Soil and Water Conservation*, **51**		Collaborative work on water erosion at field-scale started in Senegal

Box 3.2 (*cont.*)

Date	Event	Outcomes/Product	Ongoing activities commenced
Dec 1996	Publication of metadata for models, experiments and modelling programmes (GCTE Report 6)		
Early 1997	Publication of Oxford NATO workshop book in NATO-ASI Series		
Apr 1997	Network Workshop: 'Modelling water erosion at catchment scale' (Utrecht, NL)	Series of papers for publication in book or journal special issue anticipated	Editing Utrecht papers Detailed planning of GCTE presentations at Wind Erosion Congress, GCTE-LUCC Open Science Conference, Activity 3.3 Session at ISSS Congress, and Focus 3 Science Conference
Nov 1997	Network Workshop: 'Erosion Thresholds' (Almeria, Spain)	Erosion thresholds Core Research Project (CPR2) launched. Detailed plans for GCTE Erosion Network input to GCTE-LUCC Open Science Conference and GCTE Special Session at ISSS Congress; and proposals for input to Focus 3 Conference	
Mar 1998	GCTE-LUCC Open Science Conference (Barcelona, Spain)		Focus 3 session to include reporting of Soils Activity results

Aug 1998 planned	ISSS Congress: GCTE Activity 3.3 Session (Montpellier, France)	Series of papers as Journal Special Issue	Editing ISSS Special Session papers
Sept 1999 planned	Focus 3 Science Conference (Reading, UK)	To include a full review and synthesis of GCTE soil erosion research	Editing Focus 3 Conference papers

3.3.2   Thematic networks

GCTE has launched several thematic networks and consortia where the emphasis is on linking similar types of research or research addressing a similar issue. Geographical spread, although common, is not always a primary objective.

The 'Elevated CO_2 Consortium' (Fig. 3.1), launched by Focus 1 'Ecosystem Physiology' in 1992, has gained considerable momentum, and a consortium of developers of dynamic global vegetation models (DGVM) has been established by Focus 2 'Ecosystem Structure and Function'. The work of these two consortia is further described in Chapters 7 and 8, respectively.

Many of GCTE's thematic networks have model comparison as a central activity. Many of the models involved in these exercises have previously been well evaluated for the conditions for which they were developed. To use them with confidence for global change studies, however, they need to be tested for robustness under conditions outside the environmental 'envelope' in which they were designed. This analysis requires a thorough and systematic evaluation using data from contrasting experimental sites. Several GCTE networks, especially those in Focus 3, are designed to deal specifically, or in part, with this issue.

GCTE is implementing much of its research on agriculture through a series of Crop Networks. Networks for wheat, rice and potato are now well established (see Table 3.1), and are formally recognized as GCTE Core Research. Similar networks for cassava, sorghum, maize and groundnut are being planned.

The Crop Networks have linked objectives:

- To refine and adapt current crop production models for use in global change studies in a wide variety of conditions.
- To design and undertake experiments to provide improved mechanistic understanding of global change impact on crop production, to aid in model development.

An initial goal is the identification of models, or modules, for application to field experimental programmes. A major aim is then to determine (by model comparisons with contrasting datasets) the robustness of well-established and new models under conditions of environmental change, and to provide a guide to the application of crop models under global change within national programmes. Model evaluation is an important step in this process, but it is not intended to be a competition. Rather, it is designed to stimulate the further development of generic crop models for global change studies by identifying strengths and weaknesses in existing models (see Chapter 9 for a discussion of results from the GCTE Wheat and Rice Networks). A further aim is to strengthen model links to 'yield reducing' factors (see Box 9.1, Chapter 9); the

Crop Networks will further integrate pest, disease, weed and soil components with the current simulation models for major crops, to allow better predictive capabilities for whole agricultural systems for changed environments.

Models have often been developed using experimental data from closed and open top chamber experiments, but will eventually be tested against datasets generated by the Free-Air CO_2 Enrichment (FACE) (and other field-scale) experiments (see Chapter 2). In addition to global change experiments, which provide ideal datasets for the analysis of model responses to, for example, elevated CO_2, many experiments with no formal global change dimension are valuable for analysing the effects of contrasting photoperiod, soils or irrigation.

GCTE's managed forest programme, also part of Focus 3, has launched a research network combining elements of both the 'site' and 'thematic' approaches. Unique within GCTE, this network has implemented detailed common experimental protocols for investigating interactions of elevated CO_2 and nutrients. This will greatly facilitate the synthesis of results across the network, while quantifying the specific environmental factors contributing to site results. Initial network sites are in boreal and temperate systems.

3.3.3 Regional networks for impact studies

There is a strong and growing need for the developing world to strengthen links both within major regions (between scientists and with the policy community), and between regional scientists and global change scientists worldwide. In response to this a series of regional networks is being developed dealing specifically with impacts of global change. These are established in close collaboration with the START (Global Change SysTem for Analysis, Research and Training) programme, and are termed Impacts Centres. The philosophy of these Centres is to conduct policy-driven studies and to provide an advisory network to the policy community. A regional approach helps identify problems common to neighbouring countries, therefore allowing for the more efficient sharing of resources and application of results.

GCTE Impacts Centres aim to assist regional scientists build their own capacity to analyse, interpret and predict the impacts of global change on terrestrial systems, including agriculture, production forests, natural ecosystems and nature reserves; and to promote planning for sustainable development and biodiversity conservation. These aims will be achieved by offering training courses, research fellowships, study visits and small equipment grants; by undertaking collaborative impact analyses with appropriate groups in the region; and by providing expert advice to the policy community and resource managers. The Impacts Centres will also facilitate regional involvement in GCTE networks, and in applying strategic research results to regional issues.

The first of the Impacts Centres was launched in 1995 with a grant from the

Australian Government. Located in Bogor, Indonesia, to serve Southeast Asia, it is designed to be self-supporting in the long-term by obtaining grants and contracts through regional partnerships. Other Centres are planned for South America, Southern Africa and South Asia.

3.3.4   Information networks

To complement the research networks described above, a fourth type of network has also been developed. 'Information networks' link together a wide group of researchers and appropriate end users. They provide a mechanism regularly to distribute information on GCTE findings and activities for application to resource management and for guiding new research directions and efforts; they do not involve a specific research programme. An example is the GCTE Pastures and Rangelands Information Network, which has over 300 members worldwide. These include users in the pastoral industries and scientists currently active in GCTE research. Importantly, it also includes groups or individuals in different regions that may not currently have active global change research programmes, but which are considering developing such programmes based on information obtained through the Information Network.

3.4   Standardizing networking 'tools'

3.4.1   Experimental protocols

Agreeing upon common experimental protocols is notoriously difficult; persuading laboratories to change well-established methods to a new standard, finding methods suitable across widely ranging conditions and different levels of training, equipping and infrastructure are all commonly cited reasons. Nevertheless, progress is often frustrated until common protocols are established. Many of GCTE's Tasks, networks and core research projects (CRPs) are tackling specific standardization issues, usually by convening international working groups.

3.4.2   Collation of metadata

Collations of metadata (metadatabases) are an efficient way of determining and communicating what data and models are available for synthesis, integration and analysis. They also serve well in enabling researchers both to 'advertise' their work, and to identify suitable opportunities for bi- and multi-lateral collaboration, without compromising their intellectual property; having little intrinsic intellectual content, metadata can be made widely – and freely – available.

GCTE has invested considerable effort in developing a series of meta-

databases because information presented clearly in a systematic way is quick and
efficient to search, and allows a good overview of a given piece of research,
whether model or experiment. By asking specific questions, the metadatabase
design also shows potential network applicants what types of research are of
interest to the network, and a 'worked example' of completed metadata indicates
the level of detail requested. GCTE compilations of metadata are published in
the GCTE Report Series (Table 3.2) and on the GCTE worldwide web site.
They are often unique references for global change research on selected topics.

3.4.3 Data requirements and the advantage of a standard file structure

One of the primary objectives of the GCTE networks is to compare, refine
and evaluate simulation models for use as tools in global change studies. This
requires that, firstly, models work well and, secondly, suitable data exists to
permit their application. To determine how well models work, data are initially
needed to evaluate them. An understanding of the quality of the data is crucial,
and close cooperation between modellers and experimentalists is required to
establish this.

Model evaluation requires a balanced coverage of datasets representing a
range of environments. Datasets need to cover all normal and, where possible,
extreme growing conditions. Crops datasets, for example, should cover a range
of cultivars and management practices (including pest, disease and weed control), and should include data on phenology, yield (and other plant components)
and total biomass, leaf area index (LAI) development and disease, pest and weed
impact. To obtain quality data, their variability must be recognized (i.e. an
adequate number of samples must be taken), as well as their consistency (e.g.
whether there are missing values, or varied ways of measuring parameters were
used), one of the main problems inherent in datasets.

Model application requires the selection of a range of representative sites
and/or regions with known cultivar characteristics and crop management. Data
on regional yields and meteorology are required to evaluate regional applications.

GCTE has made rapid progress in developing standardized protocols for the
collection and dissemination of data, most notably from agricultural experiments. The gathering of new data of a suitable type can be costly so it is essential
to use such comprehensive datasets as widely and efficiently as possible. Additionally, there are existing datasets that are inadequate for model construction
but which can provide useful datapoints for validation across multiple environments and conditions. Access to both types is facilitated by adopting a standard
format for data collation, and, for crops, an example is well documented by the
International Benchmark Sites Network for Agrotechnological Transfer

Table 3.2 *Major products from GCTE networks and consortia: books and journal special and thematic issues*

Ballantine, S. (ed.) (1996). GCTE Soil Erosion Network papers. *Journal of Soil and Water Conservation*, **51**(5) (Special Issue; 5 papers.)

Boardman, J. and Favis-Mortlock, D.T. (eds.) (1997). *Modelling Erosion by Water*. Springer-Verlag NATO-ASI Global Change Series, Heidelberg.

Cramer, W. and Woodward, F.I. (eds.) 1996. A Global Key of Plant Functional Types (PFT) for Modelling Ecosystems Responses to Global Change. *Journal of Vegetation Science*, **7**(3). (Special Issue; 14 papers.)

Hirose, T. and Walker, B.H. (eds.) (1996). Global Change and Terrestrial Ecosystems in Monsoon Asia. *Vegetatio*, **121**(1). (Special Issue; 16 papers.)

Koch, G.W. and Mooney, H.A. (eds.) (1996). *Carbon Dioxide and Terrestrial Ecosystems*. Academic Press, San Diego. 443 pp.

Körner, Ch. and Bazzaz, F. (eds.) 1996. *Community, Population and Evolutionary Responses to Elevated CO₂*. Academic Press, San Diego. 465 pp.

Nösberger, J. and Campbell, B.D. (eds.) (1997). Interactions between elevated CO_2 and water supply in grasslands. *Global Change Biology* **3**(3). (COST-GCTE Thematic Set; 7 papers.)

Powlson, D.S., Smith, P. and Smith, J.U. (eds.) (1996). *Evaluation of Soil Organic Matter Models*. Springer-Verlag, Berlin. 429 pp.

Schulze, E.-D. and Mooney, H.A. (eds.) (1993). Design and execution of experiments on CO_2 enrichment. *Ecosystems Research Report* 6, European Commission, Brussels. 420 pp.

Luo, Y. and Mooney, H.A. (1999). *Carbon Dioxide and Environmental Stress* Physiological Ecology Series, Academic Press (*In press*).

Smith, P., Powlson, D.S. Smith, J.U. and Elliott, T. (eds.) (1997). Evaluation and comparison of soil organic matter models using datasets from seven long-term experiments. *Geoderma* **XX** (Special Issue; 12 papers.)

Smith, T.M. (ed.) (1996). The application of patch models of vegetation dynamics to global change issues. *Climatic Change* **34**(2). (Special Issue; 15 papers.)

Smith, T.M. Shugart, H.H. and Woodward, F.I. (eds.) (1997). *Towards the Development of a Functional Classification of Plants*. IGBP Book Series No. 1, Cambridge University Press. 369 pp.

Steffen, W.L., Walker, B.H., Ingram, J.S.I. and Koch, G.W. (eds.) (1992). Global change and terrestrial ecosystems: the operational plan. *IGBP Report*, **21**. IGBP, Stockholm. 95 pp.

Stott, P. (1995). Global Change Interactions with Terrestrial Ecosystems. *Journal of Biogeography*, **22**(2–5) (Special Issue; 88 papers.)

Tinker, P.B., Gregory, P.J., Canadell, J. and Ingram, J.S.I. (eds). (1996). Plant–soil carbon belowground: the effects of elevated CO_2. *Plant and Soil*, **187** (Special Issue; 19 papers.)

Walker, B.H., and Steffen, W.L., (eds). (1996). *Global Change and Terrestrial Ecosystems*. IGBP Book Series No. 2. Cambridge University Press. 619 pp.

Walker, B.H., Steffen, W.L., Canadell, J. and Ingram, J.S.I. (eds). (1999). *The Terrestrial Biosphere and Global Change: Implications for Natural and Managed Ecosystems. A Synthesis of GCTE and Related Research*. IGBP Book Series No. 4. Cambridge University Press. 432 pp.

(IBSNAT) and GCTE (Box 3.3). The International Consortium for Agricultural Systems Applications (ICASA), in collaboration with GCTE, is currently refining these data standards.

Box 3.3   Standard file formats

Standard datafiles should be machine-readable and use only ASCII characters, which are easily understood, transferred and edited. Datasets should be self-contained and include comments on quality, authorship and variable names. A suitable data structure is: *dataset title; !comment; @header with variable names/ abbreviations; and with 'space' used for delimiters.

A worked example is thus:

```
*EXP.DATA(A) : WABO8201 NITROGEN
!EXAMPLE FOR GCTE
@TRNO ADAT CWAM
1 83172 14000
2 83174 14500
3 83180 15000
```

The objective of designing standard files is to simplify the transfer, storage and multiple use of datasets. It is not an attempt to constrain data reporting, but merely to avoid a large number of conversion programmes being required. A few file-naming conventions help to promote universality, but it is important to maintain scope for local names. Standard files need defined codes, but also need scope to add new codes and to define them within the file, which applies both for variable names and for inputs. The structure of input and output data for the models also needs to be standardized for ease of model comparison and interchange of model components, and the IBSNAT input format (Hunt *et al.*, 1994) is proving successful. It has been formally adopted by the Crop Networks, and it may also be of value to other networks.

Network members are encouraged to convert their datasets into the GCTE standard, and GCTE offers advice and some assistance in doing this. The goal is to have all network datasets thus converted to maximize the ease of transferability and model validation. Significant progress has been made for the Wheat Network, with 31 datasets now converted and stored in the standard format.

3.5   Protection of intellectual property rights

The emphasis placed on networking, and hence on sharing and exchanging data and model code, has generated much debate on the protection of intellectual property rights. GCTE has addressed this by drafting its own Data and Model

Box 3.4   GCTE Data and Model Sharing Policy

Many of GCTE's Tasks involve experimental and modelling networks that call for the sharing of data and models among network members. The 'spirit' of GCTE is to share models and data, as far as institutional or other constraints will allow. The following guidelines clarify GCTE's policy regarding ownership and sharing of data and models.

1. Property Rights. All data and models (computer code) remain the property of the research worker(s) who obtained it. No network member or coordination centre shall pass data or models on to a third party outside the network without the owner's expressed permission. The data and models shall not be used in GCTE publications, other than with the permission of the research worker(s) concerned.

2. Use within Network. There is often a long lag time between the collection of data or the evolution of a model or model component to a 'stable' state and its publication in the scientific literature. The GCTE networks aim to reduce this lag by facilitating the exchange of data and models prior to publication. GCTE will ensure that such exchanges preserve the originator's right to legitimate first use of the data and model, their right to know to whom within the network the information has been distributed, and the recognition of priority. Members of GCTE networks are obliged to observe these rights, and to release models and data to other current members of the network and to the coordinating centre under these conditions.

3. Publications. Each researcher is encouraged to publish his/her work in the normal way, and is requested to state in the acknowledgements that the work is a contribution to the GCTE Core Research Programme. Reprints of publications will be gratefully received by the GCTE Core Project Office and the relevant Focus Office.

4. Synthesis and Reviews. One of the aims of GCTE is to conduct syntheses of the data gathered from sites within a network. Once synthesised, data from specific sites will lose their individuality and become part of the GCTE regional, and ultimately global, picture. Those conducting a synthesis exercise on behalf of GCTE may feel that their work merits publication in its own right, and they should be encouraged to publish in an appropriate journal. Any such paper must include suitable acknowledgement of contributing research workers, and state that it is part of the GCTE Core Research Programme. Should published data be used in the synthesis, references must be made in the normal style of a review.

Any contributor whose data are used in a significant way and whose data collection is wholly or largely intended for the network is entitled to become a co-author. It must, however, be recognized that there is a level of participation (or data use) below which addition of a further co-author would be inequitable to major authors. In the unlikely event that concern regarding inclusion or not of a given co-author becomes an issue, the Chairman of GCTE will confer with those concerned and act as arbitrator.

These guidelines are not legally enforceable, but it is expected that members of GCTE networks will abide by them.

Sharing Policy (Box 3.4), in which the key issues are presented. A condition for membership in many of the GCTE networks is that the project's principal investigator signs an undertaking to abide by the terms of the policy.

The agreement by all members of a given network to share data and code, working under the 'protection' of the terms of the policy, has both advanced scientific progress and helped GCTE knit the wide-ranging communities into coherent and effective collaborative partnerships.

3.6   Networking benefits and products

From its inception GCTE has benefited from improved communication technologies and the establishment of scientific collaborative links between researchers worldwide. This is a major factor in GCTE's gathering momentum; over 400 scientists attended the First GCTE Science Conference held in Woods Hole, USA in 1994 (Stott, 1995; Walker & Steffen, 1996). Many new collaborative links have developed (usually between scientists from different countries – and often between different disciplines), and the strong links built between wide-ranging groups of individuals and teams have resulted in effective international partnerships to undertake global change studies.

The individual GCTE networks are also proving very effective, and have gained considerable scientific momentum. Once the research goals were established, formal networks have helped to identify, collate, organize and undertake research. An analysis of a network's contributing projects shows not only where the strengths of the network lie but also where thematic and geographic gaps exist. Networking thus provides a systematic mechanism for reviewing ongoing research and planning the next generation of experiments and model exercises.

Networking also provides a mechanism for including projects which, in themselves, might not form part of GCTE's Core Research Programme (being, for example, too site-specific). For instance, the Elevated CO_2 Consortium has identified and integrated results from many ecosystem-level CO_2 research projects from a worldwide range of natural ecosystems, and several detailed syntheses of the effects of increasing atmospheric CO_2 have now been conducted (Fig. 3.1). Similarly, the rapid progress witnessed in the highly complex issue of DGVM development (see Chapter 8) has been substantially due to GCTE bringing together the world's leading DGVM groups, particularly for model comparison exercises. A third example is the assemblages of worldwide datasets for crop, forage and tree growth, increasingly in standard format, and the rigorous and candid examination and development of models ranging from single-point, single-season crop growth models to global, long-term simulations

Past and Planned GCTE Sponsored and Co-sponsored Workshops in the area of
Ecosystem Responses to Elevated CO_2 (1992–1999)

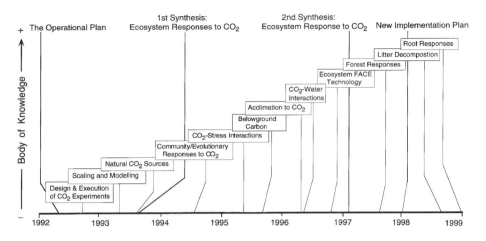

Figure 3.1   Elevated CO_2 consortium 'Time-line'.

of forest stand dynamics; these would not have been possible without GCTE
networking.

Networks of relatively small research undertakings can amount to sizeable
financial investments, particularly where the collation of long-term datasets are
concerned. For example, the estimated 'value' of the field data (i.e. the cost of
undertaking all the experimentation) included in GCTE SOMNET has been
calculated (Smith *et al.*, 1996*a*). By assuming that each long-term experiment at
Rothamsted Experimental Station, UK cost US$12 500 per year to run in 1989
(a conservative estimate), then multiplying this figure by the number of years
the experiment has run, an estimate of the 'value' of that experiment's data can
be derived. Summing similarly derived figures for all the experiments in
SOMNET yields the figure of US$35.6 million (Smith *et al.*, 1996*b*).

GCTE networks have already produced several interim products. Networks'
results are often rapidly published and disseminated in the GCTE Working
Document Series, or lead to multi-authored papers in the peer-reviewed literature. In addition to these numerous multi-authored papers, several books and
journal special issues have been published as a direct result of GCTE network
activities (Table 3.2). Network activities have also produced reports and implementation plans which are made widely and freely available in the IGBP and
GCTE Report Series, and as GCTE working documents. Finally, networks
have also assisted in developing and publicizing proposals for synthesis, experimentation and model development. This has furthered international collabor-

ation and has strengthened interdisciplinary links between ecologists, soil scientists, crop physiologists, pest, disease and weed scientists and agronomists.

The progress reported above and elsewhere in this volume has not, however, been without cost; much discussion and significant amounts of very limited GCTE resources and staff time have gone into network design, membership establishment and arranging and raising funds for model comparison and synthesis workshops. In addition to these costs, networks run the risk of loading an administrative burden upon the members. GCTE has therefore sought to minimize administrative effort on the part of the network membership while maintaining close links among them, and with the working group and the Task Leader(s).

3.7   Conclusions

The benefits and examples of achievements discussed above show how networks have helped GCTE, and highlight why networking is central to many of its implementation strategies. While networking may not always be the best approach to all research issues – and indeed many projects in the GCTE Core Research portfolio are 'stand alone' – it has often proved the *only* way to achieve success. In addition to the results from the individual networks, the effective communication and collaboration established by GCTE (as a network itself) has rapidly advanced the community's breadth and depth in the understanding of terrestrial ecosystem science; it has directly resulted in this synthesis volume.

Finally it is worth noting that, although a primary goal of GCTE is to develop a predictive capability to support long-term planning and adaptation strategies, the existence of well-established networks of experts constitutes a major asset in the event of sudden, unexpected environmental changes. The GCTE networks, and their equivalents in the other IGBP, IHDP and WCRP projects, provide a 'reaction force', capable of diagnosing unanticipated events and developing adaptation responses. This may prove to be one of the most important products of the current international global change research effort.

4 The IGBP Terrestrial Transects

W.L. Steffen, R.J. Scholes, C. Valentin, X. Zhang, J.-C. Menaut and
E.-D. Schulze

4.1 Introduction

Transects of various types and scales have been a common ecological tool for
many decades. In most cases transects are organized along a gradient of some
factor thought to be of importance for the system under study. The basic
rationale behind this approach is simple: organizing experiments and observations along a well-defined and continuous variation in an environmental factor
allows insights into how that factor influences important system processes.
Thus, a transect approach adds a further, powerful analytical dimension to
comparative research based on a network of research sites.

Recently this approach has been adapted for global change research, and is
proving to be a valuable tool for a number of additional reasons. In the global
change context, transect studies: (i) facilitate the important scale transition from
local process study through landscape effects to regional and global understanding; (ii) promote interdisciplinary research through sharing of sites and resources; (iii) provide ideal 'ground-truth' test beds for both remotely sensed
data and global models; (iv) facilitate application of global change research to
more immediate resource management problems; and (v) promote efficient use
of scarce research resources.

This chapter presents a brief overview of the IGBP (International Geosphere
– Biosphere Programme) Terrestrial Transects as tools for global change
research. First the basic transect design is outlined, focusing on the North East
China Transect (NECT) as an example of an operational transect. The chapter
then discusses the scientific rationale for the transect approach, emphasizing
those features of the transect that are specifically aimed at global change issues.
Next the important role of remote sensing in transect research is described.
Finally, the chapter focuses on the use of transects as tools for synthesis and
integration of research, and describes the challenges that must be met for the
transects to reach their full potential as a coordinated international set. The
approach in this chapter is to set out the principles briefly, and then use
examples from operational IGBP transects to demonstrate the principles.

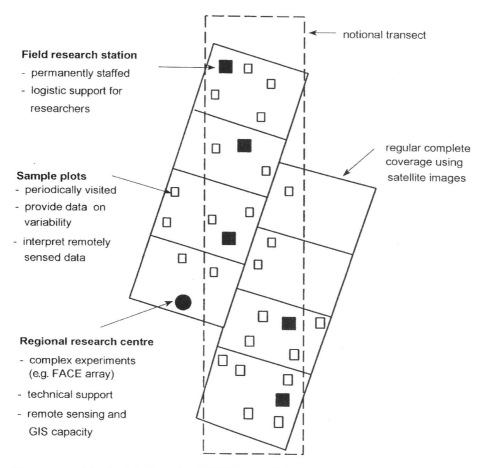

Field research station
- permanently staffed
- logistic support for
  researchers

Sample plots
- periodically visited
- provide data on
  variability
- interpret remotely
  sensed data

Regional research centre
- complex experiments
  (e.g. FACE array)
- technical support
- remote sensing and
  GIS capacity

notional transect

regular complete
coverage using
satellite images

Figure 4.1 Idealized design of an IGBP Terrestrial Transect. The notional transect is about 1000 km long and a few hundred km wide (from Koch *et al.*, 1995b).

An example of the use of the IGBP transects for an actual global change analysis is given in Chapter 12.

More detailed information on the IGBP Terrestrial Transects can be found in Koch *et al.*, 1995*a*,*b*.

4.2   Transect design and structure

4.2.1   Generic transect design

The needs of global change research have led to a generic transect design, shown schematically in Fig. 4.1. The transects consist of a set of study sites and are of the order of 1000 km in length and wide enough to encompass the dimensions of remote sensing images and several grid cells of global models (e.g. Dynamic

Global Vegetation Models (DGVMs) and General Circulation Models (GCMs)). The transects can be visualized most easily where they represent a simple gradient of a single controlling factor that varies in space, such as a gradient of temperature from tundra to boreal forest. In addition, some transects are organized around a conceptual gradient of land-use intensity; see Section 4.2.2.

As shown in Fig. 4.1, the transects include field research stations, where most of the intensive experimental work is carried out (e.g. eddy correlation measurements of water and energy exchange and trace gas fluxes), and a larger number of sample plots. The latter are important to sample various successional stages of a basic vegetation type or to form a secondary gradient, such as grazing intensity in semi-arid tropical transects organized around a primary rainfall gradient. Studies along coupled primary and secondary gradients can be valuable in determining the relative importance of multiple variables and their interaction.

4.2.2   Global change gradients

Each IGBP transect has been designed around the variation of a major environmental factor as it influences terrestrial ecosystem structure and functioning (e.g. carbon and nutrient cycling, biosphere–atmosphere trace gas exchange, and hydrologic cycling). Most are organized around a gradient of either temperature or precipitation that varies continuously in geographical space. Important ecosystem properties, such as biome type, tree : grass ratios, leaf area index, net primary production (NPP), and above-ground biomass, are often directly related to the major underlying gradient.

> *Example*: Figure 4.2 shows the variation of a number of ecosystem properties along a rainfall gradient in the Argentinian transect, one of the mid-latitude transects based on a moisture gradient. The figure includes properties that vary linearly with rainfall (e.g. root biomass), properties that are largely invariant (e.g. soil depth with 50% of root biomass), and properties that show sharp thresholds (e.g. leaf area index). These relationships give insights into how rainfall controls belowground processes of these ecosystems.

In addition to relatively straightforward gradients in which a single environmental factor varies continuously in space, a set of IGBP transects is based on an underlying gradient of intensity of land use. These are 'conceptual gradients', in that ecosystems that experience varying intensities of land use are rarely distributed continuously in geographical space.

> *Example*: A land use-intensity gradient from Jambi Province in Sumatra (Indonesia) is shown in Fig. 4.3 (M. van Noordwijk, personal communication). The figure demonstrates the analytical power of this conceptual gradient by relating agricultural (degree of management), ecological (degree of disturbance

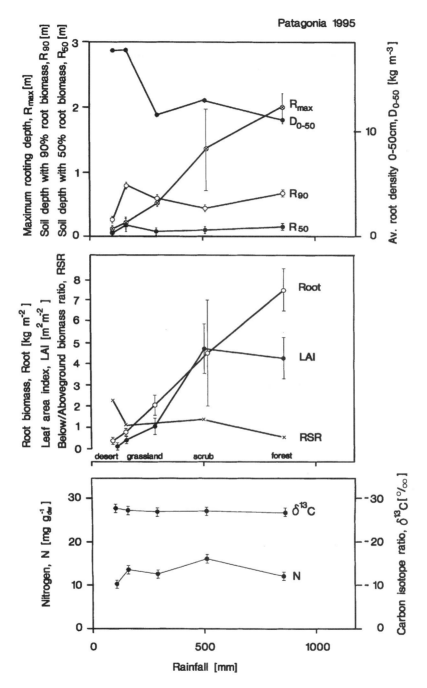

Figure 4.2 Variation of vegetation (leaf area index, above/belowground biomass, leaf nitrogen content, carbon isotope ratio) and root (biomass, rooting depth) characteristics along a rainfall gradient in Patagonia, Argentina (from Schulze *et al.*, 1997).

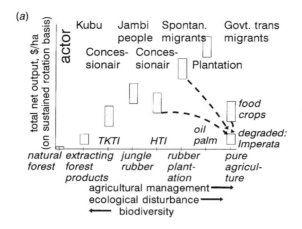

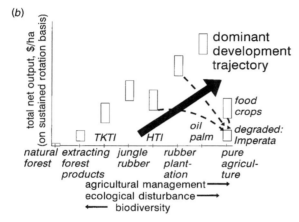

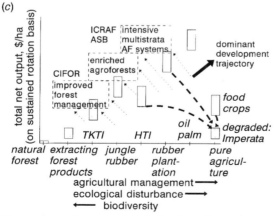

Figure 4.3   An example of a land-use intensity gradient. Tentative assessment of returns to a range of land-use options in the peneplain of Sumatra along an 'intensification' axis (TKTI = Indonesian selective logging system; HTI = industrial timber plantation); (a) indicates the main actor groups; (b) the dominant current changes; and (c) possible new developments of more productive variants of extensive systems (from M. van Noordwijk, personal communication).

IGBP Terrestrial Transects: Operational or Advanced Planning

▓ high latitude ☐ mid latitude ▒ semi-arid tropics ☐ humid/sub-humid tropics

1 = Kalahari Transect (KALA)
2 = Savanna on the Long-Term (SALT)
3 = North Australian Tropical Transect (NATT)
4 = Argentina Transect
5 = North East Chinese Transect (NECT)
6 = North American Mid-Latitude Transect (NAMER)
7 = Siberia Far East Transect (SIBE)

8 = Siberia West Transect
9 = Europe Transect
10 = Boreal Forest Transect Case Study (BFTCS)
11 = Alaskan Latitudinal Gradient (ALG)
12 = Amazon (LBA)
13 = Miombo Woodlands Transect
14 = SE Asian Transect

Note: Transects 12-14 are based on a conceptual gradient of land-use change intensity.

Figure 4.4   Locations of existing IGBP Terrestrial Transects and those in the advanced planning phase (from J. Langridge, personal communication).

of natural systems and level of biodiversity), and socio–economic (net economic output for various groups) factors to the intensity of land use. The work is part of the Alternatives to Slash-and-Burn Program, and will become a key component of the Integrated Southeast Asian Study, one of the IGBP humid tropical transect studies.

4.2.3   The initial set of IGBP Terrestrial Transects

The IGBP Terrestrial Transects are largely established in a 'bottom–up' fashion, with regional priorities and human and financial resources a major factor in their development. Thus, the international set of IGBP transects is evolving as the various regions reach the critical level of coordination and support. Figure 4.4 shows the approximate geographical location of the operational IGBP transects or those in the advanced planning stages. Table 4.1 sets out the priority regions, the land cover, the primary and secondary gradients, the contributing transects, and their current status.

These are the initial set of IGBP Terrestrial Transects; other transects will be added to the set to the extent that they (i) meet the criteria set out in Koch *et al.*, 1995*b*, and (ii) add something significant (e.g. a new environmental factor, a new transition between biomes, a globally significant region) to the existing set.

Table 4.1 *The initial set of IGBP terrestrial transects*

Region	Land cover	Primary global change gradient	Secondary gradient	Contributing transects in initial set	Status
Humid tropics	Tropical forest (humid and dry) & its agricultural derivatives	Land-use intensity	Precipitation	Amazon Basin (LBA)	Operational
				Central Africa (Miombo Network)	Advanced planning
				Southeast Asia	Advanced planning
Semi-arid tropics	Forest–woodland–shrubland–grassland (savannas)	Precipitation	Land-use intensity Nutrient status	Savannas in the Long Term (SALT) – West Africa	Operational
				Kalahari – Southern Africa	Operational
				Northern Australia Tropical Transect (NATT)	Operational
Mid-latitude	Forest–shrubland–grassland	Precipitation	Land-use intensity	Great Plains (USA)	Operational
				Argentina	Operational
				North East China Transect (NECT)	Operational
High latitude	Boreal forest–tundra	Temperature	Precipitation Nutrient status	Alaska	Operational
				Boreal Forest Transect Case Study (BFTCS) – Canada	Advanced planning
				Central Siberia	Advanced planning
				Far East Siberia	Advanced planning
				Scandinavia–Northern Europe (SCANTRAN)	Advanced planning

Box 4.1 An Example of an IGBP Terrestrial Transect – North East China Transect (NECT)

The North East China Transect (NECT, from 42° to 46 °N latitude and 110° to 132 °E longitude) extends from temperate conifer-broadleaf mixed forests in the east through meadow steppes and agricultural lands in the middle to steppes and desert grasslands in the west (Fig. 4.5). The transect is characterized by a large moisture gradient, with 709 mm of annual precipitation at the east end and only 177 mm in the west.

Figure 4.5 Vegetation at points along the North East China Transect (NECT): (a) coniferous-deciduous broadleaf forest, at the eastern (moist) end; (b) agriculture land in the middle; and (c) desert stepe, at the western (dry) end (from X. Zhang, personal communication).

Research has been carried out in the region of the NECT for more than 15 years. In 1993 the transect itself, which is coordinated by the Laboratory of Quantitative Vegetation Ecology of the Chinese Academy of Sciences, was formalized with the collation on a common GIS system of climatic, vegetation, terrain and soil data (Plate 1). A field survey was conducted during June and July of 1994 along the transect and a number of quadrats were established for routine sampling to quantify vegetation status (e.g. Fig. 4.6) and soil chemistry. Empirical models of vegetation distribution (Li, 1995) and net primary production (NPP) (Zhou & Zhang, 1995) have been developed and applied to the transect (see Plate 1 for NPP simulations and Plate 2 for a simulation of NPP by a generic biogeochemical model).

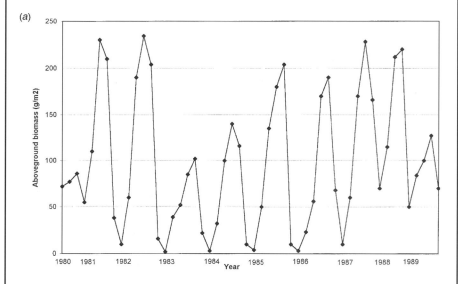

Figure 4.6   Observed (a) above ground and (b) below ground biomass for *Aneurolepidium chinese* steppe in the Xilin River basin of Inner Mongolia (from X. Zhang, personal communication).

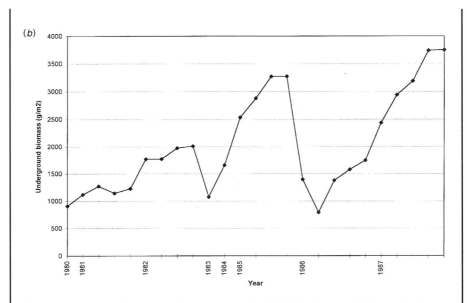

There are three major research sites along the NECT: Changbai Mountain Forest
Ecosystem Research Station (Li *et al.*, 1981) and the Inner Mongolia Grassland
Ecosystem Research Station (Chen *et al.*, 1988), both operated by the Chinese
Academy of Sciences, and the Grassland Research Station operated by the Nor-
theast Normal University in Changchun (Li & Wang, 1993). The number of
Chinese and international research groups carrying out collaborative research
on the NECT is increasing rapidly.

4.3   Rationale for the transect approach

4.3.1   Gradient-driven processes

In addition to *properties* that vary systematically along the transects, there are
also *processes* that are driven by the underlying gradient. Many of these processes are related to global change, and understanding how they might be
altered in future can best be achieved through use of the transects. An obvious
example of a gradient-driven process is the movement of water through a
catchment. Clearly changes in land–use and cover coupled with changes in
climate, especially intensity and timing of rainfall events, will affect the process.
But gradient-driven processes are not limited to the biophysical world; processes within human societies also respond to biophysical gradients, and there is
no doubt that human responses to changing gradients within the global environment will be a crucial part of global change in the next decades.

Example: An important example of a human gradient-driven process is the seasonal movement of people and cattle along the SALT (Savannas in the Long Term) transect in West Africa (Menaut *et al.*, 1995). This process links a complex of suite of socio-economic and cultural human behaviours to a gradient in a biophysical parameter that is predicted to change over the coming decades. SALT is a north/south transect based on a strong precipitation gradient extending from the wet savannas of Cote d'Ivoire in the south to the arid Sahel regions in the north. The transect has a substantial human population with a 3% increase per year. The distribution and activities of this human population are strongly affected by the underlying precipitation gradient.

Traditionally nomadic people and their cattle have moved north following seasonal rains and subsequent grass production, and retreated south during the dry periods to moister climatic zones in the middle section of the transect. More recently, cropping has become more widespread in the mesic savannas of the middle zone, and the adoption of pastoralism by the agriculturists in this area has changed a seasonal oscillation of people and cattle tracking rainfall into a net flux from the Sahel regions to the mesic savannas as both human and cattle populations increase.

In a broader sense, the predominant rainfall gradient and consequent increasing gradient of natural resources from north to south is counterbalanced by the occurrence of severely leached soils and increasing pest hazards in the southern end of the transect. As a result, people concentrate in the middle zone of the transect. This associated demographic and socio-economic gradient provides a further strong pull for the migration of people from the north to the middle zones of the transect.

4.3.2   Thresholds along a continuum

Traditional biological experiments establish two or a few levels of treatment, with several replicated experimental units at each. This type of design is powerful for demonstrating effects, but weak for revealing the location of critical thresholds along a continuum, especially where that continuum is nonlinear and may involve the interaction of several factors. Many of the issues in global change research have these characteristics. Transects which include multiple environmental gradients provide a sensitive tool for locating these thresholds. Conceptually, they are 'natural experiments' with no true replicates, but an infinite number of treatment levels and combinations. They therefore need to be analysed and interpreted using different statistical tools (nonlinear multiple regressions rather than ANOVA, for instance).

Example: One example of a property showing a pronounced threshold along a continuum comes from southern Africa. A third of the interior of this region is a vast, sand-filled basin known as the Kalahari. It stretches from South Africa in the south, where the mean annual rainfall is 100 mm, to Zaire in the north, with a rainfall of 2000 mm. There is a smooth increase in plant biomass corresponding to this gradient, but at approximately 600 mm mean annual rainfall, with no associated change in soil type, the vegetation composition changes abruptly from a fine-leafed, thorny savanna to a broad-leafed savanna. This transition is associated with a decrease in grass palatability, with major consequences for large mammal carrying capacity (Scholes, 1993) and the emissions of trace gases from the fires lit by pastoralists to burn off the uneaten grass (Scholes *et al.*, 1996a).

A hypothesis being tested on the Kalahari Transect (Fig. 4.7, Scholes & Parsons, 1997) is that the transition between palatable and unpalatable savannas is related to the ratio of carbon assimilation to nitrogen assimilation. These two processes are both influenced by rainfall and temperature (Ellery *et al.*, 1996), but in different ways. Thus, the location of the transition, which is important for local livelihoods, biodiversity conservation and greenhouse gas emissions, is predicted to migrate northwards with increasing temperature and nitrogen availability, and southwards with increasing atmospheric carbon dioxide and rainfall. Finding the transition point and its sensitivity to change would be almost impossible without the continuous gradients superimposed on a homogeneous soil offered by the Kalahari Transect.

4.3.3 Facilitation of interdisciplinary research

Interdisciplinarity is the essense of both global change research and the basis for systems-oriented resource management strategies. The transect approach promotes collaboration among different disciplines through the sharing of common research facilities and sites, the requirement for common databases, and the design of gradient-based research around multi-faceted global change or resource management questions.

Example: Collaboration between the ecological and hydrological global change research communities was a prominent feature of the HAPEX (Hydrologic Atmospheric Pilot Experiment)-SAHEL land-surface experiment. The work, which studied the land surface–atmosphere interaction of a West African savanna, was largely carried out at a site about 70 km from Niamey, Niger. The site also serves as one of the field research stations of the SALT transect, and SALT ecologists participated in the development and implementation of HAPEX. Vegetation and soils process studies, databases and associated maps (Valentin & d'Herbes, 1997) developed by the ecological community underpinned the interpretation of the land-surface experiment, which in turn provided flux measurements for important ecosystem processes at precisely the sites where intensive ecological research is being carried out.

Example: Similar collaboration is developing between the Boreal Forest Transect Case Study – BFTCS (Halliwell *et al.*, 1995; Price & Apps, 1995) – and the BOREAS (Boreal Ecosystems Atmosphere Study) land-surface experiment (Sellers *et al.*, 1995b). As for SALT, the BFTCS encompasses the land-surface experiment sites, and extends the latter in both space and time from a few sites studied intensively for a short period of time to a larger number of sites studied in less detail but over a much longer time. Thus, BFTCS facilitates the aggregation of BOREAS results to a regional scale over

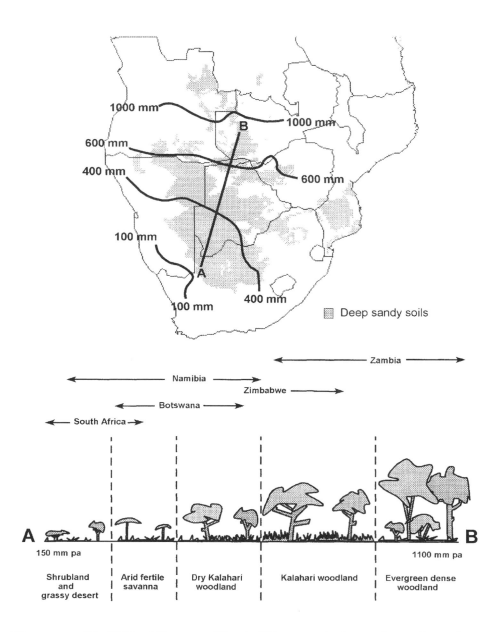

Figure 4.7   The Kalahari Transect of South Africa, organized around a north–south gradient of rainfall. The sharp transition between fertile and infertile savanna is indicated on the diagram as the transition between 'arid fertile savanna' and 'dry Kalahari woodland' (from R.J. Scholes, personal communication).

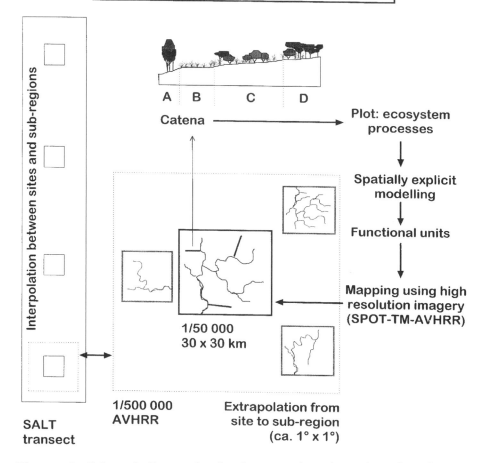

Figure 4.8 Schematic diagram showing the approach used to scale up from the catena to the region in the SALT (Savannas in the Long Term) transect of West Africa (from J.-C. Menaut, personal communication).

timeframes relevant to global change processes, by inclusion of disturbances, vegetation succession and human influences. These two examples show the value-adding of the IGBP transects to the international suite of land-surface experiments.

4.3.4   Use of transects for scaling up

A key function of the transects is to facilitate the transfer of results, understanding, data and models from one scale to another (Fig. 4.8). One of the most important scale transitions is that from a relatively homogeneous patch of a system – the scale at which many ecologists and agricultural scientists work – through landscapes to a region, about the size of a grid cell of many global models. The critical question is: when can a process or property be scaled up simply as an area-weighted average and when must the spatial pattern be taken into account in the scaling-up process?

Example: A common example where an area-weighted average approach is appropriate is the estimation of above-ground biomass from the relative areas of various vegetation types. This technique is used in the inventory method for calculating changes in the terrestrial carbon cycle, and has proven to be a valuable tool in many studies (e.g. Houghton, 1995).

Example: In general, the simple area-weighted averaging technique fails where the process has a significant horizontal transport component. An example where the spatial pattern and connectivity of the landscape is critical in scaling-up comes from the miombo woodlands of south-central Africa. Where the ancient land surfaces of the African Shield occur under a climate of hot, wet summers and warm, dry winters, the undisturbed landscape consists of two main components: sandy ridges covered by drought-deciduous woodlands with a grass understory (known locally as *miombo*), and clayey depressions covered by seasonally waterlogged grasslands known as *dambos* (Fig. 4.9). This landscape is the focus of an IGBP transect known as the Miombo Network (Desanker *et al.*, 1997). The total area is about 3 million km².

Although the dambos make up about one-fifth of the landscape, they do not appear on continental-scale soil and vegetation maps, since they are individually small (one to five kilometres wide). Global climate and vegetation models ignore them, since they cannot be represented at a grid scale of half a degree (50 km) or more. The water redistribution, primary production and trace gas emissions of the landscape cannot be correctly predicted without taking the presence of dambos into account, and furthermore, explicitly linking them to the woodlands. This is because the characteristics of these two landscape facets are totally different, and they depend on each other.

The typical global-scale modelling 'fix' to this situation would be declare every fifth grid cell in the miombo region to have the characteristics of a dambo. This is no solution, since under the prevailing climate, the dambo grid cell would never become waterlogged, even if its soil and vegetation were correctly specified. The water balance of the dambo regions is determined by drainage of water, mostly beneath the soil surface, from the surrounding woodlands. The evapotranspiration, NPP and CH4 emissions would all be underestimated by 50% or more.

This is a case where spatial context matters – it is essential to get both the proportions of the landscape facets correct, as well as their relationships. A partial solution would simulate both a miombo and dambo patch within the grid cell, and transfer water (and, in the long term, carbon and nutrients as well) from the miombo to the dambo. A spatially explicit approach would need to reduce the grid size until dambos appeared. This would allow, for instance,

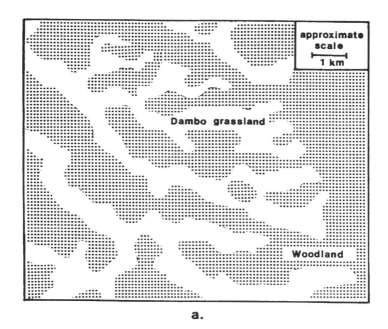

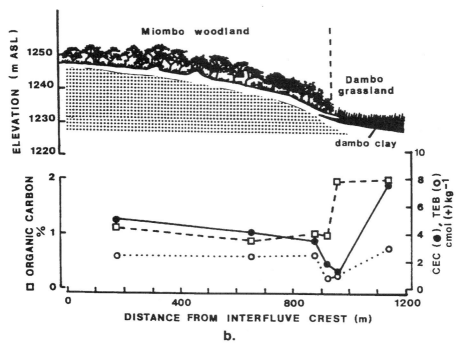

Figure 4.9 Diagram showing (a) the spatial pattern of miombo woodland and dambo grassland; and (b) the sharp difference in soil organic carbon between miombo and dambo due to the horizontal transport of water and nutrients from miombo to dambo (from R.J. Scholes, personal communication).

the supressive effect of evaporation from an adjacent moist dambo on the
vapour pressure deficit in the miombo to be simulated.

4.4   Remote sensing applications to the transects

Remote sensing technologies offer an excellent opportunity to provide globally
consistent, long time-series data on the Earth's land surface for global change
research. The IGBP transects can facilitate this application of remotely sensed
data, through provision of (i) ground validation sites, (ii) mechanisms for scaling
data from site through region to globe, and (iii) comparisons to regional and
global models. Recent initiatives by IGBP-DIS (Data and Information Systems) and CEOS (Committee on Earth Observing Satellites) are creating
effective linkages between transects scientists and the remote sensing community. The transects are also poised to establish strong links with other developing initiatives such as GTOS (Global Terrestrial Observing System) and LQI
(Land Quality Indicators).

> *Example*: Work at the Laboratory of Quantitative Ecology of the Chinese
> Academy of Sciences has led to a remote sensing-driven model of vegetation
> dynamics to probe the responses of ecosystems along the NECT to elevated
> CO_2 and climate change (Gao & Zhang, 1997). NDVI (Normalized Difference
> Vegetation Index) obtained from NOAA AVHRR data was used to parameter-
> ize and partially validate the model at present CO_2 concentration and climate,
> based on the relationship between NDVI and green biomass established by
> Tucker (1979*a*,*b*). The model was run for doubled CO_2 concentration and
> altered climate to predict vegetation structure, biomass distribution and NPP
> for a number of climate scenarios.

> *Example*: Scientists working on the Northern Australia Tropical Transect
> (NATT) are using SPOT (visible images) and AIRSAR (radar images) syner-
> gistically to improve classification of land cover, including both vegetation and
> soils (Ahmad *et al.*, 1996). The pilot study is based on the NATT site at
> Kidman Springs, and uses one SPOT and two AIRSAR images, coupled with
> ground surveys, to map soil type and key vegetation parameters such as tree
> basal area and percentage grass cover. The early results are encouraging, with
> the composite images (i) capable of distinguishing between red and black
> cracking clay soils, and (ii) showing a good correlation between radar backscat-
> ter and tree basal area ('woodiness').

4.5   Transects as tools for synthesis and integration

Much of the value of the transects for facilitating innovative, interdisciplinary
research would be lost without the inclusion of explicit synthesis and integration
activities within the overall transect research effort. Such activities are not extra

'add-ons' at the end of process studies and observations, but rather are crucial elements to the design of the overall study at the beginning and to the interpretation and application of the results at the end of major transect-based studies. Synthesis and integration activities are essential for both the practical application of transects research to management issues, and for the delivery of regional-level understanding and modelling expertise to Earth system science.

4.5.1 Regional studies and modelling-management

The IGBP Terrestrial Transects are ideally placed to transfer the results of global change research to the scale of interest to resource management authorities. Indeed, nearly all of the transects have been initially established with practical management outcomes as the primary aim. As shown schematically in Fig. 4.10, the co-location of global change research and management-oriented studies along the same transect strongly facilities the two-way transfer of understanding and expertise: (i) the implications of long-term environmental change for the development of effective shorter-term, management strategies (global change 'feedforwards'); and (ii) the scaling-up of the consequences of various local and regional scale management options for further global environmental change (global change 'feedbacks').

Example: A fascinating example demonstrating the identification of a threshold effect and its subsequent application to a management problem comes from the SALT transect in West Africa. One of the dominant features in the mid-section of this transect is a banded pattern of vegetation known as tiger bush. Tiger bush is formed by parallel strips of woody vegetation interspersed with barren zones. The run-off of rainfall from barren areas concentrates moisture and nutrients, leading to high productivity in the woody zones. In fact, the productivity of tiger bush equals that of the forest in the wet savannas to the south, and is double that of landscape systems with no tiger bush patterns.

The location and productivity of the tiger bush is controlled by two opposing trends linked to mean annual rainfall, because productivity is closely related to total water input, run-on as well as rainfall. From north to south, the rainfall increases but the barren zone/thicket ratio decreases concurrently, and thus the amount of run-on provided to the thicket also decreases. The result is that woody biomass of these banded systems reaches a maximum for 600–650 mm mean annual rainfall. When rainfall exceeds this threshold, the woody biomass of the tiger bush plummets dramatically and becomes as low as that for unpatterned vegetation. This effect, and its controlling agents, is most easily detected and understood using the gradient appoach, as embodied in the SALT transect. At the scale of West Africa, this rainfall range has been recognized as the rainfall optimum for run-off production. Under natural conditions, the run-off coefficient regularly declines while rainfall increases so that mean annual run-off amount reaches a maximum for an annual rainfall of approximately 600 mm (Valentin, 1996).

Lack of understanding of this system has led to poor management in the

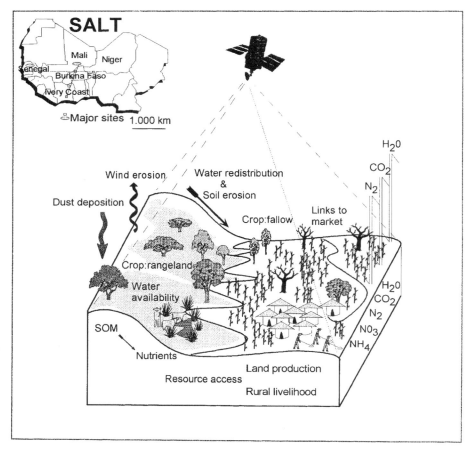

Figure 4.10   Schematic diagram showing the main processes studied along the SALT (Savannas in the Long Term) transect of West Africa. The interaction between resource management and global change issues is indicated by the interest in socio–economic factors such as rural livelihood and links to markets as well as biophysical impacts such as trace gas emissions and water and energy exchange (from C. Valentin, personal communication).

past. Foresters considered the bare interbands to be a sign of degradation, and reafforestation attempts were initiated. In Niger the main technology was the construction of half-moon shaped furrows in the interbands. The reafforestation attempts were spectacular failures. Not only did the seedlings in the interbands receive insufficient rainfall to survive, the furrows and seedlings trapped overland flow necessary for the maintainance of the downslope vegetation, which consequently began to decay. With further disturbance, such as intensive fuelwood harvesting or ephemeral cropping of sorghum or millet, the spatial pattern of water redistribution was altered, run-off increased, and serious gully erosion resulted. Breakdown of the banded patterns of woody vegetation caused a marked degradation of the landscape.

Table 4.2 *Site classification hierarchy for the Canadian BFTCS (Halliwell et al., 1995)*

Parameter/Code	Description
Species dominance	
A	Aspen
B	Black spruce
J	Jack pine
M	Mixed
Age	
D	Recently disturbed; young
I	Immature
M	Mature
Productivity	
L	Low
M	Medium
H	High

Understanding the dynamics of the system has led to more effective management regimes. Rehabilitation strategies which mimic the natural tiger bush ecosystem are now being employed; tree planting in the appropriate areas is coupled with a restoration of the soil condition through the use of mulching, which protects the surface from erosion and attracts termites. The sustainable use of tiger bush in the longer term requires the careful maintenance of this natural water harvesting system, with no attempts at afforestation in the interbands, no agriculture, and only moderate wood harvesting in the vegetation bands.

4.5.2 Comparisons across transects

Considerable benefits and efficiences, to both Earth system science and regional sustainable development studies, can be obtained by well-designed intercomparisons of results and simulations across transects of the same type. The first of these activities are now being planned, and many more are required for the transects to reach their full potential (see Section 4.6).

Example: A workshop aimed at comparing research on the Alaskan and Canadian transects was held in April 1997 (F.S. Chapin III, personal communication). The meeting focused on measurements and simulations of water and energy exchange and of trace gas fluxes, using models developed for sites on one transect (e.g. Alaska) to predict fluxes at sites on the other transect (e.g. Canada). The results from experimental campaigns and locally developed models were then compared to fluxes as simulated by DGVMs (Dynamic Global Vegetation Models). A novel feature of the workshop was use of the expertise gained on the North American high latitude transects to predict the corresponding flux rates on the two Siberian transects, which are just beginning their implementation phase. These prospective studies give indications of what fluxes to expect in Siberia, given our current level of understanding of North America.

Example: Making measurements and observations using common methodologies and protocols is a useful technique for enhancing comparative studies. Table 4.2 shows the site classification system used for the Canadian BFTCS (Halliwell *et al.*, 1995). If all of the IGBP high latitude transects adopted it or a similar common classification system (perhaps using functional type rather than species classifications), the gain in understanding of long-term system dynamics and enhancement of model development through intercomparisons across the transects would be significant.

4.5.3   Global model intercomparisons based on the transects

Regional and global scale models are a valuable mechanism for placing the current and projected future composition and performance of vegetation along the transects into a coherent, consistent global context. The transects, conversely, are an ideal test bed for developing and validating such models. They can act as a valuable 'reality check' for models, by providing long-term observations of critical ecosystem parameters, results from physical experimentation, and comparisons with process models developed for specific systems at much finer scales.

Example: An example is the use of the NECT for model comparison, testing and validation. Figure 4.8 shows a simulation by the CENTURY model (Parton *et al.*, 1987, 1989, 1993*b*) of net primary production (NPP) under current climate for the NECT. The pattern can be compared directly to that simulated by finer scale models developed specifically for the NECT (see Fig. 4.6). It should be noted that CENTURY was originally developed in North America, and has since evolved into a general biogeochemical process model used for many different systems around the world at many different scales. Thus, it has not been 'tuned' in any way for the NECT. The generally good correspondence between the patterns simulated by the two models (with the highest values in the southeast and lowest in the northwest reflecting the transect's dominant moisture gradient) gives confidence that the NPP simulations of both are reliable. Comparisons of the output of both models to measurements of NPP at various sites along the transect is a further test.

Example: The development of Dynamic Global Vegetation Models (DGVMs) can also be enhanced by using the transects as test beds. Chapter 12 uses pairs of transects of each of the four types (high latitude, mid latitude, semi-arid tropics, and humid tropics) to analyse future change in vegetation composition and structure. As a first step, the simulation of vegetation type under current climate is compared with maps of actual potential vegetation to check the performance of the models. This analysis emphasizes the need for increased standardization across models, measurements and data (see Section 4.6).

4.6   Future challenges

The individual IGBP Terrestrial Transects are making rapid progress toward becoming facilities for interdisciplinary, integrative global change research, as shown in the examples above. However, to reach their full potential they need to be implemented as a coordinated international set, yielding even further insights into global change interactions with terrestrial ecosystem functioning, composition and structure. To achieve this higher level of integration requires at least three key steps:

- Consistent basic datasets. Each transect needs to have a GIS database containing basic transect information such as soil type and characteristics, vegetation composition, climate (present and future scenarios), and land cover and use. Ideally, each transect should use the same grid cell size, same classifications of soil, vegetation and land-use/cover types (or perhaps a regionally preferred classification that can be translated into an international standard), and common climate scenarios.
- Comparative experimental and observational research. Examples include adoption of a single methodology and instrument type for measuring LAI (leaf area index), agreement on common protocols for soil decomposition studies, and use of the same instruments and techniques for trace gas flux measurements. Where such commonality is not possible, different measurement systems should be carefully intercompared so that appropriate correction factors can be applied.
- Intercomparisons of results and simulations. Workshops should be held at regular intervals among scientists from transects of the same type (e.g. high latitudes, semi-arid tropical savannas) to compare results on specific topics (e.g. water and energy exchange). The use of models developed on one transect to simulate the same process on a companion transect, and then comparison to measurements, is an especially powerful technique for building robust simulation tools.

Achieving this level of integration among the transects is not an easy process, and will require additional resources, support from the international global change research community, and good will from the coordinators and participating groups on the individual transects. If it can be achieved, the rewards in terms of rapid increases in understanding of global change processes will be great.

⑤ Data needs and limitations for broad-scale ecosystem modelling

W. Cramer, R. Leemans, E.-D. Schulze, A. Bondeau* and R.J. Scholes

*Previous name: Alberte Fischer

5.1   Introduction: Why global models and databases?

This chapter describes the data needs for global biospheric models as they are developed within the GCTE (Global Change and Terrestrial Ecosystems) science community. Current approaches for globally comprehensive and spatially explicit biosphere models are briefly reviewed, focusing on data use. The problem of spatial sampling in irregular networks is discussed, and examples for the most important data sources are given. However, in the context of this chapter, a full catalogue of available datasets is not provided. Finally, data requirements for the next generation of global models are considered.

Projections of possible future changes in the terrestrial biosphere are a fundamental task in GCTE (Walker, 1996). The Intergovernmental Panel on Climatic Change (IPCC) required such projections (Houghton *et al.*, 1996; Watson *et al.*, 1996) when they asked with respect to the terrestrial biosphere: (i) what are the likely impacts of climatic change on the structure and functioning of major vegetation complexes, such as the world's forests and grasslands; and (ii) what are the possible feedbacks of such changes to the climate system? Such projections require the synthesis of current knowledge of the main driving forces of global change as well as of the mechanisms by which these forces influence terrestrial ecosystems. Such an 'Earth system analysis' requires mathematical models containing representations of the various subsystems as well as the most important interactions between them. The models currently available for this purpose differ widely in detail and emphasis (see Chapter 6). All model development and application are, however, limited by the availability of data on past and present states of the physical environment on Earth and of the terrestrial biosphere (Cramer & Fischer, 1996).

This chapter first discusses the types and nature of the data needed for global biospheric modelling, including sources of the data, spatial and temporal sampling strategies, and methods for interpolating between data points. It then describes data availability in five critical areas: climate, soils, ecosystem properties, radiative properties of the land surface, and land use and land cover. Finally, future developments are briefly discussed.

5.2   Data considerations for global biospheric modelling

5.2.1   Globally comprehensive, spatially explicit datasets

This chapter is primarily concerned with modelling activities that are *globally comprehensive* and *spatially explicit*. While regional models are relevant for many purposes within the overall framework of global change, they are not included in this discussion as their data requirements are regionally specific. The dual requirement of global comprehensiveness and spatial explicitness for global models can be achieved over a wide range of spatial resolutions, from a few to several hundreds of kilometres. The practical limits to this resolution are set by both data availability and computational resources. Most global datasets, in their original form, do not directly provide the spatial and temporal features that are required by ecosystem models; methods to re-scale data are discussed below.

5.2.2   Data needs for model development and evaluation

Data for global models may be 'static', that is, they provide a modelling framework (e.g. topography or coastlines), or 'dynamic', that is, they are required for model application and evaluation (e.g. climate or land use). Whether a particular dataset is considered static or dynamic may change for different applications. Global comprehensiveness usually requires some static data to be provided, such as topography. Much soils data (e.g. texture) is considered static by most terrestrial biosphere models, although in future models some of these could be turned into dynamic variables as well. Most data on climate, land use and ambient CO_2 concentrations are part of the dynamic simulation.

Simulation models use data in several ways. Modellers use observations of real-world mechanisms to *calibrate* parameters in such a way that the model component produces results that are close to what has been measured, although there are important exceptions to this. Independently of whether it has been calibrated or not, each model requires *evaluation* of its performance in relation to observation (also referred to as validation, but this term is avoided since full validation of a global model is impossible, see Oreskes *et al.*, 1994). Clearly, the same observations cannot be used for calibration *and* evaluation. However, a single large dataset may be split in two subsets, one of which is used for calibration while the other one remains for evaluation. In large-scale global change studies, evaluation is generally difficult, because few global observations of critical variables exist; the global results of the model cannot be compared to an observed quantity if it is unknown to the model developer. Evaluation may be more feasible for some model subsystems than for others, and thereby contribute to the overall plausibility of the model results. In the absence of suitable validation data, the only alternative test for consistency in coupled models of the Earth system (atmosphere, oceans and biosphere) is to check whether the overall behaviour of the biospheric component produces instability of the coupled model or not.

A model that has been evaluated to yield plausible results may be applied to conditions that are different from those present during calibration and/or evaluation. Such an application is based on data which are usually generated from some scenario technique. The most common type of scenario in global modelling studies are those of future forcings of greenhouse gases (GHGs) on the atmospheric circulation, which result in predicted values for climatic variables. Values from these global circulation model simulations are used by terrestrial biosphere models as scenario data for applying the models. A general requirement for all databases (model output or observation) that are used in biospheric modelling is consistency between all different datasets that are used with the same model (Kittel *et al.*, 1995). For example, consistency is required between variables with respect to spatial and temporal resolution (see below).

5.2.3 Data sources

Despite the large number of data sources with widely varying characteristics for biospheric modelling, shortage of data remains the overriding problem. One limitation is that using data from different sources always implies considering different sampling strategies. Observation networks, such as those of the world's meteorological agencies, are an important data source. These are comprised of stations making regular observations over long periods at a large number of points – the resulting data are therefore considered more or less representative of the range of conditions that occur throughout the spatial and temporal domain of interest. Satellite remote sensing is another important data source which has the basic characteristic of wide (often complete global) coverage at regular intervals. Depending on the type of sensor, these data can be used for estimation of many different variables that are important for global models (Fischer *et al.*, 1996).

Single observations should not be overlooked in their importance for global modelling studies. Many indicators of key ecosystem properties, such as soil variables, biomass or net primary productivity (NPP), can only be estimated as aggregates from single investigations at many different locations: networking activities are underway to collect the results of such investigations in a common format that makes them suitable for global modelling. An example of this is the Global Primary Production Data Initiative (GPPDI), which, under the coordination of the IGBP Data and Information System (DIS), collects observations of NPP and other ecosystem observations throughout the major vegetation types of the world (Prince *et al.*, 1994; Olson *et al.*, 1997). Another example is the dataset of river discharges, which can validate the hydrological component in global ecosystem as well as climate models (Perry *et al.*, 1996).

5.2.4   Spatial and temporal characteristics

Spatial sampling

When collecting spatially explicit data, sampling is usually made at a series of 'points', chosen to represent areas of a more or less well-defined size, within which any heterogeneity is ignored. While this abstraction is rarely discussed, it represents a significant element in the modelling strategy. Usually, however, it is a logical step, since most models only consider the broader trends across a region and hence have to ignore local phenomena. For data gathering, this means that points generally must be assumed to be representative for their surroundings. In some cases, the point-based information may also carry elements of this heterogeneity, e.g. elevational range.

For ground-based observations of climate or other habitat variables, coverage is taken as global if most of the overall domain appears to be covered by the weather stations, allowing interpolation between them. Complete coverage of the global land surface can, however, only be achieved by some type of sampling scheme, e.g. by following a regular grid with a defined mesh size or by using polygons with clearly defined criteria. Complete coverage data may come from a device directly designed for it (such as a remote sensing instrument on a satellite), from some data assimilation algorithm (such as spatial interpolation), or as output from grid-based models. For simulations of biospheric processes, grid-based models and databases (as opposed to vector-based structures) have been particularly popular.

Conceptually, grid-based approaches have the advantage that most features of the terrestrial biosphere and its driving forces have standard spatial dimensions. Many studies of biospheric aspects of climate change, for example, have been made using grid mesh sizes of 0.5 or 1.0 degrees longitude and latitude. These cells have edge lengths of approximately 55 or 111 km, respectively, at the equator, with E–W edges shortening as latitude increases. Whether this resolution is fully adequate for the biospheric models remains an open issue. An advantage of grid-based approaches is that the resolution and hence the size of the smallest spatial features are clearly defined and can be considered in the evaluation of models. Non-rectangular polygon-based approaches have the capability of defining a minimal area size, but this feature is usually less obvious to the user than it is with a grid-based approach. There is a risk that the limitations of the original point dataset become obscured by the gridding procedure. Despite its continuous appearance, a grid-based map cannot automatically be considered to represent a continuum or regular raster. Rather, it is seen as a regularly spaced number of points, each of which is more or less representative for the average conditions in their grid cell. In a careful analysis, different grid cells may well be supplied with adequate information about the degree to which local data density was sufficient for producing the value at the grid location.

In conclusion, due to the conceptual and practical advantages as well as to the steadily improving availability of large computer storage devices, grid-structured databases are most frequently used as a fundamental spatial data structure of biospheric models.

Temporal sampling

There is considerable interest in modelling the *temporal* dynamics of the biosphere to increase our understanding not only of the changing global environment, but also of the inherent changes that are characteristic of many biospheric subsystems. This results in a demand for series data over time which cover the important global change variables. For time series, the immediate questions are 'for how long?' and 'at what intervals?', and the answers to both involve trade-offs against spatial coverage. For instance, while there is a large array of climate stations with coverage of several (and sometimes many) decades and intervals down to a day or less, attempting to create a global dataset that samples most climatic regimes is problematic; if one requires frequent sampling of short time intervals, then spatial coverage is low and often differs from time to time, and if one accepts long averaging intervals (such as climatic normals), then the suitable network is more dense but temporal variability less covered. It is important to specify the use for which the dataset is required, since the acceptable trade-offs between coverage, duration and interval may vary greatly between different models.

Due to the nature of the different systems being monitored, the requirement for temporal consistency may also vary greatly between databases. Some soil variables (e.g. texture) do not change significantly over years or even decades, and the temporal sampling of the points in a global database is less critical than in a weather database. Similarly, climatic information (i.e. long-term averages) is less sensitive to consistency between different observation periods than instantaneous weather data, and this sensitivity changes greatly between regions.

Interpolation in space and time

The sparse and often irregular sampling of data within both the temporal and spatial domains necessitates the 'assimilation' of data, that is, procedures that produce more or less continuous fields of data for the application or evaluation of models. Interpolation is the generic term for methods that fill gaps between data points in such a way as to approximate the conditions between them. The underlying assumption is that parameters usually vary smoothly from one point to the other. The main difference between different interpolation techniques is the mathematical method of smoothing that is applied. A further consideration is whether additional information beyond the spatial coordinates or the times of the observations is used. Spatial interpolation techniques, for example, now include the capability of accounting for vertical, in addition to horizontal, space dimensions (Hutchinson & Gessler, 1994), and as a result the quality of the interpolated fields has increased considerably for many climate variables.

Another, more advanced, type of gap-filling is required when the variability between certain points in time and/or space is an issue. This may be the case, for example, when a realistic distribution of weather conditions is required and only climatic information is available. Numerous 'weather generator' techniques for this purpose exist in the agrometeorological literature, such as the popular WGEN algorithm (Richardson, 1981). The key problem for global modelling is that most of these weather generators require parameters that are difficult or impossible to collect at the global scale (Friend, 1997). Work is therefore underway to analyse the global applicability of these approaches and, where possible, reduce their complexity (Hutchinson, 1995).

5.2.5 Modelling approaches: equilibrium versus dynamic models

The first approaches to simulating responses of the biosphere to global change were all based on models that assumed equilibrium between atmospheric chemistry, physics and the state of the biosphere. Despite the time-lags that are to be expected in the response of the actual Earth system, it appeared to be an appropriate first approximation for developing equilibrium biogeochemical models such as the Terrestrial Ecosystem Model TEM (Melillo *et al.*, 1993), equilibrium plant biogeography models such as BIOME (Prentice *et al.*, 1992) and satellite-driven production efficiency models such as CASA (Potter *et al.*, 1993). Although involving short-term dynamics of the daily and seasonal cycles in many processes, the biogeochemical and biogeography models are in equilibrium with longer-term climatic conditions. The data needs for all of these models are relatively straightforward. Essentially, they require climatic (temperature, precipitation, radiation), soils (texture) and sometimes vegetation data at relatively high spatial resolution and global coverage (some models require more data, such as air pressure, wind speed, humidity and sometimes the satellite-derived Normalised Differential Vegetation Index, NDVI), and they have been developed in such a way as to provide plausible results even with the limitations of the global datasets. Most of these models now use the climate database developed by Leemans & Cramer (1991), as well as various derivations from the FAO soil map (FAO/UNESCO, 1974) which are interpreted for (static) soil texture. Dynamics are considered only in so far as they occur within the daily or seasonal cycle. Therefore, they normally use long-term monthly averages of climate variables (and sometimes satellite data) to describe seasonality. The daily cycles of radiation or temperature are usually not supported by data, but they are instead computed using solar angles and simple assumptions on the general relationships between day-time and night-time conditions.

Dynamic Global Vegetation Models (DGVMs, models that simulate long-term vegetation dynamics at the global scale) were first proposed by a work group at IIASA in 1988 (Prentice *et al.*, 1989), but their development has only recently gained momentum (see Chapters 6 and 8). In contrast to equilibrium models, DGVMs need globally comprehensive time series of climate input data, and such data are still in poor supply at the resolutions that now have become typical for simulating the biosphere. Currently, GCM output is used as a surrogate for historical data, even for the present century (see Chapter 8).

5.3   Data availability for broad-scale ecosystem models – examples for GCTE-related datasets

5.3.1   Climate data

A database of monthly means of temperature, precipitation and cloudiness, averaged for the years 1931–60, was produced within the framework of development for the BIOME model (Leemans & Cramer, 1991). A recent (unpublished) update of this dataset is based on the use of an interpolation technique (thin-plate smooth splines) that explicitly uses elevation as a predictor variable (Hutchinson & Bischof, 1983). A broad range of applications for global and regional ecosystem modelling has shown that altitudinal gradients now are better reflected than when only linear corrections for elevation are applied. This type of climatic information is essential for biospheric models, and although similar datasets exist (e.g. Legates & Willmott, 1990), the dataset described here is the most widely used in the GCTE community. For development and testing of dynamic biosphere models, however, climatic normals such as the Leemans & Cramer (1991) data are not sufficient, because they do not include historical trends. Some datasets are available that cover parts or all of the period since 1900 (Jones *et al.*, 1985, 1986; Bradley *et al.*, 1987; Piper & Stewart, 1996; Xie *et al.*, 1996), but their spatial resolution or temporal coverage is inadequate for use in modelling the dynamics of the terrestrial biosphere. A GCTE-GAIM workshop, held in May 1996 at the Potsdam Institute for Climate Impact Research, prepared a work plan for the production of a 0.5 × 0.5 ° longitude/latitude historical climatology that is designed to fill this critical gap.

Despite the development of global climate datasets, it is still not possible to simulate all desirable ecosystem properties at the global scale by their relationships to climate, even for the equilibrium case. One of the most crucial problems in this respect is the shortage of data to support essential water balance calculations, (e.g. average vapour pressure deficit, wind speeds or soil texture). The poor spatial resolution of most datasets is also a problem because many processes vary strongly at the landscape to regional scale, and most global databases are not detailed enough to overcome this. The consequence is that the mechanisms described by the models must be generalized to such a degree that the need for data to calibrate particular relationships is minimized. A classic example, which illustrates both the potential improvements and the limitations for them, is the development of algorithms to estimate potential evapotranspiration (PET) beyond the popular Thornthwaite equation (Thornthwaite & Mather, 1957), which used only temperature observations. More recent methods are the simplifications of the Penman–Monteith approach, such as the Priestley–Taylor method (Priestley & Taylor, 1972), which require data to estimate net radiation and therefore achieve a more realistic model response to water shortage situations. The appearance of remotely sensed radiation data (such as the Earth Radiation Budget / SRB (ERBE/SRB), Barkstrom, 1984; the Total Ozone Mapping Spectrometer Photosynthetically Active Radiation, TOMS/PAR, Dye, 1992; or Whitlock et al., 1993; or the International Satellite Land Surface Climatology Project / Surface Radiation Budget (ISLSCP/SRB), Meeson et al., 1995; Sellers et al., 1995a), as well as globally interpolated cloudiness data (Leemans & Cramer, 1991), allowed this development, but the absence of other parameters such as wind speed are currently limiting further improvements.

5.3.2   Soils

Climate, hydrology, trace gas and ecosystem modellers all depend on reliable soil data; however, the assembly and use of data on soil properties at the global scale presents unique difficulties. Although there are many thousands of soil profile descriptions and analyses from around the world, these have not been collected using standard methods, and are not available in a single location and format. The primary soil attributes found in the profile descriptions – such as the fractions of sand, silt and clay – are seldom the properties needed in the models, and essential information about the depth, the organic layer or the local vegetation is mostly missing. An intermediate translation stage is required, the algorithms for which may not be available, or may not be universally applicable.

Soil properties can vary within a few metres, whereas even the most highly resolved global models work at a minimum scale of tens of kilometres; the variation in soil properties is invisible to them. A simple concept such as soil depth change, when expressed at the scale of a GCM grid cell, becomes a model parameter instead of being a measurable and verifiable number. The high degree of covariance and nonlinearity in soil properties does not allow the simple averaging or spatial interpolation of soil attributes within a grid cell. The average of a clay and a sand, for instance, is not a loam. When nonlinearities and covariances are involved, the only rigorous way of representing subgrid scale processes is to model them separately for each significantly different type within the grid, and then proportionately sum or average the results. Almost all existing climate and ecosystem models fail this test. They either select one type to represent the grid or they create a synthetic type (which may have no naturally occurring representative) by averaging the properties of different types within the grid.

Despite its known deficiencies, there is only one globally consistent soil map at the resolution needed for global change research (FAO/UNESCO, 1974; FAO, 1991*b*), and all global spatial soil products are directly or indirectly based on it. Its map units, like those of all broad-scale soil maps, are not uniform soil types, but associations of soil types. The FAO map is not explicit regarding the fractional soil type composition of each association, nor does it have quantitative soil analytical data explicitly associated with the soil types. There are currently several global-scale soil data products, all with significant limitations for biospheric modelling. The soil property database assembled by Zinke *et al.* (1984) includes a large number of profiles, but the primary data are incomplete and possibly analytically inhomogeneous. The USDA has made nearly 20 000 highly standardized soil profile descriptions available in the public domain, but they are classified in the USDA system and the vast majority are from US territory. Global fields of water holding capacity, soil carbon, soil nitrogen and other properties have been derived from the FAO soil map (e.g. Zobler, 1986; Meeson *et al.*, 1995), but the algorithms used are seldom traceable to the base data, and the methods used to determine the spatial distribution of properties obscure the small-scale variability which exists. More sophisticated methods exist but have not been applied at a global scale (Webb & Rosenzweig, 1993).

Since soil data are needed in several IGBP core projects, the IGBP-DIS framework activity undertook the task of developing an improved global soil data set (Scholes *et al.*, 1995). The World Inventory of Soil Emissions (WISE) project (Batjes & Bridges, 1994) had already combined the most complete, geo-referenced and compatible contents of three quality controlled international soil databases (FAO, USDA and ISRIC), so by reclassifying the USDA-derived pedons in the FAO system, they could all be indirectly linked to the FAO soil map. A series of 'pedo-transfer functions' were derived and globally tested for properties such as soil water retention and thermal capacity. Finally, a system of spatial linkage was developed which preserved the variation of soil properties within a grid cell. The resultant data products are available from the IGBP-DIS web page.

5.3.3 Ecosystem properties

Many models require both habitat data and input concerning functional properties of the ecosystem. These are needed for either the parameterization of processes within a pre-defined ecosystem or for the derivation of relationships that are then used in the model formulations. At the broad scale, most ecosystem properties are defined in relation to specific plant functional types (see Chapters 2 and 8) or vegetation forms. Thus, they are comprehensive as long as there are measurements available, but they are spatially explicit only to the point level (a number of reviews have been published on key plant parameters although they are mainly related to the plant/atmosphere interface).

As an example of a functional property of an ecosystem, the nitrogen concentration in leaves can characterize the metabolic activity of plants because this parameter is related to the amount of photosynthetic enzyme. The nitrogen can either be exposed as a thin layer in a broad leaf or it may be sheltered (concentrated) in a mass of leaf structure. Thus, the relation between N concentration and specific leaf area (SLA) connects the main parameters of plant processes and structure. A linear relation between N-concentration and specific leaf area holds for a broad range of plant species ($N = 1.57$ SLA, Schulze *et al.*, 1994*b*, with appendix of individual data). Although N concentration links metabolic activity (e.g. photosynthesis) to nutrition, it does not correlate with NPP because plants produce big or small as well as many or few leaves of similar N concentration. The link between NPP and nutrition is best described by the N content (amount of N per m² ground area or per individual leaf (Schulze *et al.*, 1994*c*; Bauer *et al.*, 1997)).

The main process dependent on leaf nitrogen is photosynthesis (Field & Mooney, 1986). The responses of CO_2 assimilation to climatic factors (Schulze & Hall, 1982) are known, but the maximum performance of species or plant functional types is not. Körner (1994) and Woodward & Smith (1994*a*) published a database on maximum rates of photosynthesis as well as the maximum levels of stomatal conductance for different plant life forms. Schulze *et al.* (1994*b*) added to this information a database on maximum levels of canopy conductance. Although this database provides general relations between plant nutrition and gas exchange, it is not yet possible to relate these functions to habitat conditions because slow growing and fast growing species may share the same habitat. The projection to a global scale was only possible via the global map of plant functional types or vegetation forms. Kelliher *et al.* (1995) and

Schulze *et al.* (1994*b*) provide a model for maximum canopy conductance in relation to LAI (data see Schulze, 1982), which allows a scaling of atmospheric transport processes (soil and plant water loss) with respect to vegetation types.

Rooting depth is a crucial characteristic of all terrestrial ecosystems with importance for fluxes of water and nutrients. It is a function of soil texture, dominant plant functional types and climate. A recent global review of point-based studies shows that the vegetation of the tundra, boreal forests and temperate grasslands in general has the shallowest rooting profiles, with 80–90% of roots in the top 30 cm of soil; deserts and temperate coniferous have the deepest profiles with only 50% of the roots in the upper 30 cm (Jackson *et al.*, 1996*a,b*). The same review extrapolated numerous measurements from about 80 published references to the land cover classification map (on a grid with $1.0 \times 1.0°$ longitude/latitude) of Wilson & Henderson-Sellers (1985). An accompanying study investigated the same literature for maximum rooting depths of all major vegetation types of the globe (Canadell *et al.*, 1996*a*). The root dataset has been extended to fine root biomass and surface area (Jackson *et al.*, 1997), and the information on root distribution has been successfully used to model optimal carbon allocation for different vegetation types at the global scale. This innovation has improved significantly the global pattern of plant transpiration in a coupled model of global hydrology (Kleidon & Heimann, 1998).

5.3.4   Radiative properties

The seasonal variation of the fraction of absorbed photosynthetically active radiation (fPAR) allow a rather simple estimation of terrestrial net primary productivity through production efficiency models (e.g. CASA, Potter *et al.*, 1993). Such models have become popular since the Advanced Very High Resolution Radiometer (AVHRR) sensor, carried on the NOAA-series satellites, provides a continuous and global coverage of optical measurements. The AVHRR measurements are often expressed as the Normalized Differential Vegetation Index (NDVI), which approximates leaf area or photosynthetic activity and is thus commonly used as a measure of general vegetation activity. A monthly global fPAR dataset at 1° resolution exists for the years 1987–1988 (ISLSCP, Sellers *et al.*, 1994), but because data from this group of satellites have been collected and archived for nearly two decades, several NDVI time series with different spatial and temporal resolutions, and different levels of processing, are available (e.g. GVI, Tarpley *et al.*, 1984; AVHRR-NDVI Pathfinder, Agbu & James, 1994; CESBIO-NDVI, Berthelot *et al.*, 1994). They are particularly useful for monitoring medium-term changes in the activity of the vegetation that can be related to events like the El Niño Southern Oscillation (ENSO) or the eruption of Mt. Pinatubo (Maisongrande *et al.*, 1995). The seasonal pattern of NDVI largely defines the phenology of the vegetation (Reed *et al.*, 1994) and indicates the length and other characteristics of the growing season. Start and end of the growing season can be defined by assuming a threshold value (Moulin *et al.*, 1997). A recent study using ten years of NDVI has shown an increase of the length of the growing season in the northern hemisphere, assumed to be related to the late winter/early spring warming of these areas (Myneni *et al.*, 1997). These NDVI datasets do not allow accurate estimation of the leaf area index (LAI) directly; nevertheless, global LAI data sets have also been produced together with the fPAR (ISLSCP, Sellers *et al.*, 1994).

5.3.5 Land use and land cover

Definitions

The distinction between land use and land cover is important for defining properties of natural and managed ecosystems and their responses to global change. Land cover refers to the actual character of the surface of Earth's land and is normally defined by the upper soil and the vegetation, while land use refers to the purpose to which humans put land and its land cover (Turner *et al.*, 1995). Land cover thus includes both natural and modified ecosystems as well as human-made landscapes, such as arable fields, parks, infrastructure and mine pits.

Human activities have modified and will continue to modify both land cover and land use. This is having a major impact on terrestrial ecosystem functioning and is a major component of global change in its own right (see Chapter 1). Currently, humans either use, modify or destroy at least 40% of the estimated 100×10^{12} t of organic matter produced annually by terrestrial ecosystems (Vitousek *et al.*, 1986). Approximately 11% of the land has been converted to cropland, 25% is grazed by livestock (FAO, 1991*a*), but only 6% is legally protected against direct human activities (Morris, 1995). Together, these changes have led to a decrease of 19% of the natural forest area globally. Almost half of the remaining forests and woodlands are more or less managed, logged with secondary growth or plantations. Besides these obvious changes in land cover, much of the functioning of land cover is degraded or at risk of degradation. Recent estimates range from 7.5×10^6 km² (light degradation), 9.1×10^6 km² (moderate degradation), to 3.1×10^6 km² (severe degradation) (World Resources Institute, 1992; Oldeman, 1993).

The potential distribution of land use is relatively well understood and can be modelled for different types of land evaluation studies (e.g. Brinkman, 1987; Leemans & Solomon, 1993; Rosenzweig & Parry, 1994). Broad-scale natural land cover patterns defined by the physiognomy of vegetation and its potential species composition can also be modelled (Woodward, 1987a; Cramer & Leemans, 1993). The combination of natural vegetation and human land use thus establishes the actual land cover pattern in any region. Not all land use changes, however, necessarily lead to apparent changes in land cover (e.g. change from natural forest to managed forest). A different land management may only alter some of the properties of land cover (e.g. enhance plant growth through fertilization), without changing the overall structure or land-cover type. The consequences of such local modifications are neglected in regional and global databases (Leemans *et al.*, 1996), which emphasize major land-cover classes and conversions between them, but the cumulative effects of such local modifications could be significant when determining greenhouse gas fluxes between the atmosphere and the biosphere.

Available land-cover datasets

Many compilations and datasets of global land cover have been developed. The most widely used for climate change impact assessments are the geographic databases by Matthews (1983) and Olson *et al.* (1985). Climatic classification schemes (Cramer & Leemans, 1993), which classify land *potentially* suitable for human land use, and the tabular databases provided regularly by FAO (e.g. the production yearbooks), which provide statistics on a country-by-country basis, are alternative approaches. The classification of land-cover patterns can be based on various structural attributes (e.g. trees versus shrubs), on phenology (e.g. deciduous versus evergreen), on floristics (e.g. oak–hickory forest), or on mixtures of these; the terminology is not always unequivocal (such as 'woodlands') or consistent between regions. Many classifications include indices that are not directly related to land cover but to environmental variables such as climate (e.g. tropical rain forest). Such differences are not problematic when a unique classification is applied throughout, but many global datasets are mixtures from different sources, using similar terminology but largely different criteria (Leemans *et al.*, 1996).

The only globally comprehensive global classification with appropriate resolution is the hierarchical UNESCO classification (UNESCO, 1973). This hybrid classification (physiognomic and structural characteristics at the top levels, species composition at the lower levels) was developed for the description of potential vegetation at a climax stage, which is probably the reason why it has never resulted in a global assessment of land cover; land cover types resulting from land use are difficult to classify. Some UNESCO vegetation maps have, however, been produced for single regions (e.g. Africa: White, 1983). Some of the newer approaches have already led to more comprehensive regional estimates of deforestation (e.g. Skole & Tucker, 1993) and to land cover maps (De Fries & Townshend, 1994). Several international organizations, including

UNEP and FAO (e.g. UNEP/GEMS, 1994), have now concluded that the implementation of an improved scheme should be prioritized in order to allow for consistent land-use evaluations, biodiversity assessments and environmental change studies.

There are large differences between global land cover databases. Plate 3 presents an aggregation of the different land-cover classes in each of a range of different databases into broad, but comparable classes. Use of different approaches, such as statistical, gridded and geo-referenced datasets and the inclusion or exclusion of regions with extreme climates (e.g. Greenland), has led to large differences in total land surface (between $121–148 \times 10^6$ km²). A reliable estimate of the total extent of land is 136.3×10^6 km², excluding Antarctica (Bartholomew et al., 1988). The most striking observation of this simple comparison is the wide range of recorded cultivated lands, ranging from 0 to 55×10^6 km². The model-based databases without cultivated land focus on potential vegetation only (e.g. Neilson et al., 1992; Prentice et al., 1992), while the largest values (e.g. Solomon et al., 1993) stem from evaluations of the land with potential for cultivation. The estimate for cultivated land (14.5×10^6 km²) from the FAO (1991a) is one of the most authoritative because their data come from national surveys with relatively well-accepted reporting protocols. In several databases, some classes are based on poorly defined categories and the global compilations are based on a multitude of sources with varying uncertainties (Buringh & Dudal, 1987; De Fries & Townshend, 1994; Leemans et al., 1996). Nevertheless, many of these datasets are frequently cited and used in global change studies.

Global land-cover databases are used in connection with models to determine possible impacts of climate change on global land-cover patterns (e.g. Emanuel et al., 1985a; Alcamo, 1994), for global carbon cycle modelling (e.g. Smith et al., 1992a; Melillo et al., 1993), and for parameterizing land surface properties in climate models (e.g. Hummel & Reck, 1979; Matthews, 1983). The data uncertainties, for example with respect to the spatial pattern, clearly limit the robustness or realism of such studies. By linking several geo-referenced environmental databases, some of the inconsistencies between the existing land-cover databases can now be removed (Loveland et al., 1991; Running et al., 1995). These and other improvements are imminent, but there are still many hurdles to be overcome, especially with respect to a unified classification and comprehensive global coverage (Townshend, 1992; UNEP/GEMS, 1994).

Statistical data on land cover and land use
The FAO Agricultural Production Yearbooks (published annually since the 1970s) contain extensive tables of land-use statistics on crop and meat productivity, extent and trade for all nations. Similar data are produced for forestry

products (see Chapter 9, Tables 9.2 and 9.3). Since the early nineties, these data are available in digital format (FAO Agrostat database, see also `http://www.fao.org`). They are used by several international organizations in their annual assessments (e.g. World Resources Institute, 1992). The FAO land-use data are based on the best statistical data available. Unfortunately, they are sometimes still only approximations, as the quality of the data depends on the reporting capabilities of individual countries. The FAO only collects, harmonizes and assembles the final datasets; if data is not available, the FAO will, through expert judgement and other sources, try to fill the gaps. This results in some imperfections, but generally the quality of the database is reliable for the crop and wood productivity statistics. The quality is lower if forest and agricultural extents are considered (see Chapter 9). From comparison of the total areas it appears that there are overlaps among classes; however, this disadvantage does not limit the use of Agrostat, as there are (or soon will be) more detailed databases on land cover. The largest drawback of the Agrostat data for global change assessments is that the scope is national and that the data are not geo-referenced. This results in large differences among countries due to differences in size as the data cannot be located to a specific locality (or grid cell).

Datasets derived from modelling land use and land cover

The IMAGE2 model (Alcamo *et al.*, 1996*a*) contains a dynamic land-use and land-cover change model, which reconciles regional demands for food, wood and fuel with the local potential to produce these products. The output is a dynamically changing land cover that is the basis for carbon flux calculations between the biosphere and the atmosphere. The demand component is calibrated with the Agrostat data for the period 1970 to 1990. A reliable initial geo-reference gridded database of land cover for 1970 was obtained by calculating potential crop patterns based on soil and climate patterns (Leemans & Van den Born, 1994). This was achieved by an iterative process linked to several regional databases of cultivation intensity, land cover and urbanization to create a realistic global pattern of agricultural land (both arable and pastures). The remaining land was filled in with potential natural patterns. This approach yielded a global land-cover database at a resolution of $0.5 \times 0.5°$ longitude/latitude, which was consistent with the national summary tables of FAO. The approach has also been used to create historic land-use and land-cover datasets for the period 1890–1990 (HYDE, Klein Goldewijk & Battjes, 1995). However, the reliability of these datasets is lower due to limited availability of statistical and environmental data to create the land-cover patterns.

New developments

The most promising global land-cover dataset is currently being developed by IGBP-DIS (Belward & Loveland, 1995). This dataset has a resolution of 1 km^2

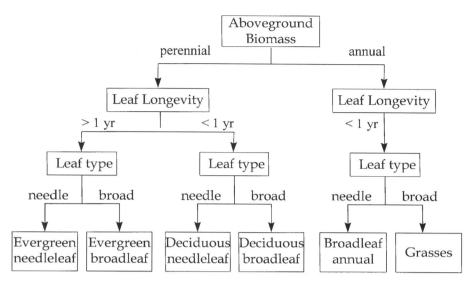

Figure 5.1   A schematic diagram of the IGBP-DIS 1 km² land-cover classification (after Running *et al.*, 1995).

and is derived from the AVHRR satellite data. The classification used is based on three primary attributes of vegetation structure: (i) permanence of aboveground live biomass (annual versus perennial); (ii) leaf longevity (evergreen versus deciduous); and (iii) leaf type (needle versus broad) (Fig. 5.1). Six combinations of these three vegetation attributes are possible but these occur across a variety of soil types and a range of climate types (Running *et al.*, 1995). All attributes can be derived from remotely sensed data. The classification is too limited for many applications – for example, the distinctions between woody and herbaceous (Schulze, 1982) is critical for most models. The scheme can be extended using data from different sources to include additional distinctions in the classification (e.g. climate for tropical, temperate and boreal or topography for lowland, highland, alpine, nival).

The approach is to use the seasonal path of the NDVI. For this dataset continental NDVI composites of 10 to 14 days were used to determine the characteristics of the growing seasons. Different NDVI thresholds in the non-growing season distinguish forests, shrublands, grassland and bare soil. For example, a very low NDVI value (less than 0.1) denotes bare land and, if observed for part of the year, is a good indication of arable land. The seasonal amplitude of NDVI allows distinction of evergreen versus deciduous canopies. Needleleaf, broadleaf or grass is distinguished by evaluating different reflectance patterns. The addition of the thermal channels of the AVHRR data improved the different distinctions (Table 5.1).

The approach has been implemented successfully for the conterminous USA

Table 5.1 *The 17 classes of the IGBP-DIS landcover dataset*

Evergreen needleleaf forest
Evergreen broadleaf forest
Deciduous needleleaf forest
Deciduous broadleaf forest
Mixed forest
Closed shrublands
Open shrublands
Woody savannas
Savannas
Grasslands
Permanent wetlands
Croplands
Urban and built-up areas
Cropland/natural vegetation mosaic
Snow and ice
Barren and sparsely vegetated
Water bodies

(Loveland *et al.*, 1991) and much progress has been made with a global implementation (for more information see http://edcwww.cr.usgs.gov.land-daac/glcc). This activity is coordinated by the Earth Resources Observation System (EROS) Data Center of the US Geological Survey with contributions from other individuals and institutes. The first implementation spans a 12-month period from April 1992 to March 1993 (Belward & Loveland, 1995). Classified land-cover data are already available for North America, Latin America and Africa; data for the rest of the world should become available in late 1997 or early 1998. A new data collection activity for the year 1995 has started and will lead to a new land-cover database. Comparisons between the two periods could present the first global analysis of global land-cover change based on observational data.

The selected classification and resolution could also facilitate the development or improvement of geo-referenced land-use databases based on the FAO Agrostat (1991*a*). The resolution of the IGBP-DIS land-cover dataset is high enough to allow fine-scale tuning between land use and cover. The use of these datasets developed for well-defined periods will accelerate global change model development and enhance understanding of land-use and land-cover change by giving empirical insights into the dynamics of land-cover conversions. The effectiveness of these future databases for assessing more complex land-cover modifications is still open to debate.

5.4 What will be needed in the future?

The development of models and databases are highly interdependent. Data gathering activities are increasingly driven by model requirements, whereas model development is limited by the availability of data. Potential new sources of high quality global datasets have recently been summarised by a special issue of the IGBP Newsletter (No. 27, September 1996), covering the activities of IGBP-DIS. Most prominently featured is the 1 km² AVHRR global land dataset, as well as the derived land-cover data (Belward & Loveland, 1996). Both of these will be of immense value for the modelling of biospheric processes. Due to their unprecedented high spatial resolution, they are likely to stimulate new modelling approaches, for example, for spatially distributed hydrological models with global coverage (using the connected digital elevation database which is also under development, see Hastings, 1996).

Model comparisons have shown that a significant limitation to global model development is the lack of readily available site data for basic fluxes such as net primary and ecosystem productivity (NPP/NEP) (Cramer *et al.*, 1996), as well as vegetation structure and disturbance regimes. Although many observations exist from different sources, a compilation of them, with appropriate checks for quality and consistency, has only just begun through the GPPDI (Olson & Prince, 1996; Olson *et al.*, 1997). This database also will be essential for future assessments of, for example, carbon fluxes between the atmosphere and the terrestrial biosphere.

Dynamic modelling of transient processes requires, however, two additional data elements that are currently missing:

- A historical climate dataset with sufficiently high resolution in time and space. This dataset needs to reflect climatic conditions at the canopy level as much as possible. Therefore, the climatological re-analysis datasets that are currently under development (Trenberth, 1995) are likely to provide only a first-order solution to the problem.
- The history of human land use and the resulting land cover has been assessed by various studies resulting in useful databases (Esser, 1989; Klein Goldewijk *et al.*, 1994). There is consensus, however, that a much improved source of this data is needed.

Model development will continue with or without these two global datasets, though the difficulties of model testing and limitations of model application will remain. Projections of the future of Earth's terrestrial vegetation will be more credible if the trends that have occurred during recent decades can be simulated accurately.

6 Use of models in global change studies

J. Goudriaan, H.H. Shugart, H. Bugmann, W. Cramer, A. Bondeau*,
R.H. Gardner, L.A. Hunt, W.K. Lauenroth, J.J. Landsberg, S. Linder,
I.R. Noble, W.J. Parton, L.F. Pitelka, M. Stafford Smith,
R.W. Sutherst, C. Valentin and F.I. Woodward
*Previous name: Alberte Fischer

6.1 Introduction

Computer models are important tools for gaining a better understanding of how terrestrial ecosystems interact with a rapidly changing environment. Models can be used to synthesize existing knowledge in a common framework and to explore the consequences of known or hypothetical mechanisms at higher levels of integration. Quantitative models of the dynamics of vegetation under current and altered environmental conditions can be tested by attempting to reconstruct the response of terrestrial ecosystems to past changes and used to project future changes.

Ecological modelling in the context of global change addresses many basic issues of ecosystem structure and functioning, for example, the fundamental nature of succession, the importance of competition in structuring communities, and the role of internal feedbacks in ecosystem dynamics. A critical factor in these phenomena is scale – i.e. the need to define carefully the time and space scales of relevance to the issue and to ensure an appropriate match of the sets of observations being used to test theory. For example, within seasons periods of limitation by nitrogen may be unavoidable and important in understanding the dynamics of ecosystems on annual time scales, but on the century time scale the nitrogen supply may well be strongly linked to carbon cycling rates and can be assumed to be non-limiting. Ignoring processes that are present in reality but not relevant to the scale of inquiry is not a weakness of models, but a necessary step in the scientific process leading to an improved understanding of the behaviour of ecosystems.

6.2 Model types and assumptions

Ecological models are mathematical expressions developed to be analogous with an ecosystem or a part of an ecosystem. The models of principal interest here are those that are used to integrate information and to produce predictions of responses of ecosystems to environmental change. The capabilities for predicting the responses of terrestrial ecosystems to global change and to the natural fluctuations in the Earth's climate have developed to a remarkable degree in the past decade. The ability to predict the consequences of change derives from an array of models, each with different underlying assumptions and with different data requirements.

There is a wide variety of ways to categorize ecological models. In this chapter, a simple categorization is adopted, starting with a review of models aimed for managed terrestrial ecosystems, i.e. crops, grasslands, and forests. Biogeochemical models, which describe the fluxes of water, carbon and nutrients between the vegetation and the atmosphere, have often been used to study aspects of ecosystem functioning in global change research and are considered next. These models typically ignore changes in community structure and composition. Models that deal with these issues (but in turn largely ignore ecosystem functioning) are considered in a section of its own. Finally, recent model-based attempts to unify structural and functional dynamics of ecosystems are reviewed.

6.2.1 Models for managed terrestrial ecosystems

There are few terrestrial ecosystems that are not influenced by humans in some way or another. Selective removal of trees in a natural forest may gradually cause a shift in species composition, for instance towards an oil palm grove, as is happening in some parts of the wet tropics in Africa. Models for managed systems cover the whole range from extensive to intensive management. In this review some specific models are mentioned for illustrative purposes, but in no way is this intended as an exhaustive review. More comprehensive inventories of agro–ecosystem models exist, such as that compiled by Plentinger & Penning de Vries (1996).

Agricultural crop models

Global change presents considerable challenges to crop growth models. First, models are now required to be more robust over a broader range of environmental conditions than those for which they were developed; extending understanding into conditions that do not currently exist becomes pre-eminent. Second, additional physiological functions, such as the effects of increasing atmospheric CO_2 concentration on growth and yield (see Chapters 7 and 9), need to be modelled explicitly. Third, generic crop models applicable over broad areas need to be developed; the current generation of crop models are plot-specific and thus less suitable for global change studies simulating food production over regions, continents or the globe.

Table 6.1 *Crop production situations (Rabbinge, 1993)*

Production Situation	Limiting or reducing factors	Dry biomass at the end of the period of growth ($g\ m^{-2}$)
Potential yields	Radiation and temperature	1000 – 2000
Attainable yields	a Water	500 – 1000
	b Nitrogen	300 – 500
	c Phosphorus	200 – 500
Actual yields	Weeds, pests and diseases	Less than either potential or attainable

Current plot-specific crop growth models simulate the development of a crop plant through its growing season as a function of weather and of the uptake of water and nutrients from the soil. Thus, the models have the physico-chemical environment as input, and biomass with its components as output, with an obvious emphasis on yield of edible components. Most of them simulate a 'potential' crop growth level that is achieved when the crop is free from pests and diseases and when it has an optimal root environment (Rabbinge, 1993). For potential growth, only the atmospheric climate determines the growth. When there is a shortage of belowground resources such as water or nutrients, production decreases, resulting in the so-called attainable yield. These factors are designated 'growth-limiting factors' (not to be confused with the 'reducing factors' pests, diseases and weeds; see Table 6.1 and Box 9.1). The distinction between potential and attainable yield is made on practical grounds, as it is agronomically possible to improve the belowground factors water and nutrients, while the climatic factors are outside control except in greenhouses. Some models directly use water or nutrient efficiency; in other models the processes of uptake and conversion are included so that the resulting efficiency is a model output.

Phenology (the time course of the development of the plant) is the main determinant of the length of the growing season for crops. Even the simplest crop growth models have some kind of phenological calendar to determine the timing of onset and termination of growth. Phenological development is primarily driven by temperature, with photoperiod (daylength) and cold requirement (vernalization) as secondary modifiers, although moisture availability also affects phenology, especially in the tropical zone. The scale of phenological development of a crop is an ordinal scale with arbitrary stage indications, but through practice it has been further developed such that a practically linear scale with degree-days could be constructed (Zadoks *et al.*, 1974). In rice and wheat growth models the effect of temperature is usually represented by an accumula-

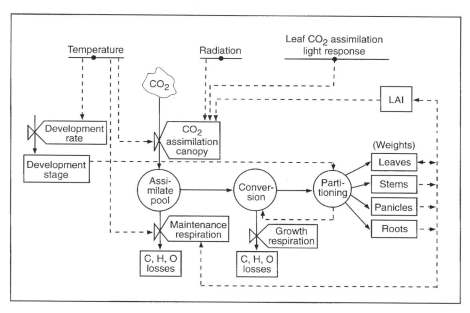

Figure 6.1 Structure of a typical crop growth model (e.g. SUCROS, Van Laar *et al.*, 1992; and AFRCWHEAT, Porter, 1993).

tion of degree-days after germination (the temperature sum, or sum of heat units). The correspondence between the temperature sum and phenological stage is strongly variety-specific, which is reflected in the need for extended databases for variety parameters (Hunt *et al.*, 1990; Hunt & Pararajasingham, 1995; Hunt & Boote, 1997). Also, the precise interaction of temperature, photoperiod and vernalization on rate of development is handled in various ways in the different models, presumably because of real differences between species and even cultivars (Yin, 1996).

Allocation of assimilates to the various plant organs is directly related to the phenological stage of the plant. In the beginning of its life cycle much of the assimilates is allocated towards the root system. After flowering, this downward flow is much smaller, and the seed-filling stage will begin so that most assimilates will flow to the reproductive organ. The precise quantitative formulation of these relations is quite variable among the different models, although they agree on the major characteristics. Examples of crop models (Fig. 6.1) are SUCROS, CERES, AFRCWHEAT, CROPSIM, SINCLAIR, Sirius, DE-METER, SOYMOD, SWHEAT, ECOSYS (Ingram, 1996; Plentinger & Penning de Vries, 1996). This list is not exhaustive, and is continually growing.

The conceptual level of crop models is mostly biophysical, but obviously there is also interest in outputs in terms of concepts such as money flows, job opportunities, landscape effects, and the like. To reach such a high level of

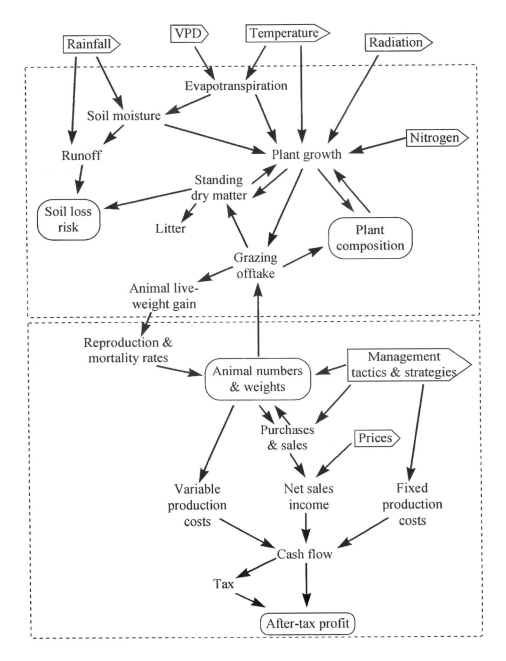

Figure 6.2 Structure of a generic rangelands model for global change research.

abstraction, the outputs of the biophysical models have to be linked into integrated, farm-level system models which include management decision-making, and processes even further along the chain of decision-making (e.g. effects of decisions at the agri-business level). Interestingly, grassland models are further developed in this direction than crop growth models.

Grassland (pasture and rangelands) models

Grassland models have been used to improve understanding of the responses of grassland ecosystems to global change and to simulate the impact of grazing management systems (Fig. 6.2). The range of available management treatments is large (stocking rates, nutrient amendments, fire or cutting frequencies, etc.) and there are a variety of environmental conditions that may be imposed on the system. Models are essential to interpolate and extend experimental results and explore the implications of rare events such as particular sequences of rainfall, which are important in determining the production of pastures and rangelands.

Examples of different types of grassland models that have been used in global change research are GEM2 , SPUR , CENTURY and GRASP (McKeon *et al.*, 1993). The grassland ecosystem models GEM (Hunt *et al.*, 1991) and GEM2 seek to predict ecosystem level effects of CO_2 enrichment and climate change on primary production and biogeochemical cycling. SPUR and GRASP simulate the ecosystem in somewhat less detail, but carry pasture production on to secondary production in terms of animal growth. GRASP has been linked to a herd dynamics and economics model, Herd-Econ, to simulate the interactions between the biophysical system and management decisions (e.g. about stocking rates), economic effects (e.g. differing levels of product demand) and policy (e.g. changes in impacts such as taxation), Campbell *et al.*, (1997). The CENTURY model has been used at the site, regional and global scales for different climatic change scenarios and for the interaction between climatic change and increasing atmospheric CO_2 concentration. Reviews of the use of grassland models for developing management strategies and for exploring global change issues are presented by Parton *et al.* (1994*b*), Hanson *et al.* (1985) and Campbell *et al.* (1997).

Forest yield models

Traditional forest models used by managers are empirical descriptions of forest stand development over time, based on long-established mensuration procedures. These models use concepts such as 'site index' (often expressed as the height of dominant trees in even-aged stands, as a measure of a site's 'growth potential' or 'quality'), which depends on some form of calibration, and are essentially statistical descriptions of the growth of stands with specified stem populations. While the nutritional status of the soil is reflected in the site index, these models almost invariably lack any physiological basis and do not take site water balance into account (except in so far as average water conditions are reflected in the long-term average tree growth rates and thus parameterized in the site index). Thus, they cannot be used to estimate forest growth rates or carbon fixation under changed environmental conditions. However, from the point of view of yield prediction under an unchanging environment in managed areas where there is a reasonably long history of forest measurements, these conventional forest models are currently more useful than process-based models and provide the basis for most operational forest management decisions. Landsberg & Waring (1997) have recently produced a simplified process-based model (called 3-PG) specifically designed to predict growth rates and yield of evergreen forests, where basic information on climate and soils is available, but given the widespread use of forest yield models, adoption of a model such as 3-PG, even when it has been extensively tested, will probably take a long time.

Several process-based forest productivity models of varying levels of complexity have been developed in recent years (McMurtrie *et al.*, 1990*b*; Ryan *et al.*, 1996*a,b*). These can provide useful estimates of carbon fixation by forests and stand growth rates in terms of biomass production. Some of these models, in particular FOREST-BGC (Running & Gower, 1991) and the so-called e-model (see Landsberg & Hingston, 1996), although point models dealing only with homogeneous plant communities, are being used in association with remote sensing measurements to estimate forest productivity over large areas. Coops *et al.* (1997) produced a version of the 3-PG model (3-PG S) adapted for use with remote sensing measurements. Initial tests against data collected in forests in Australia and New Zealand have shown that the model can simulate observed mean annual increments with considerable accuracy. Other models, such as G'DAY (Comins & McMurtrie, 1993), include nitrogen uptake and soil organic matter turnover, and are emerging as powerful tools for the analysis of ecosystems and their probable responses to perturbations imposed by management or caused by global change (Fig. 6.3).

Yield–reducing processes: pests, diseases, weeds

A wide range of analytical tools have been used to estimate the impact of climate change on pests, diseases and weeds (see Chapter 9 and Box 9.1). Here, examples of some of the different approaches are briefly described.

CORRELATIVE MODELS.   There are large numbers of species of pests, diseases and weeds to consider when evaluating the likely impact of global change on terrestrial ecosystems. In practice there is inadequate data on all but the most thoroughly studied species to be able to develop a realistic, dynamic

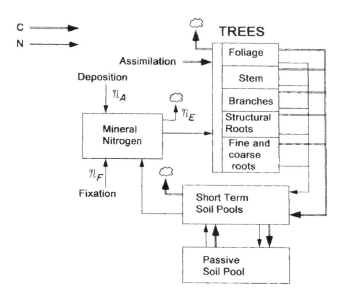

Figure 6.3   Schematic representation of the G'DAY model, an example of a process-based plant–soil model used to simulate the responses of forest productivity to global change (from Comins & McMurtrie, 1993).

population model. An alternative approach is to use a simplified model – such as the CLIMEX model (Sutherst *et al.*, 1995) – to describe the essential features of a species' climatic requirements and then to run the model with a geographical database of climate data containing current climate data and climate change scenarios. The response of a species is related to those seasonal conditions that favour population growth and to those that ensure survival until the next growth season. These 'growth' and 'stress' indices are combined into an annual 'ecoclimatic' index to give a measure of the potential performance of a species in a given location or year in relation to climate. Output can be linked with other data in a GIS to carry out an evaluation of the combined response to climate and other environmental variables such as soil type or host availability.

The parameter values are inferred from the species' geographical distributions, taking care to ensure that the limits are determined by climate rather than some other variable like host availability (a significant problem for any modeller using geographical data to estimate species parameters). Additional information of value in parameter estimation is the seasonal phenology of the species and experimental data, which add confidence to the model estimates. Numerous analyses of climate change impacts on plants and animals have been carried out using CLIMEX (e.g. Sutherst, 1991; Sutherst *et al.*, 1995).

POPULATION DYNAMICS MODELS.   Simulation models that describe the population dynamics of pests, disease and weeds, using climatic driving variables (e.g. as temperature, moisture, light or pollution) to take account of the different climates in different geographical areas, are often useful tools for global change studies. In each case there are varying needs to incorporate both the direct and indirect effects of changes in atmospheric composition, particularly CO_2. Such models are usually derived using data from a limited number of point studies, with evaluation using independent data from different areas. They can be run in different locations, using locally available data. A common limitation of most pest dynamics models is their limited ability to predict the size of the population accurately, although they can give reliable estimates of the timing of events that depend on temperature, in particular.

Simulation of weed population dynamics is based on an understanding of the competitive interactions between the weed and the surrounding vegetation. Large parts of these models, such as the canopy process modules, are similar to the crop, forest and grassland models described earlier. Weed models require additional information on management practices such as planting times, crop varieties and prior use of the land, as they must simulate the growth of two different plants, one of which is strongly influenced by management.

Migration is frequently a significant process when dealing with the impacts of global change on pests, diseases and weeds. Climate change can result in a geographical shift in the area suitable for propagation of a noxious species; concurrently, wind patterns may shift leading to changes in the migration routes of those species. To address these issues, spatially explicit models have often been used so that the effects of immigration and emigration can be simulated explicitly. For example, Heong *et al.* (1995) used a model to investigate the impact of climate change on the migratory patterns of the brown plant hopper (a major pest of rice in Southeast Asia) and its predators. Plant migration in general is discussed further in Chapter 8.

PEST AND DISEASE IMPACT MODELLING.   Assessment of the impacts of pests and diseases on vegetative growth and agronomic production requires integration of the separate models for plants and pests/diseases (Boote *et al.*, 1983; Teng *et al.*, 1996). Examples of such coupling include the work of Bastiaans (1993) for the effect of leaf blast on rice; that of Van Oijen (1991) and Rossing *et al.* (1992) for the effect of late blight on potato; and that of Xia (1997) for the effect of cotton aphid and its biological control on cotton.

The mechanisms of pest and disease damage vary enormously: reduction of light interception due to removal of foliage, reduction of leaf photosynthesis due to increased senescence, removal of assimilates and nutrients; hormonal interference with effects of growth and morphogenesis; and deterioration of product quality.

Pests and diseases involve a higher trophic level than do weeds, which are plants themselves competing for the same resources. Thus, the modelling of the impacts of pests and diseases is an order of magnitude more complex than the modelling of weed impacts. The reliable inclusion of the effects of global change into these coupled models of plant and pest/disease is still in the early development phase (e.g. Goudriaan & Zadoks, 1995). The unpredictability and short time frame of disease outbreaks, in particular, necessitate careful, long-term observations of epidemiological factors. Modelling requires sound knowledge of the causal factors of diseases outbreaks and their impacts, and is just one tool in developing effective strategies to combat disease outbreaks.

6.2.2 Biogeochemical models

Several types of models have been used to investigate the effects of global change on the functioning of terrestrial ecosystems in terms of the fluxes of carbon, water, and energy. Here, models are categorized according to whether they calculate the full biogeochemical cycle, including belowground dynamics and fluxes ('ecosystem models'), whether they consider the fluxes of CO_2, water and energy from the canopy only ('canopy models'), or whether they concentrate on parts of those fluxes only ('models of partial ecosystem processes').

Ecosystem models

The fluxes of energy, water and basic elements (C, N, P) between the terrestrial biosphere and the atmosphere have received great attention during recent years due to the debate about possible feedbacks between atmospheric and biospheric change, and particularly the potential role of terrestrial vegetation as a major sink of atmospheric carbon (see Chapters 7 and 10 and Section 12.2.1). For the improvement of estimates of this sink and of its behaviour over time, it is necessary to simulate the relationship between the state of the ecosystem (such as its structure), climatic conditions and the ambient concentration of CO_2 in the atmosphere. A broad range of models have been developed for this purpose. Such models capture basic ecophysiological processes, such as the energy balance at the leaf or canopy level, the photosynthetic assimilation of carbon to carbohydrates, their allocation within the plant to different tissue pools or to autotrophic respiration, the production of litter, and the heterotrophic respiration of the litter pool through various pools of soil organic matter. Ågren *et al.* (1991) and Ryan *et al.* (1996*a,b*) distinguish ecosystem models based on the level of resolution, ranging from detailed physiological models to simplified ecosystem models. The physiological models simulate the response of individual plants to environmental factors and

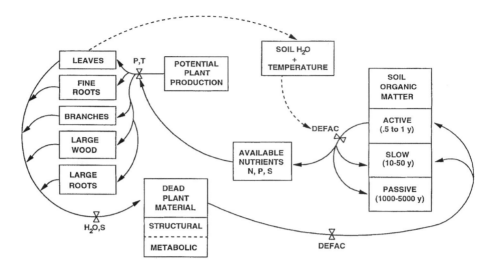

Figure 6.4   General flow diagram for the CENTURY biogeochemical model (from Parton *et al.*, 1994*a*).

incorporate considerable ecophysiological detail on such processes as photosynthesis, respiration and transpiration. These models are most appropriate for examining ecosystem responses at specific sites and over relatively short periods of time (days to decades) because of the extensive input data and computational complexity of the models. An example of such a model is ECOSYS (Grant *et al.*, 1995*a,b*).

Simplified ecosystem process models simulate the fluxes of elements such as carbon, water and nitrogen among plant and soil pools within the ecosystem. They tend to aggregate plant compartments into relatively few classes but have more soil compartments compared to the detailed plant physiology models. The more simplified ecosystem process models generally use longer time steps (month to year), require less input data, and have been used at the regional and global scale for most major ecosystem types. Examples of this type of model are TEM (Melillo *et al.*, 1993) and CENTURY (Parton *et al.*, 1987, 1993), both of which have been used to evaluate possible global change impacts at regional and global scales (Fig. 6.4).

The processes determining growth and their environmental interactions in managed ecosystems are basically similar to those in natural ecosystems. These processes underly biomass formation in both natural and managed ecosystems, and the description of the processes of photosynthesis, stomatal regulation and transpiration is similar. The difference comes with the representation of phenological processes, allocation and senescence. Grassland models generally represent the functional processes less precisely than crop models, but they are more complex in that they also represent grazing and mowing, and often also the secondary production by the grazing animals. In all ecosystems primary productivity can be reduced by pests and diseases; in crops, competing weeds may take their share of water, nutrients and radiation.

Applications of biogeochemical models to global change issues are described in Chapters 7 and 10.

Canopy models

A number of canopy process models have been developed in recent years, dealing with carbon fixation and forest growth at various levels of complexity and with emphasis on different processes and interactions. A representative sample of these models is reviewed here to illustrate the principles involved and the capabilities and shortcomings of the approaches, especially with respect to evaluating the impacts of global change. The principles embodied in them, at least those relating to PAR (Photosynthetically Active Radiation) interception and CO_2 uptake by canopies, are essentially the same across most models, at varying levels of detail. The belowground factors are assumed to be expressed through the functional properties of the photosynthesizing leaves and through growth characteristics such as partitioning and specific leaf area, so that the attention here is focused on the canopy processes themselves.

In this context, two model types can be distinguished: models that are formulated at the canopy scale (designated here as 'canopy-based models'; Fig. 6.5); and models that are formulated at the leaf scale, followed by integration across many leaves to scale up to the canopy level (designated here as 'leaf-based models'; Fig. 6.1) (Table 6.2).

In canopy-based models net primary production is immediately calculated from the total amount of radiation intercepted by the leaf canopy. The radiation use efficiency is expressed in grams of dry matter per MJ of short-wave radiation intercepted by the plant canopy (photosynthetically active radiation and near infrared radiation taken together), typically about 0.8 g MJ^{-1} for leguminous species, about 1.0 g MJ^{-1} for C3 non-leguminous species, and 1.5 g MJ^{-1} for C4 species. This ratio is an output variable in the more detailed leaf-based models, but here it is an input. Reduction factors for non–optimal temperature or shortage of water or nutrients may be used, as well as reduction of intercepted radiation due to competition. A direct CO_2 effect can be simulated as a multiplier. Respiratory processes are included in the value of the radiation use efficiency.

Leaf-based models, on the other hand, represent the wide diversity in irradiance conditions for the individual leaves, with its repercussions for photosynthesis, stomatal conductance and transpiration. These models are usually applied to basically homogeneous canopies, but sometimes also for canopies

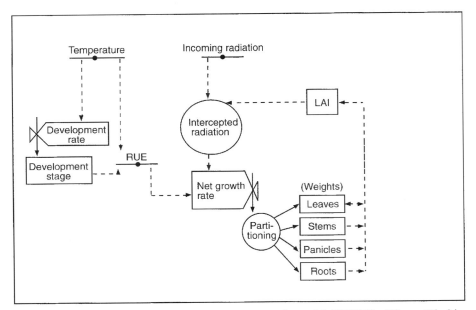

Figure 6.5   Structure of a canopy-based crop growth model (CERES – Wheat, Ritchie & Otter, 1985). (Figure 6.1 shows the structure of a leaf-based crop growth model.)

with a crown structure (with a consequent need to integrate across all leaves).

Both leaf-based and canopy-based models can be used to separate the fluxes associated with CO_2 uptake by canopy photosynthesis from other fluxes such as soil respiration. Photorespiration is normally parameterized through the value of the CO_2 compensation point, which rises with increasing temperature. This process operates on the leaf scale and is a major component of the positive effect of ambient CO_2 on canopy photosynthesis (see Chapter 7).

Canopy-based and leaf-based models both generate a net carbon assimilation rate on a daily basis. The allocation of these assimilates to generate growth of leaves, roots, stems and reproductive organs is affected by phenology as expressed as flowering, maturation, senescence, and onset of dormancy. Whereas canopy processes are basically similar for forests, grasslands and crops, their developmental and growth processes are not. These differences lead to separate descriptions in models of growth processes for specific biome types.

The seasonal time course of leaf area index, or canopy-soil cover, is one of the most important determinants of radiation interception and crop growth. It is often possible to drive the models with remote sensing data so that the estimate for LAI is improved and does not drift away from real-world values. Some models such as BIOME-BCG were specifically designed for this application.

Table 6.2 *Canopy-based and leaf-based models of primary production*

Canopy-based models		
Model name	Reference	Registered in
LINTUL	Spitters & Schapendonk, 1990	CAMASE
SINCLAIR	Amir & Sinclair, 1991	GCTE
Sirius	Jamieson *et al.*, 1998*b*	GCTE
SIMRIW	Horie et al., 1995	GCTE
CROPSIM	Hunt, 1994*a*; Hunt & Pararajasingham, 1995	GCTE
WOFOST	Van Diepen *et al.*, 1989.	GCTE/CAMASE
CERES	Ritchie & Otter, 1985	GCTE
EPIC	Williams *et al.*, 1990.	CAMASE
BIOME-BGC	Running & Hunt, 1993	

Leaf-based models		
Model name	Reference	Registered in
SUCROS	Van Laar *et al.*, 1992.	GCTE/CAMASE
ORYZA	Drenth *et al.*, 1994.	CAMASE
AFRCWHEAT	Porter, 1993	GCTE
WTGROWS	Aggarwal & Kalra, 1994	GCTE
DEMETER	Kartschall *et al.*, 1996	GCTE
ECOSYS	Grant *et al.*, 1995a,b	GCTE
MAESTRO	Wang & Jarvis, 1990a,b	
BIOMASS	McMurtrie *et al.*, 1992*b*	
ForGro	Mohren *et al.*, 1993.	CAMASE

Models of partial ecosystem processes

WATER BALANCE. Shortage of water is the most common factor that limits the formation of biomass. In models for this situation the water content of the soil is included as a balance of precipitation and drainage, evaporation and transpiration. The soil water balance is often calculated by the Penman–Monteith equation (Penman, 1948; Monteith & Unsworth, 1990) and daily precipitation; rooting zones are estimated; and the water balance is used to constrain stomatal conductance values, which are also constrained by atmospheric vapour pressure deficits (see also Section 7.4).

The calculation of drainage can be done in a complex manner, using the Richards equation (Richards, 1931; van Genuchten, 1980; Feddes *et al.*, 1988) and the soil hydraulic properties, and in a simpler manner with the so-called bucket approach (e.g. as in the models SUCROS and CERES). The latter approach has simplified drainage to an event that occurs only if the overlying soil layer exceeds field capacity. A so-called drainage coefficient may be included as a multiplier that reduces the rate of drainage of excess water within the next time interval. A proper allowance for surface run-off requires the actual topology of adjacent plots. This is usually not done; instead run-off and run-on are characterized by parameterization (e.g. Williams *et al.*, 1990).

In these models, climate operates primarily through the direct effects on transpiration and evaporation, which usually are expressed by the Penman–Monteith equation. However, increasing CO_2 also has an effect through stomatal conductance (see Chapter 7). To date, most crop growth models have parameterized this indirect CO_2 effect on transpiration, rather than building it in to interact with the assimilation rate. The work of Leuning (1995) provides a methodology to include the effect in the detailed leaf-based models.

NUTRIENTS.   Within GCTE's work on managed ecosystems, global change effects on nutrient availability have not yet received much attention. Some models such as CERES, AFRCWHEAT2, SWHEAT (Van Keulen & Seligman, 1987) do simulate the interaction between altered potential growth and nitrogen availability, but there has not been a concerted model comparison for this specific issue yet. It is fair to say that in this field the existing traditional knowledge of fertilization has not been fully applied.

However, for natural ecosystems, both modelling and experimental work on the effects of altered nutrient availability on ecosystem functioning are underway within GCTE (see Sections 7.3 and 7.5). Some ecosystem modelling approaches are now incorporating the interactive effects of CO_2, temperature, precipitation and nitrogen availability (Section 7.8).

SOIL ORGANIC MATTER (SOM) AND SOIL EROSION.   Most ecosystem models have components that simulate SOM decomposition. For example, Comins & McMurtrie (1993) have produced a model called G'DAY (Generic Decomposition and Growth), based to some extent on BIOMASS but incorporating SOM decomposition modules based on CENTURY (Parton *et al.*, 1993) and a dynamic N-uptake algorithm. Carbon partitioning remains empirical. G'DAY has been designed to produce simulations for periods up to hundreds of years and provides a tool for investigating the interactions between N-uptake and utilization and carbon fixation. The use of G'DAY for such long-term simulations is described in Chapter 7.

Whereas models that simulate SOM decomposition do not have to be spatially heterogeneous, soil erosion models are heterogeneous by necessity. They can be categorized by the main erosive agent (Table 6.3) and by spatial scale (field scale and catchment (< 10 km^2) scale). Daily time steps are used in many current models, and these are then aggregated to assist with the validation of coarse-resolution models.

Table 6.3 *Soil erosion models*

Name of model	Reference
Water erosion WEPP	Lane & Nearing, 1989; Flanagan & Nearing, 1995
Wind erosion WEPS	Hagen, 1991
Integrating crops (CERES model) & soil erosion EPIC	Williams et al., 1990; Skidmore & Williams, 1991

The European Soil Erosion Model (EUROSEM, Morgan *et al.*, 1998) is a dynamic distributed model for simulating the erosion, transport and deposition of sediment over the land surface by inter-rill and rill processes. It is designed as an event-based model for both individual and small catchments. EUROSEM describes catchments by dividing them into a simplified cascading network of elements. The Limburg Soil Erosion model (LISEM, De Roo *et al.*, 1997) is a physically based hydrological and soil erosion event model for drainage basins. It is one of the first examples of a physically based model that is completely incorporated in a raster Geographical Information System, which facilitates the use of remotely sensed data from airplanes or satellites.

Expert systems may be an effective alternative to the highly mechanistic, data-intensive models discussed above (Boardman *et al.*, 1990). The rules generated by the expert system can be used to predict erosion rates at different scales across the landscape under varying climatic conditions.

Integration between crop growth and erosion models has been facilitated by interfaces such as CLIGEN (Nicks & Lane, 1989), SPUR (Wight & Skiles, 1987) and SWRRB (Arnold *et al.*, 1990). The effects of CO_2 on plant growth and water use efficiency have also been added (Stockle *et al.*, 1992; Arnold *et al.*, 1993). Ecosystem and crop models can benefit from erosion models by building the run-off subroutine of the erosion models into their water balance submodels. Run-off is rarely adequately predicted and often is simply overlooked. Similarly the prediction of rill and gully networks could also be included. In addition, soil erosion is important from the viewpoint of decreasing fertility (SOM and nutrients); thus, it amplifies a limiting factor for plant growth. The impact of deposition either by water (sediments) or by wind (dust) could also be included in ecosystem and crop models. These deposits induce a marked change in soil properties (physical, chemical and biological). Conversely, soil erosion models could use the outputs from ecosystem and crop models in terms of canopy development, plant colonization, plant residues and organic matter production.

By and large, there are no real interactions yet between landscape models developed by ecologists and soil erosion models, even at the watershed level, although they do share some common approaches (e.g. GIS, cellular automata).

6.2.3   Models of community dynamics
Patch models of community structure and composition

In these models vegetation dynamics are modelled at the scale of a 'patch'. Ideally, the size of a patch is chosen so that it can be assumed that (i) every plant on the patch fully interacts with every other plant, and (ii) a mature individual occupies the whole area of the patch. The interactions considered include shading and the competition for water and nutrients. On such small patches the detailed horizontal relationships between individual plants can thus be ignored, reducing model complexity and saving much effort in parameter estimation and computation. This assumption has been found to work effectively for many ecosystems, particularly forest ecosystems, where patch size is in the order of 0.01–0.1 ha. The suite of models originating with JABOWA of Botkin *et al.* (1972) and further developed by Shugart and colleagues via the FORET family of models (Shugart, 1984) is a successful application of the concept. These models have been used widely to investigate the dynamics of forest stands over periods of decades to centuries (Solomon *et al.*, 1981; Solomon & Tharp, 1985; Solomon, 1986; Pastor & Post, 1988) .

Another approach has been developed to avoid the assumption that each individual plant in a patch interacts fully with every other individual (assumptions (i) and (ii) above). These models take the full spatial position of each individual into account in evaluating competition and other interactions between individuals (e.g. Pacala *et al.*, 1993). These are discussed further in the section on landscape models below.

The most common patch models used in global change studies are gap models (Fig. 6.6). Other kinds of models that fall into the same category but have been less commonly used are Markov models (e.g. Horn, 1975) and state-and-transition models (e.g. Starfield & Chapin, 1996). There are several comprehensive reviews of gap models in the literature (Coffin & Lauenroth, 1990; Prentice *et al.*, 1993; Bugmann, 1996a); here only a brief overview with specific emphasis on their use in global change studies is given.

Gap models simulate population dynamics of individual plants over time by considering establishment, growth, and mortality. The models keep track of the size and age of each individual on the plot. Models of forests typically are based on the assumption that light is the most critical resource, while shrubland and grassland models focus on belowground resources, i.e. either water or nitrogen, or both.

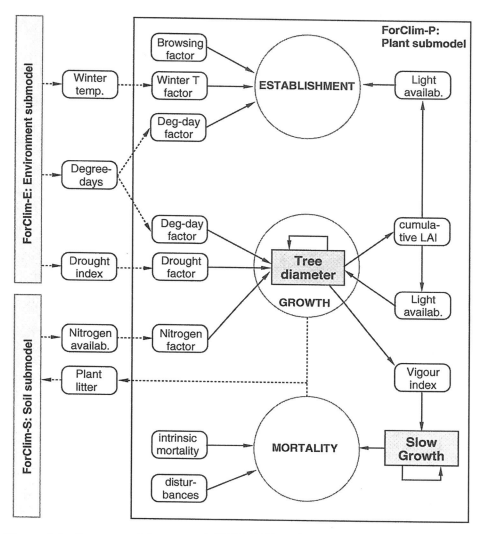

Figure 6.6   Structure of the gap model FORCLIM, with an emphasis on the plant submodel. Square boxes denote state variables, boxes with rounded corners are auxiliary variables. Arrow from box *x* to box *y* indicates that $y = f(x)$, and broken lines denote the calculation of input/output variables. Redrawn from Bugmann (1994).

 The establishment of new individuals on a patch is controlled by the availability of propagules, environmental constraints, and competition from existing individuals. Environmental constraints are most often represented as the temperature and soil water conditions required for establishment. New individuals are added to the plot by randomly drawing individuals from a pool determined by the above constraints. Traditionally, growth is calculated as a potential biomass increment based on the current size of each individual relative to its maximum size. From this potential increment the actual increment is calculated based on temperature and light, water, and nitrogen availability. Light availability is influenced by vertical canopy geometry, its modelling ranging from simple approaches (e.g. JABOWA in Botkin *et al.*, 1972) to more detailed formulations of crown shapes (e.g. FORSKA in Prentice & Leemans, 1990). In the early models, the effect of temperature on tree growth was usually based on the concept of growing degree–days, with zero growth occurring at the northern and southern limits of current distribution of each species and a parabolic response function between these points, assuming maximum growth in the centre of the range.

The problems associated with this approach have been recognized (Bonan & Sirois, 1992) and various alternative approaches have been formulated (Coffin & Lauenroth, 1990; Prentice *et al.*, 1993; Cumming & Burton, 1996; Bugmann & Solomon, 1998). Soil moisture effects are typically represented by a drought index, using a soil water submodel that calculates the annual course of potential and actual evapotranspiration. Maximum growth is then reduced depending on the drought sensitivity of the individual. Direct representations of the effect of water availability on growth require an estimate of soil water and a water-use efficiency parameter to translate water availability into an amount of biomass or carbon that can be supported. Nitrogen has been represented in an analogous fashion to water – either as an index or as a direct resource. Direct representation of nitrogen requires a calculation of nitrogen-use efficiency to estimate the amount of biomass that can be synthesized by the available nitrogen. Improved modelling of the effects of water and nitrogen on tree growth is an important area for future work on gap models to enhance their applicability for global change studies. The information generated by gap models about the composition of material that is available for mineralization as simulated by spatially homogeneous biogeochemical models may increase our ability to understand the interactions between ecosystem structure and biogeochemical processes in responses to global change.

Mortality is one of the most difficult processes to represent adequately in gap models because there are few data on tree mortality rates in forests. Two aspects are often represented: an intrinsic mortality that is parameterized from the maximum longevity of the trees, to which a resource-related mortality rate is added. For the base mortality, only 1% of a tree cohort is usually assumed to survive to maximum tree age, resulting in a negative exponential curve of survivorship. Typically, the resource-related mortality rate is based on the actual growth performance of each individual relative to its potential performance. If during several years an individual cannot obtain sufficient resources to support a minimal growth rate, it is subjected to increased hazard of mortality. Because of lack of understanding of factors accounting for mortality in long-lived individuals like trees, the mortality rates in some models have been adjusted so that the composition and structure of the simulated stands agreed with observed data for the particular system. Given the likelihood that mortality rates will change significantly as a result of rapid environmental change, research on tree mortality is another critical area for improvement of this class of models for global change studies.

Landscape models

Ecological dynamics at the landscape scale are usually described in three different ways:

(i) Non-interactive patches: the models simulate a landscape unit that is composed of discrete, non-interactive elements (individual patches). At the landscape scale, the response of interest to environmental change is independent of the landscape configuration and spatial effects. In this case, the patch models described above can be applied.

(ii) Spatially connected patches: the models are constructed with a recognition that the patches comprising the landscape interact in ways that alter each other's dynamics. Typical examples are lateral flows of water and lateral shading effects. In this case, the landscape-scale response to environmental change has a pronounced spatial component that either amplifies or attenuates the local response.

(iii) Spatially connected individual plants: the landscape is composed of individuals with known location and extent in space. In this representation the response of the landscape to its environment again has a strong spatial component, but the models are computationally more demanding than in the case of spatially connected patches.

Non-interactive patch models have been reviewed in the preceding section; below, an overview of the other two categories of models is given.

SPATIALLY CONNECTED PATCHES.   In these models, a landscape is usually represented by a regular grid of cells. In each of those cells, the dynamics of the vegetation are simulated by a patch model that is internally non-spatial, but which interacts as a unit with the neighbouring cells (e.g. a Markov model as in Gardner *et al.*, 1996; or a rule-based model as in Noble & Gitay, 1996). The cells interact with each other via the exchange of materials (e.g. run-off), propagules, or the propagation of disturbances such as fire. Some modellers have tackled larger spatial units by linking together a series of gap models with a selected range of inter-patch interactions (e.g. Urban *et al.*, 1991). Most patch models deal with an area of less than 1000 m^2, which means that a landscape as small as 10 × 10 km would still require > 10^5 patches. The heavy computational load of this 'brute force' approach implies that the processes and interactions that emerge at landscape scales via such models have been little explored.

It is possible to explore the behaviour of landscapes by using much simpler models to describe the behaviour of the plants within the cells of the grid. In some cases the biology has been reduced to very simple rule systems (e.g. a 'game of life' cellular automaton used by Thiery *et al.*, 1995). These allow the exploration of behaviours that arise from interactions and exchanges between the elements of the landscape but it is nevertheless important to ensure that the biological representation of the cells is sufficient to produce sensible responses to more subtle landscape interactions. Hochberg *et al.* (1994) have used a more complex biological representation in a cellular automaton to investigate the interaction between fire regimes and the spread of woody plants in savanna systems. They found that fire is essential to understand the clustered presence of trees in these savannas. Noble & Gitay (1996) have used the vital attributes model (Noble & Slatyer, 1980) as the basis of a model incorporating complex successional pathways, species migrations and disturbance by fire to explore the differences in composition predicted by point and landscape representations.

The more mechanistic approaches will increase in their applicability as the impacts of atmospheric and climate change (temperature, moisture and direct CO_2 effects) on forest succession increase. Markov representations, on the other hand, in which the transition probabilities are estimated from a knowledge of forest dynamics in the past (Gardner *et al.*, 1996), may become less reliable. In general, given the importance of land-use change over the next few decades as a component of global change, models that can simulate the overall, net effects on ecosystem structure and functioning of simultaneous changes in landscape pattern and in climate will be the most useful for global change research.

SPATIALLY CONNECTED INDIVIDUAL PLANTS.   In some situations, the assumptions of horizontal homogeneity and lack of interactions between patches made in gap models clearly are invalid. For example, in open woodlands and savannas trees occur both in clumps and as scattered individuals, but there is no continuous forest cover. These systems are also characterized by the dominance of mixed life-forms or plant functional types. In such systems, even small changes in driving factors can affect the balance between plant functional types (e.g. grasses and trees). Savannas also experience recurrent, but irregular disturbances that directly affect structural attributes and consequently functional features. Modelling savannas and their development under global change therefore poses a specific challenge.

Under such conditions, the landscape-scale response of the system may be

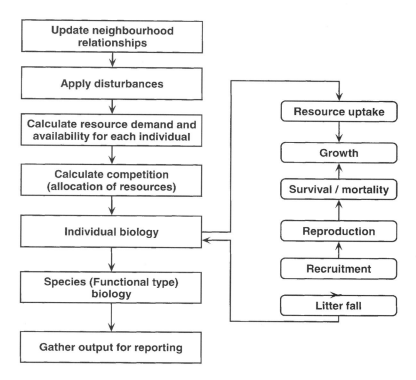

Figure 6.7   Structure of the spatially explicit stand model, MUSE. The flow diagram shows the sequence of processes dealt with within the MUSE shell. The biological detail in each step is flexible and can range from simple Lotka–Volterra models of population dynamics to spatially explicit, physiologically based plant models.

simulated by explicitly taking into account the spatial location of each individual relative to its neighbours. A model of this type is MUSE (Multistrata Spatially Explicit model, Gignoux *et al.*, 1995, 1997, Fig. 6.7), which is an individual-based model that includes interactions between plants depending on their position in space and their shape. Using an object-oriented language, MUSE allows users to modify assumptions on the interactions between plants, change the description of the major biological properties according to chosen life-forms, insert these independently into the main code (i.e. without affecting the core of the model), and examine the model output in a comprehensive manner. MUSE thus behaves as a modelling environment able to incorporate a series of different submodels, and to run them at the different time steps and spatial ranges at which the relevant processes (function or dynamics) occur.

The model couples a water budget submodel (soil water reserve in two horizons), an evapotranspiration submodel, and a production model simulating the seasonal variation of grass bio- and necromass, together with that of living and dead leaf area. The soil water and evapotranspiration models are coupled to the primary production model via phenology, on a daily time step. The effective coupling of water availability and primary production via an accurate modelling of LAI dynamics (including feedbacks) is a major feature of the model.

Being able to couple a wide range of biological models (functional processes) to a large range of spatial representations (vegetation structure), MUSE can simulate a large variety of ecosystem types, from boreal or tropical forest to grasslands (Gignoux *et al.*, 1995, 1996). MUSE thus appears as an efficient tool to explore management options (man-made disturbances) or the response of ecosystems to global change (stress and disturbances), or combinations of both.

The related model FRENCH aims to explore tree population dynamics as influenced by competition between individuals and by grass biomass (itself modified by tree density and distribution), which together determine fire spread and intensity (Menaut *et al.*, 1990). FRENCH was developed on the basic assumptions characterizing MUSE, and uses its complex spatial formalism. It includes little physiology (competition is not for explicit resources but is conveyed by a neighbourhood/overlap index), and focuses on the demographic aspects of tree dynamics. Tree spatial patterning is the result of fire occurrence and intensity, and of the interaction between tree species with contrasting fire resistance strategies (Gignoux *et al.*, 1997). Tree density can only be regulated in the long term by aperiodic, extreme events.

6.2.4   Global vegetation models

Equilibrium models of global vegetation structure

Over longer time-scales, e.g. centuries to millennia, most modellers assume that there is an equilibrium between the distribution of climate zones and of vegetation (Huntley *et al.*, 1997). Because the broad-scale climatic constraints for the distribution of major plant types are relatively well understood, it is possible to design simple models that allow the prediction of equilibrium vegetation changes as a function of climate change. An early example for these models was the 'life zone' scheme proposed by Holdridge (1947), initially with the aim of recovering the climatic zonation in a landscape from the observation of its vegetation zones. The Holdridge scheme was introduced into global biogeographic modelling by Emanuel *et al.* (1985*a*) and has since been used for a variety of assessments of climate change impacts on the broad distribution of biomes, as well as the associated changes in terrestrial carbon storage (Prentice & Fung, 1990; Smith *et al.*, 1992*b*). Climatic variables defining these life zones are 'biotemperature' (an index closely related to the annual number of growing degree-days) and total annual precipitation.

Despite its apparent success in simulating the major belts of global vegetation at a grid scale of 0.5 degrees longitude and latitude (Emanuel *et al.*, 1985*a*), there are several significant improvements over the Holdridge scheme in the model developed by Box (1981). First of all, vegetation is no longer considered to consist of homogenous entities, i.e. biomes, but to be composed of a combination of plant life forms drawn from as many as 90 different types. Although this number in itself may be open to debate, this work showed that climatically determined combinations of these life forms may show a great variety between different regions, much as real vegetation does. Secondly, the relations between life forms were expressed by a simple dominance rule, based on the assumption that plants that are near their distributional limits are likely to be less competitive than those in the centre of their potential range. Third, the climatic determinants themselves were defined to reflect ecologically more meaningful climatic indices than annual rainfall and temperature, e.g. by using a moisture index that was derived from estimates of potential evapotranspiration. An important aspect is that these climatic indices reflect differences in seasonality between different climates, while the Holdridge approach relied on annual averages alone. However, the model still required that the climatic range of each life form was determined as a model parameter.

As a direct precursor and later as a part of GCTE activities, a new model was constructed that contains a similar philosophy to the model of Box, but differs by using a more consistent physical and ecophysiological rationale (BIOME1 by Prentice *et al.*, 1992). It is based on only 13 plant types and five bioclimatic variables (see Table 6.4). One of these variables is the Priestley & Taylor moisture index (Priestley & Taylor, 1972), which is estimated using monthly temperatures, precipitation totals and cloudiness percentages, as well as a global soil texture database. The distribution limits for each plant functional type in relation to the climatic variables are defined as the minimum necessary set only, i.e. in contrast to Box's model the table listing the plant types and their climatic limits is largely empty. Furthermore, to a considerable extent the climatic limits used in BIOME are based on ecophysiological considerations, such as the importance of minimum temperatures earlier investigated by Woodward (1987*a*).

In addition to BIOME1, other global biogeographical models exist, e.g. MAPSS (Neilson, 1993; Neilson & Running, 1996). All of them are difficult to validate against observations, due to several circumstances (see also Section 6.3). First, their development is usually not carried out strictly independently of the current distribution of global vegetation (although different levels of process understanding are used for different models, see above). Second, the current distribution of the world's vegetation is determined by both climate, soils and land use and may therefore not be expected to be simulated adequately by a model that uses climate and soil data alone. Third, the distribution of biomes

Table 6.4 *Derivation of the bioclimatic indices in the BIOME1 model from monthly means of climatic data*

Tolerance/ requirement	Ecophysiological mechanism	Bioclimatic index	Climatic variable used
Cold tolerance	Killing temperature during coldest period of the year	T_{min} temperature of the coldest month (lower limit)	Temperature
Chilling requirement	Winter chilling period required for budburst of woody plants	T_{min} temperature of the coldest month (upper limit)	Temperature
Heat requirement	Annual growth respiration requirement	GDD growing degree days above 0 °C and 5 °C	Temperature
Moisture requirement/ drought tolerance	Soil moisture availability	AET/PET (annual) actual evapotranspiration/ potential evapotranspiration	Temperature, precipitation, cloudiness

is not mapped equally well on all continents (some satellite-based datasets are currently under development, but not complete yet, see Chapter 5). Nevertheless, in a systematic comparison between several observed vegetation datasets (such as Matthews, 1983; Olson *et al.*, 1983) , the distribution predicted by the BIOME model was shown to be in better agreement with observed vegetation patterns than that of any other equilibrium model (Leemans *et al.*, 1996) .

Global models linking vegetation structure and functioning

While the earlier generation of global models of terrestrial biospheric functioning relied on a vegetation structure which is prescribed from a vegetation map or from using a biogeography model such as BIOME1 (Prentice *et al.*, 1992) off-line, more recent models now include structure as an integral part of the simulation. The Frankfurt Biosphere Model (FBM, Kindermann *et al.*, 1993; Lüdeke *et al.*, 1994) provided an early partial solution for this, since it prescribes the vegetation type based on a mapped distribution, but every other aspect of structure, such as total biomass or the leaf area index, are simulated dynamically. More recent models, such as DOLY (Woodward *et al.*, 1995) or BIOME3 (Haxeltine & Prentice, 1996*b*; Haxeltine *et al.*, 1996) , no longer rely on any prescription of vegetation structure but simulate it as an essential ecosystem state description (which therefore can be compared with the map-

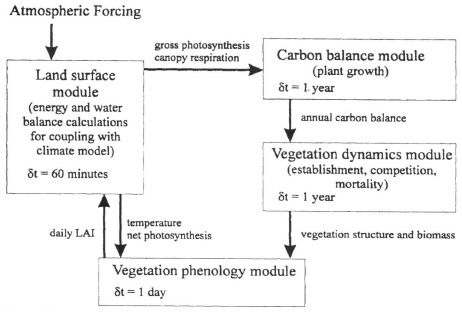

Figure 6.8   Basic modules in a Dynamic Global Vegetation Model (IBIS, adapted from Foley *et al.* 1996).

ped distributions, as long as the impact of human land use is considered adequately).

However, the simulation of structure in these models is essentially static, while the real world's terrestrial vegetation structure obviously changes, driven by changes in atmospheric composition and climate, as well as by human land use. A new class of global biosphere models, known as Dynamic Global Vegetation Models (DGVMs), has recently emerged within the pantheon of global ecological models. DGVMs are intended to overcome two fundamental problems associated with the previous generation of global biosphere models, namely: (i) the lack of explicit integration between biophysical, ecophysiological, biogeochemical, and biogeographical processes within models; and (ii) the absence of explicit vegetation dynamics mechanisms, which give rise to transient changes in ecosystem structure, and consequently also ecosystem functioning, over time.

In a hierarchical scheme of process descriptions, covering time-scales from hours to days and years (Fig. 6.8), DGVMs simulate growth of vegetation similarly to the biogeochemistry models above (i.e. using a full carbon balance simulation). The additional feature is realistic temporal behaviour under non-equilibrium conditions. For this, population level processes, such as establishment, competition and mortality, are included, using a philosophy similar to that of patch models (see Section 6.2.3) but with the added requirement of simplicity with respect to the diversity of species/functional types. Competition is modelled for the most basic resources only, such as light and water, and the plant functional types are defined so as to make simplified formulations of resource use possible (Foley et al., 1998). A limitation of current DGVMs is their very limited ability to simulate disturbances such as fire, wind storms, and pest attacks. More research is required to achieve globally applicable formulations of these processes as a function of the environment.

The potential applications of DGVMs include the following: (i) characterize the linkages between the dynamics of ecosystem structure and ecosystem functioning (e.g. water balance, net ecosystem carbon exchange, and biogeochemical cycling) in the terrestrial biosphere; (ii) provide a continental-scale, consistent description of the transient changes in global vegetation patterns under future atmospheric composition, climate change and land-use change scenarios; (iii) estimate the transient behaviour of terrestrial carbon reservoirs, including the potential effects of CO_2 fertilization, climatic variability and climatic change, and decadal-scale ecosystem dynamics (e.g. disturbances, see Section 12.2.1); and (iv) provide a fully interactive representation of terrestrial ecosystems within atmospheric general circulation models, to examine the potential for vegetation–climate interactions and feedback mechanisms. The use of the first generation of DGVMs to simulate future global vegetation structure and functioning is described in Chapter 8; their use in providing continental-scale comparisons of vegetation pattern based on the IGBP Terrestrial Transects is described in Chapter 12.

6.3   Model evaluation and intercomparison

6.3.1   Evaluation of ecosystem models: issues and difficulties

The critical evaluation of the structure and behaviour of models, which are abstractions of reality and necessarily ignore many components of reality, is a prerequisite before they can be used and applied with confidence. Strict 'validation' of models is never possible (Oreskes *et al.*, 1994) – models represent hypotheses about reality, and hypotheses can only be falsified. The situation is even more peculiar for models of global phenomena because there is not a single independent set of observations to test the models against, since this would require a second planet with comparable features. Careful evaluation, either by comparison against partial observations or even by comparison against each other, may nevertheless help to assess the applicability of models.

In the context of global change, careful and thorough model evaluation is a critical process for at least three reasons: (i) in many cases there are several ways to describe a single ecological process, and because the choice of a particular formulation is not straightforward, it may have strong effects on the simulated dynamics; (ii) the models may be used for projecting the behaviour or performance of ecosystems under new and possibly unprecedented environmental conditions; and (iii) the model may eventually be used by the policy community to guide the decision-making process on global change issues.

Model evaluation consists of two phases: internal and external evaluation. It is often assumed that the phase of internal evaluation (verification) has been successfully passed once a model is published, but the experience in the model comparison workshops run by GCTE (see Chapter 3) has shown that this assumption is not always justified. There are many different techniques for model verification, each checking a different aspect of model quality. They may include checks on consistent use of units or checks for conservation of energy and matter during transformations and transport of elements among model compartments. The more tests to which a model has been subjected, the more confidence one can have in its quality, but certainty can never be assured.

Subtle errors may slip in when definitions within different modules are not consistent, such as leaf area being used for leaf blade in one place and for total leaf plus sheath area in another. These kinds of errors are usually found during discussions between model authors and not by standard techniques. Model comparison workshops are instrumental for such discussions, especially when the authors find that their models behave differently. The GCTE model comparison exercises have shown that use of artificial input data is extremely valuable, even more so than use of real data. Stochastic fluctuations in the real world may mask differences in model behaviour that will show up with smooth model input. Sensitivity analyses, in which the effects of isolated environmental characteristics or of single model parameters are examined by systematically varying them and simulating the system response, are the next step in internal model verification.

6.3.2 Model testing against independent data

The testing of ecological models against observational or experimental data often poses serious problems because of the long-term nature of the processes being simulated. Except for those systems dominated by annual or short-lived perennial plants, it is rarely possible to conduct experiments that can be used to test long-term predictions of ecosystem development. Instead, ecologists have to rely on various types of often incomplete observational datasets that were not necessarily collected for the purpose of model testing. For example, modellers of ecosystem structure can sometimes utilize palaeoecological data and run their models to see if they can replicate the composition of past plant communities. Modellers of ecosystem functioning and biogeochemistry often rely on data collected over relatively short periods of time on a few key processes. For example, data on standing biomass, annual litterfall and the chemical composition of throughfall, soil, soil solution or stream run-off can be used to test model components. Data are prone to errors just as models are, and the process of data testing is as rigorous as that of model testing (Hunt, 1994*b*; Jones & Kropff, 1996; Hunt & Boote, 1997). In addition, it is rarely possible to document all environmental influences on the behaviour of existing or past plant communities. For these reasons, it is often difficult to know whether a good match between model predictions and available data is fortuitous or actually indicates that the model properly incorporates the critical processes.

In recent years, global change ecologists have begun running models of ecosystem structure or functioning for large geographic regions as opposed to single sites or experimental plots. While this scaling-up activity is certainly required to address the large-scale impacts of global change on ecosystems, it has presented new challenges for obtaining validation data and for conducting model comparisons. Typically, models run for a continent or the whole globe utilize a fairly coarse spatial grid. Current grid size may be, at best, 10 km × 10 km when models are simulating relatively small regions, whereas global models are still often run for grid scales of several degrees. This means that the models are simulating the average conditions (e.g. community composition or ecosystem productivity) existing over an area of hundreds to thousands of square kilometres.

How does one test a model at large scales when the average conditions as simulated by the model may, in fact, not exist at any one locality? A possible approach is to test the model at a series of sites along an environmental gradient (Solomon, 1986; Kienast, 1991) or, preferably, along continuously varying environmental conditions (Bugmann, 1996*a*). The availability of data from the IGBP Terrestrial Transects around the globe (see Chapter 4) will provide invaluable data for testing current models (see Section 12.3 for an example of this). Another solution, especially for models of ecosystem functioning, involves the application of recently developed techniques (e.g. eddy correlation measurements) that can measure gas fluxes integrated over fairly large areas (Raupach *et al.*, 1992; Grace *et al.*, 1996, see also Chapter 2).

Finally, model results may agree with observational or experimental data because model errors sometimes cancel each other under current climatic conditions but become visible under a changed climate (Martin, 1992; Bugmann & Martin, 1995; Fischlin *et al.*, 1995). This implies that (i) models should be tested against climatic conditions that differ from those of today, and (ii) additional tests of models are required to reveal structural deficiencies, e.g. model comparisons.

Testing models against satellite data

Due to their global coverage and continuous operation, satellite observations can directly infer land surface properties or provide input data for models (see Chapter 5), but they are also suitable to test model behaviour at scales where ground data are very incomplete. The variables that can be evaluated by satellite data are not often the ones that the model simulates as final output variables. Nevertheless, evaluation against satellite data may be more feasible for some model subsystems, and thereby contribute to the overall plausibility of the model results.

Available since the early 1980s, the National Oceanographic and Atmospheric Administration/Advanced Very High Resolution Radiometer (NOAA/ AVHRR) observations can provide regional to global estimates of leaf area index, vegetation phenology, radiative activity of the canopy, and surface temperature. Some examples of studies where satellite data have been used to evaluate predictive models are given here as illustrations:

- *Vegetation structure.* Nemani & Running (1989) have validated their simulation of a theoretical climate–soil–leaf area hydrologic equilibrium for forests at regional scales against LAI derived from NOAA/AVHRR observations. Woodward & Smith (1994*b*) have used the similar data from observations over 128 reserves in Africa to test the spatial variations of the LAI estimated by a model of equilibrium between precipitation and canopy structure.
- *Seasonal variations of the CO_2 fluxes.* In order to make realistic predictions of CO_2 fluxes, biogeochemical models must be able to simulate correctly the seasonal variations in the phenology of the vegetation canopy. The phenological submodels currently used to describe summergreen or raingreen ecosystems, for example, are rather empirical. Therefore, testing the simulated seasonality against NOAA/AVHRR time series (Lo Seen *et al.*, 1995), sometimes using phenological indicators (Lüdeke *et al.*, 1996), provides a better evaluation of the model's capabilities. In addition, using satellite data as input, a simple diagnostic model of net primary productivity can validate the CO_2 fluxes simulated by a predictive biogeochemical model (E.R. Hunt *et al.*, 1996).
- *Spatial pattern of evapotranspiration.* The surface temperature derived from the AVHRR thermal channels has been used to evaluate the evapotranspiration simulated by a hydrological model (Mauser *et al.*, 1997).

Other ecosystems properties can be derived from microwave measurements (e.g. Kerr & Njoku, 1993, Kasischke *et al.*, 1997). These, and the applications mentioned above, will benefit significantly from more advanced or future sensors, such as the POLarization and Directionality of the Earth's Reflectances sensor on board the Advanced Earth Observing Satellite (ADEOS/POLDER) since 1996 (Deschamps *et al.*, 1994) or the Moderate-Resolution Imaging Spectrometer sensor on board the Earth Observing Satellite (EOS/MODIS), which will be launched in 1998 (Running *et al.*, 1994). These new sensors will achieve a considerable increase in accuracy of measurement and hence in the potential to validate crucial components of ecosystem models. Biomass, height, vegetation water content, soil moisture, etc., are all variables that could be tested at large scale and in a continuous manner.

6.3.3   Model comparisons

For the reasons outlined above, organized model comparisons have become an important approach for evaluating ecological models. A variety of problems are encountered in conducting such exercises (see Chapters 8 and 9). A major objective of model comparisons is to identify the basis of differences in the predictions of different models. The challenge is to determine whether differences are consequences of dissimilar formulations and assumptions of the models, or whether they are due to the implementation of different temporal or spatial scales, or the use of different input datasets. Because it is important to minimize the effect of factors other than those internal to the models themselves, the availability of common spatial data on current climate, soil, vegetation and other key environmental factors that can be utilized to drive models having dissimilar data requirements is critical. However, several problems arise in obtaining such datasets. First, for many parts of the world the needed data may not exist at the required temporal or spatial resolution; this is particularly true for climatic data. Various methods, including the use of interpolation techniques and weather generators, can be employed (Kittel *et al.*, 1995). A second problem often arises when models use different time steps. For instance, some ecological models require monthly data, while others require daily data. In conducting a comparison of such models, the daily and monthly climate data for any particular location must represent the same mean conditions, yet the daily data must also include variability that is characteristic for the actual weather. Datasets for different environmental factors must be spatially consistent as well. Thus, for example, both climate and vegetation should properly reflect topographic features of the landscape at the resolution of the analysis.

Three GCTE-sponsored model comparisons illustrate the value of these exercises for model development. The first example is a comparison workshop for wheat growth models, held in Lunteren, Netherlands in 1993 (see Chapter 9). That comparison revealed an unexpected (at that time) difference in model outputs, which made the modellers aware of how arbitrary some model constructs are and helped to identify and categorize model processes in which differences were found. The comparison was based on the relatively simple situation of ample supply of water and nutrients and absence of weeds, pests and diseases. In this situation there are only three major groups of processes that govern crop growth: (i) development and phenology; (ii) morphogenesis and leaf area formation; (iii) dry matter formation. To simplify the model comparison, the first two groups (development and morphogenesis) were made uniform, so that any differences would be exclusively due to the modelling of dry matter formation.

This well-constrained, highly focused comparison revealed a surprising number of differences among the models. Most models formulate a dry matter formation rate on the basis of photosynthetic properties and respiratory losses. The net dry matter gain as calculated for a fully active unsenesced canopy, at an optimal temperature, was used as a point of reference. This rate is then modified for the effects of temperature and senescence. These multipliers, plotted as a function of day of year, clearly exposed significant model differences. Also, canopy-based models (Fig. 6.5) and leaf-based models (Fig. 6.1) handle biomass respiration differently, through different parameterizations for characteristics such as optimal temperature; effect of temperature change; slope, intercept and maximum level of the radiation response curve; and radiation extinction coefficient. Even some operational definitions appeared to be different, which only emerged after close examination of each model's results. Interestingly, total dry matter integrated over the season was very similar in the three models CERES-wheat, AFRC-wheat and SUCROS, probably because of inadvertent parameter calibration. The conclusion of the exercise was that an ongoing and well-structured process of continued comparisons, both model–model and model–data, is necessary to further improve model quality.

The second example arose from an initial model comparison designed to help to evaluate the applicability of ecological models along environmental gradients (Bugmann, 1996a). In this study, the realised niches of the tree species dominating the natural forest vegetation of central Europe were simulated with two forest gap models. The analysis revealed deficiencies in both models; particularly, it showed that one of the models contained a number of unrealistic threshold effects and did not simulate realistic species distributions along environmental gradients.

To address these issues in more detail, systematic model comparisons are planned for forest patch models along three of the IGBP Terrestrial Transects: BFTCS, SCANTRAN, and NECT (see Chapter 4). The results of a preparatory meeting held in December 1996 in Potsdam (Germany) suggested that the execution of quantitative model comparisons is neither scientifically nor logistically straightforward with this specific class of models (mostly gap models). Before and during this workshop, a number of pilot simulation studies along the transects were conducted by modelling teams from around the world. These activities showed that considerable work is required to homogenize input datasets (climate, soils), to derive the model-specific parameters characterizing the tree species, and to execute the comparative simulation studies themselves.

The preliminary simulation studies showed that each model falls short with respect to predicting correct species composition and stand structure along the extensive IGBP transects, because none had been developed to cover such large variations of environmental conditions. Much can be learned from these experiences in terms of model improvements, leading to more reliable assessments of the effects of climatic change on forests.

The third example arose from the combined activities of GCTE and the IGBP task forces GAIM and DIS. These activities have created an established history of comparing global biospheric models. In two workshops, 17 global models of net primary productivity have been compared, and both the advantages and the limitations of the employed strategies have been illustrated (Cramer, 1997; Fischer, 1997). Within the GCTE synthesis project, a first intercomparison of DGVMs was undertaken, with four participating groups providing their prototype models (at a later stage, more models will probably be included). It was critically important to standardize climatic driving data and spatial resolution as much as possible for all models; in the DGVM comparison, output from a transient simulation of a coupled atmosphere–ocean GCM was chosen for the test period of 1860–2100. It became apparent during the comparison that DGVMs are still in an early stage of development. They are clearly capable of capturing responses of potential natural vegetation to climate change with a strong orientation towards population processes, and are hence promising for future scenarios of realistic temporal behaviour. However, the quantitative results are still unstable (more details of the comparison are presented in Chapter 8).

6.4   Future challenges

The recent experience of model development and comparison in GCTE and elsewhere suggests a number of specific challenges that must be met over the next several years to improve the capability to simulate global change interactions with terrestrial ecosystems:

Linking management and natural successional processes. An increasing number of terrestrial ecosystems are simultaneously subject to both natural successional processes and human management interventions. Thus, models incorporating both of these processes, and their interactions, need to be developed if system responses to global change are to be simulated realistically. An example of such models is the latest generation of rangelands models (see Chapter 9), which include as drivers of system change (i) completely abiotic drivers such as drought; (ii) management interventions such as grazing intensity; and (iii) drivers dependent on both management and the abiotic environment, such as fire frequency.

Ecological complexity, functional types and ecosystem functioning. Given the increasing concern for the impacts of decreasing biodiversity and ecological complexity on ecosystem functioning (see Chapter 11), developing and improving models simulating the interaction of complexity and functioning is an immediate challenge (e.g. Heal *et al.*, 1996). As an early example of how models are becoming valuable tools to explore these relationships, a gap-type model has shown that the simulated response of a forest to elevated CO_2 is 30% greater if the community is composed of a suite of tree species of differing CO_2 response rather than a single species with the average CO_2 response of the more complex community (see Bolker *et al.*, 1995, and Fig. 11.6). In another study using a gap model of a forest, a large number of temperate and boreal tree species were grouped into six functional types, which were then tested with species composition and overall biomass production used as indicators of model performance (Bugmann, 1996*b*).

Sensitivity studies of global change impacts. Studies of global change impacts are moving away from the linear scenario–impact–consequence logic to more complex studies involving system response surfaces and multiple, interacting drivers. Thus, there is an urgent need to develop modelling tools capable of handling these complex sensitivity studies. The use of a suite of rice models to explore the sensitivity of yields to temperature increase (see Chapter 9) is an example of a simple sensitivity study.

Models coupling structure and functioning. Ecosystem physiology models do not simulate the impact of changes in available resources on stand structure and composition. On the other hand, current patch models simulating the dynamics of ecosystems largely ignore physiological mechanisms and processes by using highly simplified descriptions of ecophysiology. Over the past years, considerable evidence has accumulated that implies that changes in the structure and in the functioning of ecosystems cannot be considered independently (see also Chapter 8 and Section 12.2.2). Thus, a major challenge in global change research is to develop models that link structure and functioning. An early example is the linkage of trace gas emissions and forest successional dynamics under a changing climate (Martin & Guenther, 1995). At a global scale, the DGVMs described earlier in this chapter are an excellent example of models coupling structure and functioning.

Landscapes processes. Inclusion of landscape-scale processes, where the horizontal movement of organisms (e.g. migration of plants), elements (e.g. movement of nutrients from upland areas to coastal zones following land-cover conversion) and energy (e.g. spread of fires) is crucial, is often identified as a major lacuna in global change modelling. Such effects are obviously important for resource management, where global change effects will increasingly need to be taken into account. Even at a global scale, horizontal processes are important. For example, the inclusion of horizontal transport of carbon via rivers had a significant effect on a simulation of the dynamics of the global carbon cycle (B. Moore III, personal communication).

As more and more models are developed and published, their outlook appears confusing. Scepticism about model quality remains widespread, among both modellers and experimentalists. However, there is no other way to make theories operational in the real world, nor is there any other way to project quantitatively what might happen to terrestrial ecosystems in the future under a rapidly changing environment. Increasing application of the methodologies for model testing described in the previous section will lead, over time, to improved model quality. The ambitious goal of making realistic projections of the multiple and interacting effects of global change on terrestrial ecosystems will always need to be addressed with caution (see Section 12.5), but quantitative models are becoming increasingly important tools for analysing and interpreting past and current trends, and for making realistic projections into the future.

⑦ Ecosystem physiology responses to global change

H.A. Mooney, J. Canadell, F.S. Chapin III, J.R. Ehleringer, Ch. Körner, R.E. McMurtrie, W.J. Parton, L.F. Pitelka and E.-D. Schulze

Box 7.1   Summary

- Most ecosystems exposed to double ambient CO_2 show higher peak season net carbon uptake than those growing at current-ambient CO_2. For grasslands, above-ground biomass increased by an average of 14%, although individual responses for a given system and year range from negative to +85%. The wide range of the biomass response shows the highly interactive nature of the CO_2 response with other environmental factors, including water and nutrient availability, and temperature. For instance, low-temperature dominated systems, such as alpine grassland, Arctic tundra, and cool climate coniferous forest trees, are among the least responsive to elevated CO_2, showing in some instances no growth response and complete acclimation of peak season gas exchange after a few years. Annual Net Ecosystem Productivity (NEP) has not been reliably measured under elevated CO_2. Faster growth in juvenile trees does not indicate whether forests as a whole will sequester more carbon or not.
- The general predictions of the degree of responsiveness of plant growth to elevated CO_2 for some of the most important functional groups (e.g. insensitivity of C_4 species relative to C_3 species; larger biomass response of species with N-fixing symbionts) are not consistently realized in complex species assemblages exposed to elevated CO_2 under natural growth conditions.
- Contrary to early predictions, litter of high CO_2-grown plants does not necessarily decompose more slowly. The basis for this result is the finding that the ratio of carbon to nitrogen in naturally senesced litter grown at elevated CO_2 is normally no higher than in the same tissue grown at ambient CO_2, despite the higher C : N in green tissues. A great deal of variation, however, occurs among species. This finding has major implications for long-term ecosystem responses.
- Herbaceous plants exposed to elevated CO_2 show a reduction in stomatal conductance, which commonly results in increased soil moisture. This increase in water availability is the dominant driver for the increased net carbon uptake in water-limited grassland systems. There is also a reduction in stomatal conductance in some tree seedlings exposed to elevated CO_2, but this does not seem to be the case for mature trees (forests) given our current experimental

datasets. If the latter is confirmed, some of the feedbacks of the terrestrial biosphere to the climate system may not occur at the intensity currently predicted.

■ Tropospheric ozone has negative effects on ecosystem Net Primary Production (NPP), but elevated CO_2 will ameliorate plant ozone injury for those species that decrease stomatal conductance at elevated CO_2. There is also the potential for UV-B to negatively impact NPP.

■ Direct effects of increased air temperature on plant growth will be smaller than is often expected because of thermal acclimation. However, there will be developmental acceleration and stimulation of litter decomposition. Indirect temperature effects are mainly associated with warming of permafrost in subpolar biomes, which will cause substantial changes in species composition, litter quality, and nutrient availability. These effects have caused, and will cause, changes in energy and carbon balance that are large enough to provide significant positive feedbacks to global warming.

■ Nitrogen deposition is associated with increased NPP in some forests, and appears to be further enhanced by the fertilization effect of elevated CO_2. However, continuous N loading will lead, in the long term, to changes in species composition that may not be associated with increased carbon sequestration at the ecosystem level. Continuous N loading, together with other associated pollutants, could lead in many instances to soil acidification with a subsequent decrease in NPP.

■ Model predictions show that the combined effect of elevated CO_2 with higher air temperature and N deposition increases N mineralization and NPP, while soil C storage is decreased by increasing soil temperature. However, little is known about the overall carbon balance of terrestrial ecosystems under elevated CO_2.

■ In the short term (years to decades), changes in plant physiology within species will dominate the response to altered climate and atmospheric composition. Species show a wide range of responses (sensitivity) to changes in atmospheric CO_2, air temperature, UV-B, tropospheric ozone, and nitrogen deposition. In the long term (decades to centuries), changes in ecosystem physiology will be dominated by changes in the relative abundance of species, and the physiological properties associated with them.

Box 7.2   Future needs

There is a need to:

■ understand rhizosphere, mycorrhizal and soil processes under elevated CO_2 and temperature in undisturbed systems, and assess belowground carbon pools (e.g. root biomass, soil carbon fractions);

■ understand physiological developmental processes in mature trees under changes in climate and atmospheric chemistry, and develop new conceptual sights to guide the interpretation of short-term experiments (FACE Experiments) to long-term predictions (> 10 years);in

■ understand nitrogen dynamics with respect to elevated CO_2, particularly of litter and decomposition processes, and the implications for carbon sequestration in ecosystems;

■ resolve the effects of elevated CO_2 on evapotranspiration in general, and for
mature forest trees in particular (i.e. on stomatal and canopy conductance and
LAI responses);

■ undertake more experimental work to determine the interactive effects of increased CO_2, air temperature, UV-B, and atmospheric pollutants, including
ozone and N-compounds, under natural conditions;

■ continue efforts to establish manipulative experiments (e.g. elevated CO_2) in
woody ecosystems, particularly in tropical savannas and forests;

■ understand soil carbon dynamics with respect to elevated temperature, particularly the controls of soil respiration;

■ understand the influence and impact of land-use changes on rooting patterns
and ecosystem water-use dynamics, especially in tropical and savanna ecosystems where forest-to-grassland conversions are common;

■ understand the capability to scale leaf- and stand-level isotopic discrimination
patterns to global carbon models determining regional and interannual source
and sink strengths;

■ establish stronger linkages with existing ecosystem eddy covariance analyses of
water and CO_2 fluxes to help provide mechanistic explanations of how ecosystem physiology constrains biosphere–atmosphere gas exchange dynamics.

7.1 Introduction

In this chapter the impact of environmental change on the physiology of
ecosystems is assessed, taking into consideration long-term hydrological and
biogeochemical feedbacks. The chapter principally examines the environmental factors that are predicted to change in the coming decades to a century: atmospheric CO_2, temperature, water availability, N deposition, UV-B
and tropospheric ozone. Major emphasis, however, is placed on atmospheric
CO_2 since there is no controversy that the concentration of CO_2 in the atmosphere has changed substantially since the onset of the industrial revolution
and that it will continue to rise into the near future as nations struggle to
control their carbon emissions. Various modelling exercises are also presented
to show the complex nature of the interactions among multiple global change
drivers, which can lead to feedbacks, both positive and negative, on ecosystem
responses.

Ecosystem physiology is a relatively new field that has arisen from the need to
understand and predict how ecosystems will respond to the changing global
environment. It is concerned with the interacting physiological processes that
determine exchanges of carbon, energy, water and nutrients of whole ecosys-

tems, and it recognizes that information derived from physiological measurements on one type of tissue and at one time scale may not be extrapolated to a whole system level nor to longer time scales because of interactions within the system. For example, increased carbon gain at the canopy level due to elevated CO_2 may not necessarily result in a proportional increase in biomass accumulation because of altered processes belowground, such as increased root turnover.

The historical development of the study of physiology has been principally based on experiments on individual organs or organisms, and there have been very few studies on the metabolic responses of whole ecosystems at any timescale. Thus, the database is limited. GCTE has therefore made a deliberate attempt to promote whole ecosystem physiological studies, particularly on the impact of elevated CO_2. This has led to the initiation of a dataset comparable across ecosystems.

The capacity of organisms to adjust to changing environmental conditions by redeploying resources in the short term (days to weeks) is well known. Plants also adjust their metabolism in response to environmental change. For example, they may adjust their temperature optimum upward as their growth temperature increases (Billings *et al.*, 1971). Such adjustments must be taken into account when building models of responses to a changing climate. In the longer term, over generations, whole-ecosystem adjustments occur not through metabolic acclimation but through shifts in population abundance and in species composition. Although these changes are demographic and not physiological, they have metabolic consequences at the whole-ecosystem level and must be considered in ecosystem physiology (see also Chapter 8). Since the life-cycle of organisms may vary from days (in the case of microbes) to centuries (in the case of trees), the dynamics of the adjustment response to environmental change are complex. Because all experiments are confined in space (less than a few m²) and time (less than 10 years), there is a challenge to provide meaningful metabolic information that can be used in models for a wide range of spatial and temporal scales.

7.2   Effects of elevated atmospheric CO_2 on terrestrial ecosystems

7.2.1   The conceptual model

By the early 1990s considerable information existed on the response of plants to elevated CO_2, but the preponderance of studies had been conducted in laboratories or greenhouses. This was a problem for two reasons. First, it was suspected that some of the responses that were seen at lower levels of integration, such as with leaves or single plants, would be dampened at the stand level.

Second, the plant-level studies could not provide information of direct use in documenting the indirect impacts of elevated CO_2 on, for example, feedbacks to the atmosphere or to the soil. For these reasons GCTE established a consortium of elevated CO_2 studies (see Chapter 3) to facilitate the rapid exchange of information among the projects and to periodically synthesize and evaluate the accumulating data. For syntheses of the consortium's main results, see Schulze & Mooney (1993), Koch & Mooney (1996), Körner & Bazzaz (1996), Tinker *et al.* (1996) and Luo & Mooney (1998).

Most of the paradigms of plant responses to elevated CO_2 have been built on the study of agricultural species. In this chapter we focus on the responses of natural systems, including extensively managed pastures and forests. We must know more about CO_2 responses of natural and seminatural systems to predict impacts on the global carbon cycle and biodiversity, since together these systems occupy over three-quarters of the global land surface.

A summary of knowledge at the beginning of GCTE was given graphically in the GCTE Operational Plan (Steffen *et al.*, 1992; Fig. 7.1). This summary figure indicates not only the direct effects of elevated CO_2 on vegetation, but also feedbacks to biogeochemistry and the hydrological cycle. This figure provided a stimulus for ongoing and proposed studies to include processes that had not generally been considered in CO_2 research at that time. There were a number of areas where there was little, if any, information, for example, fate of nutrients under elevated CO_2 and impacts on the soil water balance. Although many gaps remain, substantial progress has been made in reducing uncertainty on many key issues. It was initially thought that the major feedback effects of elevated CO_2 would be through biogeochemical cycles. However, it is now known that the indirect effects on the hydrological cycle are equally or even more important, especially for water-limited systems.

What follows is a general review, mainly but not exclusively, of the results of the GCTE Elevated CO_2 Consortium projects (see Chapter 3). Data from published papers were the primary source, but responses to a questionnaire sent to all the participants in the Consortium (Tables 7.1 and 7.2) were also used.

7.2.2 Effects of elevated CO_2 on ecosystem carbon processes

For practical reasons most elevated CO_2 research has been conducted on vegetation dominated by short statured and short-lived plants (e.g. grasslands) and on young trees. The projects discussed here encompass a wide variety of unmanaged and managed systems, including salt marsh, annual Mediterranean and temperate grasslands, tallgrass prairie, alpine grassland, Arctic tundra, chaparral, oak-woodland, coniferous forest, deciduous trees, and tropical trees (see Tables 7.1 and 7.2). The experiments used a variety of CO_2 enrichment

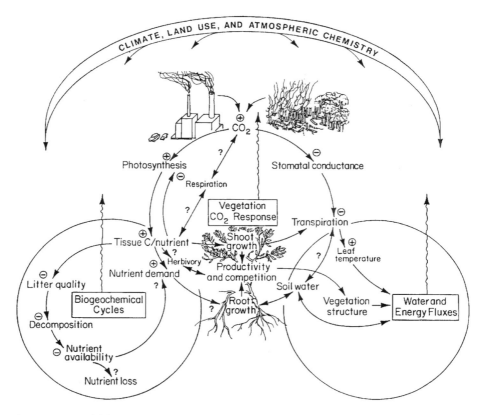

Figure 7.1   Initial research paradigm for the GCTE CO_2 research programme (from Steffen *et al.*, 1992).

technologies, including FACE, OTC, screen aided CO_2 enrichment (SAC), closed chambers (CC), ecocosms and other similar field enclosures, which are discussed in Chapter 2. Studies carried out on vegetation growing around natural CO_2 springs, which provide a valuable understanding of long-term and evolutionary responses to elevated CO_2, were also included.

Net CO_2 exchange

Increased photosynthesis at the leaf-level and net canopy CO_2 uptake (net daily or seasonal CO_2 *uptake per unit ground area*) are *probably the most consistent responses to elevated atmospheric CO_2* (see Tables 7.1 and 7.2). The current data on peak season CO_2 fluxes in ecosystems exposed to elevated CO_2 suggests that initial responses are always positive (increased net CO_2 uptake), and most often remain positive for some years. Although the degree of stimulation of photosynthesis has been observed in many studies to decline after extended exposure at elevated CO_2, a phenomenon known as 'down-regulation' (see reviews by

Gifford, 1992; Sage, 1994; Gunderson & Wullschleger, 1994; Amthor, 1995; Curtis, 1996), net CO_2 uptake during peak season at the plot level is almost always higher at elevated CO_2 than at ambient CO_2 (Drake & Leadley, 1991). For instance, Drake *et al.* (1996) showed a sustained 36% increase in peak season net canopy CO_2 uptake for a salt marsh over a six-year period, with no indication of a decreasing trend. Similarly Stocker *et al.* (1997) reported increased net canopy CO_2 uptake over two years of CO_2-enrichment in calcareous grassland (+ 34% at peak season and + 22% on a full season basis). Some studies showed net CO_2 uptake responses to have strong interactions with moisture, such as for water-limited Mediterranean annual grassland and tall-grass prairie. They showed little or no stimulation under moist conditions, but a relative enhancement of seasonal net biomass accumulation during dry years or at the end of the rainy season when soils start drying out (C. Lund & C. Field, personal communication; C. Owensby, personal communication). Low-temperature dominated systems, however, have consistently shown complete down-regulation after an initial stimulation of seasonal net CO_2 uptake. This occurred in the Arctic tundra within three years of exposure (Grulke *et al.*, 1990; Oechel & Vourlitis, 1996). Likewise, an alpine grassland, which showed a persistent enhancement of net CO_2 uptake for the first three seasons, exhibited full adjustment by year four (Körner *et al.*, 1997).

Downward adjustments per unit land area may be physiological or morphological. One example of physiological adaptation would be a decrease in rubisco concentration, while a morphological adaptation would be a reduction of leaf area index (LAI).

Another uncertainty as to how elevated CO_2 affects net CO_2 uptake in the long term is the lack of a mechanistic understanding of plant respiration. Mature leaf respiration has been reported to decrease as a consequence of the direct inhibitory effects of elevated CO_2 due to effects on non-respiratory metabolism (see Amthor, 1995 for review). Lambers *et al.* (1996) showed that while elevated CO_2 quite consistently suppresses leaf respiration, results from greenhouse and laboratory experiments showed no clear pattern for root respiration under such conditions. At the ecosystem level respiration either remained unchanged or increased under elevated CO_2.

Table 7.1 *Herbaceous systems responses to elevated CO_2*

Project no.	1	2*	3*	4	5*	6*	7*	8*	9	10	11*	12*	13*	14*
System	Arctic tundra	Alpine turf	Calcareous grassland	Limestone grassland	Tallgrass prairie	Pasture(s)	Pasture	Pasture	Pasture(s)	Pasture(s)	Annual grassland	Mediterran. grassland	Med. herbac. old field	Wetland
Location	Alaska	Switzerland	Switzerland	Lancaster UK	Kansas	Canberra	Zurich	New Zealand	Quebec	Clerm.-Ferr. France	California	Siena Italy	Montpellier France	Maryland
Contact	W. Oechel.	C. Körner	C.Körner	J. Wolgenden	C. Owensby	R. Gifford	J. Nösberger	P. Newton	C. Potvin	J. Soussana	C. Field	F. Miglietta	J. Roy	B. Drake
Number of species	15	10	25	34	100	1	12	10	10	2	50	12	12	3
Technology	CC	OTC	SACC	Mesocos. (I)	OTC	Mesocos. (I)	FACE	Mesocos. (I)	OTC	Mesocos. (O)	OTC	CO$_2$ spring	Mesocos. (I)	OTC
Yrs CO$_2$ expos.	3	4	4	3	7	4	4	1.2	4.3	2.2	4	many	3	8
Soils	natural	natural	natural	natural	natural	subsoil	natural	natural	natural	natural	natural	natural	natural	natural
Fertilization/Water	no/no	no/no	no/no	yes/yes	no/no	yes/yes	yes/no	yes/yes	no/no	yes/yes	no/no	no/no	yes/no	no/no
Annual prec. (mm)	n.a.	1600	900	n.a.	835	n.a.	1050	n.a. (874)	–	n.a.	600	700	n.a. (800)	–
Net Ecosyst. Prd.	=	+/=	+	+	+/=	+	+	+	–	+	+	–	+/=	+
Ps downreg.	canopy–total	canopy–some	canopy–some	some	none	.	leaf–some	canopy–some	leaf–none	leaf–some/no	canopy–some	leaf–some	canopy–some	canopy–some
Abovegr. biomass	=	+/=	+	+/=/–	+/=	+/=	+/=	+	+/–	+	+	=	=	.
Sps. biomass rep.	≠	≠	.	≠	≠	n.a.	≠	≠	≠	≠	≠	.	≠	≠
R : S biomass ratio	.	–	–	+	=	–	+	+	.	≠/sps	=	.	+	.
N$_2$ fixer/non-N$_2$	.	=	=	=	.	n.a.	+	+	+/–	+	+	.	+/–	.
Total plant N	–	–	–	.	+/=	–	+/=	.	.	=/–	+/=/–	–	+	+
Green tissue C:N	+	+	+	≠/sps	+/sps	+	+(small)	.	.	+/=	+(small)	+	+/=	+/sps
Abovegr. carbohydr.	+	+	+	≠/sps	.	+	+	.	.	+/=	+	+	+	+
Litter C:N	.	=	=	.	=	+	=	.	.	.	=	.	.	+/sps
Above-litter prod.	m	=	=	.	+/sps	+	+/=	.	.	.	+	.	.	.
Litter decomp. delay	m	yes	=	.	=	no	.	.	.	.	yes	.	.	yes

Leaf Area Index	=	=	=	+	=/−	+/−	+	=	+	=		=	+/=
Evapotranspiration	=	=	=		=/−	−	=	−		=/−		−/=	
Soil moisture	=	=	+	+	+	=	+	+		+	+	+	n.a.
Root turnover		+	+			+	+	+		+			
Soil respiration		=/+	=	+/=		+	+	+	+/=	+	+/=	+	
Microbial bio/act.		=		+/=	+/=	=	+	+	=	+	+	=	
Microrrhizae infect.		≠/sps	+	=	=		=	=		+	+	=	
Microflora changes				yes	yes	+		yes			yes	yes	
N mineralization		=	=	+/=			=	+	−		+		
System N losses		=	=	=/−	+	+		−	=/−				
Total soil carbon		=		+	+	=	+	=	+	=		+/=	
Leafing date		=	=	−	=		−	=			≠/sps	=	
Flowering date		=	≠/yr	=	=		earlier	=			≠/sps	=	
Number of flowers	+/=	≠/sps	≠/yr	=	−		+	=			≠/sps	+	
Seed production	+/=	≠/sps	+/−/sps	=	−						≠/sps		

*GCTE-Core Research Project; (s), summary of more than one experiment; CC, closed chamber; OTC, open top chamber; FACE, Free air CO₂ enrichment; SACC, screen aided CO₂ control; (I), indoor; (O), outdoor; Mesocos, Mesocosm (artificial ecosystem); n.a., not applicable; (=), no change; (−), decrease; (+), increase; ≠/sps, different species respond differently.

Table 7.2 *Woody systems responses to elevated co_2*

Project no.	1*	2*	3*	4*	5*	6*	7	8*	9*	10*	11*	12*	13
System, species Deciduous	Deciduous trees	Conif. tree(s) trees	Conif. forest *Pinus taeda*	Conif. trees *P. taeda*	Conif. trees *P. ponderosa*	Conif. sitka spruce	Conif. trees forest	Conif./decid. trees	Scrub oak trees	Chaparral(s)	Oak forest	Tropical	trees (s)
Location	Tennessee	Michigan	USA	N. Carolina	California	Scotland	Switzerland	Oregon	Christchurch N.Z.	Florida	California	Italy	Switzerland
Contact	R. Norby	P. Curtis	B. Strain	G. Hendrey	T. Ball	P. Jarvis	S. Hättensch.	D. Tingey	D. Whitehead	B. Drake	W. Oechel	F. Miglietta	J. Amone
Number of species	2	1	1	15	1	1	4	1	2	14	3	5	7–15
Technology	OTC	OTC	OTC	FACE	OTC	OTC	Mesocos. (I)	Mesocos. (O)	OTC	OTC	CC/FACE	CO_2 spring	Mesocos (I)
Yrs. CO_2 expos.	3.5	2	4	2.5	5	3	3	3	4	1	0.9	many	0.3–1.5
Soil	natural	mixed substr.	mixed substr.	natural	natural	natural		natural	natural	natural	natural	natural	mixed substr.
Fertilization/Water	no/no	no/no	yes/yes	no/no	yes/no	yes/yes	no/yes	no/yes	yes/yes	no/no	no/no	no/no	yes/yes
Annual prec. (mm)	1360	900	1240	1150		1200	1200	n.a. (1000)	600		490	–	n.a. (3250)
Net Ecosyst. Prod.	+	+	+	–	+		+	+	+	+	+	+	+
Ps downreg.	=	leaf–some	leaf–some	leaf–some	canopy–some	leaf–some	leaf–some	leaf–some	leaf–some	leaf–some	leaf–some	leaf–some	leaf–some
Abovegr. biomass	+		+			+	=				+		=
Sps biomass rep.	=		n.a.			n.a.	≠	≠		≠	≠		≠/=
R:S biomass ratio	=		–		+	=	+			+			+
N2 fixer/non-N2	n.a.		n.a.		n.a.	n.a.					yes		
Total plant N	+		+		+	=	–	+	+	+			=
Green tissue C:N	=	+	–	=			+	+	+	+	+	+	+/=
Abovegr. carbohydr.			+	+		+	+			+	+	+	=
Wood tissue C:N	+		–				+					+	
Litter C:N	=		–		=				=				+/=

	1	2	3	4	5	6	7	8	9	10	11	12	13
Litter carbohydrate	.	.	=	.	.	.	.	.	.	.	.	.	.=
Above-litter prod.	+	=	+	+	.	.	+	.	.	.	.	.	+/=
Litter decomp. delay	=	.	=	.	.	.	.	.	.	.	.	.	=
Leaf Area Index	+	=	+	+	.	.	−	=	+	+/=	+	+	=
Evapotranspiration	=	.	−	−	+	.	.	−	+	−	−	−	=
Soil moisture	.	.	n.a.	=	.	.	.	+	.	.	=	=	=
Root turnover	.	.	+	.	−	.	.	+/−	=	+	.	.	+
Soil respiration	+	=	+	+	+	.	+	+	+	.	=	.	+/=
Microbial biomass	.	.	+	+	+	.	.	.	.	.	=	.	.
Microrrhizae infect.	+	.	+	.	=	.	.	.	.	.	=	.	=
Microflora changes	=	.	.	.	yes	.	.	.	.	yes	yes	.	=
N mineralization	.	.	.	.	=	.	.	.	.	.	=	.	+
System N losses	.	.	+	.	=	.	.	=	.	.	.	.	+/=
Total soil carbon	+	.	+	.	+	.	.	=	=	.	.	.	+/=
leafing date	=	=	earlier	=	earlier	=	≠/year	=	.	.	earlier/=	.	.
Seed production	.	.	.	.	.	.	.	.	.	.	+/=	.	.

*GCTE–Core Research Project; (s), summary of more than one experiment; CC, closed chamber; OTC, open top chamber; FACE, Free Air CO_2 Enrichment; (I), Indoor; (O), Outdoor; Mesocos., Mesocosm (artificial ecosystem); n.a., not apply; (=), no change; (−), decrease; (+), increase; ≠/sps, different species respond differently.

Box 7.3   Biomass, NPP, NEP and NBP

When considering responses of ecosystem physiology to global change, direct effects on processes (e.g. growth) and indirect effects via changes in species composition (biodiversity) need to be distinguished. Among the direct responses related to the carbon cycle, three need to be clearly identified: (1) seasonal plant biomass accumulation (often termed 'production'); (2) annual net primary production (NPP, dry matter fixed through photosynthesis minus losses due to respiration and construction costs; losses of newly built biomass to herbivores and litter must also be accounted for. True NPP has rarely been estimated because belowground production and litter recycling usually remain unknown.); and (3) annual net ecosystem productivity (NEP), i.e. the net change in carbon pools per unit land area. It is particularly important to note that seasonal biomass accumulation, the response most commonly investigated, is not a measure of NEP (carbon sequestration). If large areas and long time-frames are considered, both the changes in species composition and in disturbance regimes need to be included in evaluating carbon sequestration. Here the concept of Net Biome Productivity (NBP) is most appropriate. See Sections 10.4 (especially Fig. 10.2) and 12.2.2 for a more detailed discussion of NPP, NEP and NBP in relation to the terrestrial carbon cycle.

Biomass responses

Aboveground biomass accumulation in whole ecosystems exposed to elevated atmospheric CO_2 in the field has been mostly studied in grasslands and crops. These experiments have shown a mean increase of 14% at twice ambient CO_2 when averaged across systems ($n = 9$ studies; 16% when averaged across systems and years, $n = 28$; see Fig. 7.2). This value is much lower than previous values derived from growth chamber and laboratory experiments, which were mostly conducted on single plants; Poorter (1993) reported for 156 species an increased plant growth at a double current CO_2 of 41% for C_3 species and 22% for C_4 species; Idso & Idso (1994) reviewed 342 published papers and found that plant dry weight was 24% higher at double CO_2 compared to ambient CO_2 when water was not limiting and 48% higher when water was limiting. They found that dry weight increased by 53% at double CO_2 in nutrient sufficient conditions and by 48% when nutrients were limiting. Note, however, that a large percentage stimulation by elevated CO_2 does not necessarily mean a large biomass response since often the largest differences between treatments occur during dry years (or treatments) when productivity is low.

It is important to note the highly variable nature of aboveground biomass production response when compared across systems and across years within the same system. Individual values in the CO_2 Consortium results range from negative responses to 85% (see Fig. 7.2) depending on the system and year, which reflects the interactive nature of the elevated CO_2 response with other

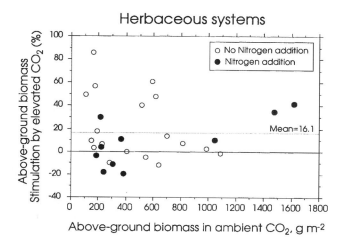

Figure 7.2   Above-ground biomass stimulation by elevated CO_2 of herbaceous ecosystems, with and without N additions.

variables (e.g. water and nutrient availability, species composition, temperature, etc.).

The average aboveground biomass response to double CO_2 of + 14% is of smaller magnitude than some of the measurements of system-plot net CO_2 uptake, which indicates the existence of significant carbon sinks other than aboveground plant growth. In some instances, changes in biomass are too small to be detected, even when net CO_2 uptake is significantly higher at elevated CO_2. This was the case in the Arctic tundra during three growing seasons of elevated CO_2 exposure before the system down-regulated (Tissue & Oechel, 1987), in an alpine grassland (Schäppi & Körner, 1996; Körner *et al.*, 1997), in a tallgrass prairie (Owensby *et al.*, 1996*a*), and in a Mediterranean grassland (Navas *et al.*, 1995) during wetter than average conditions.

No area-based biomass response data exist for mature forest trees exposed to elevated CO_2. When individual trees were investigated, positive biomass responses were found in most cases (Ceulemans & Mousseau, 1994). Seedlings or young saplings of five species in 16 different CO_2 experiments had an average biomass increase of 26% in response to elevated CO_2, with individual values ranging from nonsignificant increases to 58% increase (Jarvis, 1995*b*). However, some of the above studies and others have been conducted under rather fertile conditions and actual responses range from negative (Mousseau & Enoch, 1989) to no change (e.g. Norby *et al.*, 1992) to a several-fold increase under completely unlimited growth conditions (e.g. Idso & Kimball, 1992; Pettersson & McDonald, 1992). The reported responses were strongly dependent on nutrition, competitive environments, experimental duration, and the age of the experimental plant (Eamus & Jarvis, 1989; Loehle, 1995). Tree saplings grown in competitive assemblages in model ecosystems showed no biomass response to elevated CO_2 when nutrient supply was moderate to low (tropical: Arnone & Körner, 1995; boreal: Hättenschwiler & Körner, 1996) but a slightly positive response under more fertile settings (tropical: Körner & Arnone, 1992). Zero biomass responses have also been reported for model communities with young tree seedlings by Williams *et al.* (1986, temperate deciduous), and Reekie & Bazzaz (1989, tropical).

Seedling responses are expected to be quite different from those of adult trees but few experimental data are available for the latter. Cross-continental surveys of tree ring and community dynamics suggest that the most likely response of adult trees to current atmospheric CO_2 enrichment in the boreal zone is close to zero (Schweingruber *et al.*, 1993), while tropical trees may be stimulated (Phillips & Gentry, 1994). Analysis of tree ring data from Mediterranean oaks growing around two separate natural CO_2 springs indicates that initial positive responses are gradually reduced as trees grow in size and disappear after about 30 years (Hättenschwiler *et al.*, 1997). Such a response pattern could resolve the discrepancy between positive seedling responses commonly seen in CO_2-enrichment experiments and actual responses of mature trees. As a consequence, initial tree growth may be accelerated under elevated CO_2, but tree carbon stocks over longer time periods may not necessarily become greater. Whatever the effects on long-term carbon storage, forest dynamics are likely to become enhanced, which is of great significance for forest ecology and is of interest to commercial forest production.

Litter quantity, quality, and decomposition

Changes in litter amount and its chemical composition (quality) can alter the fluxes of carbon and nutrients within the ecosystem, and hence affect plant productivity and carbon sequestration. In young expanding systems litter quantity commonly increases under elevated CO_2 because of increases in leaf biomass and total aboveground biomass (Norby *et al.*, 1996; Jarvis, 1995*a*). Since steady state leaf area index (LAI) achieved at canopy closure in well-watered systems does not seem to increase under elevated CO_2 (Arnone & Körner, 1995; Hättenschwiler & Körner, 1996), enhanced leaf litter production may be a transient effect. Elevated CO_2 typically causes an increase in $C:N$ ratio of green leaves due to starch accumulation (Field *et al.*, 1992). However, this starch disappears at leaf senescense, so the quality of natural senesced litter, with some exceptions (e.g. Cotrufo *et al.*, 1994; Table 7.1) is often indistinguishable from litter produced under ambient CO_2 (Curtis *et al.*, 1989; Kemp *et al.*, 1994; O'Neill, 1994; Franck *et al.*, 1997; Hirschel et al., 1997). Consequently, litter of plants grown under elevated CO_2 in most cases does not differ in decomposition rate from litter grown under ambient CO_2. There is, however, a great deal of variance in decomposition rate among species (Franck *et al.*, 1997) and the long-term consequences of elevated CO_2 on litter pools may be mostly driven by changes in species composition.

Belowground carbon fluxes and soil carbon sequestration

Although persistently increased peak season net ecosystem CO_2 uptake is a common response to elevated CO_2, changes in overall biomass are usually smaller than expected, and sometimes even nil. A growing body of information indicates that CO_2 enrichment leads to enhanced carbon allocated to below-ground functions (Van Veen *et al.*, 1991; O'Neill, 1994; Körner, 1996; Canadell *et al.*, 1996*b*). Although responses are not consistent across all systems, the increased carbon flow to the belowground compartment is largely driven by accelerated root turnover (Norby, 1994; Pregitzer *et al.*, 1995; Tingey *et al.*, 1995), rhizodeposition (Cardon, 1996), mycorrhizal development (O'Neill *et al.*, 1987; Rygiewicz & Andersen, 1994; Dhillion *et al.*, 1996), and N_2-fixation (Soussana & Hartwig, 1996).

One of the most important goals of research on elevated CO_2 is to investigate whether increased net CO_2 uptake in a high CO_2 world will result in increased long-term carbon storage in ecosystems, in particular in the soil component where carbon residence times are longest. Given that most soils contain 10–20 kg C m^{-2}, it is difficult to detect C sequestration of less than 300 g m^{-2} by direct determination (Hungate *et al.*, 1996). Most often soil C sequestration has to be inferred. However, a few field studies have shown a trend, in some instances nonsignificant, towards increasing soil carbon storage in elevated CO_2 plots (see review by Canadell *et al.*, 1996*b*). Among field studies suggesting soil carbon accumulation are a wetland in Maryland (USA) after six years of CO_2 enrichment (Drake *et al.*, 1996), a Kansas (USA) tallgrass prairie after five years of CO_2 enrichment (Table 7.1), and two experiments with conifer trees (USA) after 4–5 years of high CO_2 (Table 7.2). Isotope-labelling experiments with reconstructed soils have also shown increased carbon deposition under elevated CO_2 (Lekkerkerk *et al.*, 1990), although in deep natural soils such signals are likely to be greatly diluted, and reconstructed soils show effects of disturbance for several years.

7.2.3 Impact of elevated CO_2 on water balance

The effects of elevated CO_2 on ecosystem water balance can be seen at different spatial and temporal scales. The challenge for the GCTE-CO_2 programme has been to interrelate and to model these various spatial scales and temporal dimensions. The spatial and temporal scales are to a certain degree intrinsically interconnected. For example, short-term responses are most often studied at the leaf level, while canopy-level responses are most often studied seasonally.

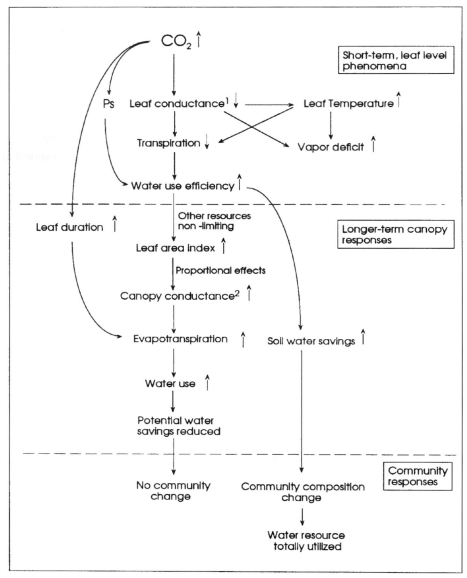

[1] Short term sensitivity to water stress enhanced under high CO_2

[2] Sensitivity of canopy conductance to evaportranspiration
depends on community roughness.

Figure 7.3   Water-mediated responses to elevated CO_2 at the leaf-, canopy-, and community-level.

Interannual responses are often seen at the community level, at least for communities with short-lived organisms. Here we describe the basic CO_2 responses, and their interrelationships and ecosystem consequences (Fig. 7.3).

Leaf level

In the short term and at the leaf level, elevated CO_2 commonly reduces stomatal conductance in herbaceous plants and tree seedlings (Morison & Gifford 1984; Field *et al.*, 1995) and its sensitivity to water stress (Owensby *et al.*, 1996*a*). However, not all species respond equally strongly and some do not respond at all, depending on the degree of environmental stress experienced (Curtis, 1996). With the current experimental data available, adult trees do not seem to exhibit the responses seen in the majority of herbaceous plants or tree seedlings. Several studies in which leaves or branches of tall canopy trees were exposed to elevated CO_2 did not reveal a change in leaf conductance (Barton *et al.*, 1993; Dufrène *et al.*, 1993; Jarvis, 1995*b*; Teskey, 1995; Körner & Würth, 1996), and first results from FACE studies in *Pinus taeda* are in line with these findings (Ellsworth *et al.*, 1995). Thus, at present, there is no experimental basis for assuming that overall forest transpiration will be reduced under elevated CO_2. Analyses of stomatal densities on leaf surfaces of plants grown under approximately doubled current ambient CO_2 show a mean decrease of 9% (Woodward & Kelly, 1995) but variation is very large and often no response is seen in the field.

Any reduction in leaf conductance would reduce transpiration. In general, this would increase leaf temperature and thus increase the vapour pressure gradient between the leaf and the surrounding air, partially offsetting the effect of the reduced conductance. Substantially higher leaf temperatures may cause the stomata to close further, resulting in a decrease in transpiration. Thus, even at the leaf level, there can be many counterbalancing interactions. However, even if no reduction of vapour loss occurred relative to water lost during the same period, water use efficiency (WUE) would increase. Data for a variety of species indicate that leaf-level WUE is proportional to atmospheric CO_2 concentration (Morison, 1993).

Canopy level

Leaf-level responses commonly do not directly translate into canopy responses (Morison, 1993; Field *et al.*, 1995). Factors that counteract or dampen the effect of leaf responses are (i) canopy conductance, (ii) feedbacks from improved plant water status via enhanced leaf area production (transpiring surface), (iii) prolonged availability of soil moisture and thus less temporal restriction of transpiration, and (iv) contributions of soil evaporation and understory evapotranspiration to canopy scale water balance.

Canopy conductance is perhaps the most important factor and is controlled

by canopy height and LAI. Tall and narrow-leafed plants such as conifer trees
are well coupled to the atmosphere, and hence stomatal conductance dominates
overall conductance. This is less true in broad-leaf forests and particularly in
low-stature vegetation such as grasslands, where aerodynamic canopy conductance may be smaller than stomatal conductance. The resistance to water vapour
movement through the canopy, rather than the stomatal conductance, may limit
the transpiration loss (see also Section 7.4). In such cases evapotranspiration
will not change greatly in response to elevated CO_2 (Field *et al.*, 1995). *In situ*
measurements of evapotranspiration in vegetation treated with elevated CO_2
have rarely been made, but the few observations available suggest relatively little
effect of elevated CO_2 on evapotranspiration (Grant *et al.*, 1995*a*; Stocker *et al.*,
1997; McConnaughay *et al.*, 1996).

Soil level

Reductions of evapotranspiration so small that they cannot easily be detected
experimentally may still accumulate to produce measurable increases in soil
moisture, particularly during prolonged dry periods. Such effects of elevated
CO_2 have been measured in Mediterranean grassland (Field *et al.*, 1996; see also
Tables 7.1 and 7.2) and in temperate grassland (Zaller & Arnone, 1997). Such
moisture savings appear to be the most influential consequence of CO_2 enrichment in periodically dry vegetation (Jackson *et al.*, 1994). As an example, the
tallgrass prairie in Kansas showed no significant increase in aboveground
productivity under elevated CO_2 during wet years (1992, 1993), an increase of
40% in an average rainfall year (1990), and about an 80% increase during a dry
year (1989) (Owensby *et al.*, 1996*a*). However, the highest elevated CO_2 response occurred during low biomass production years (dry years), and therefore
the absolute biomass change (g m^{-2}) may be small.

Increased soil water availability can enhance ecosystem productivity both by
extending the growing season and through positive effects on microbial activity
and turnover, and hence decomposition and nutrient availability. The extension
of the physiological active period is an important mechanism. The gross rate of
NH_4^+ mineralization in two California grasslands exposed to elevated CO_2
increased as a consequence of increased soil moisture resulting from decreased
stomatal conductance at elevated CO_2 (Hungate *et al.*, 1997). In the Swiss
calcareous grassland project, water savings by the majority of species seem to
enhance soil activity during dry periods, in particular the activity of earthworms, which exerts a chain reaction on nutrient relations, seed burial, and
microgap dynamics in the canopy (Zaller & Arnone, 1997). Hence, even small
improvements in water relations associated with CO_2 enrichment appear to
exert marked effects on community dynamics and organism interactions. The
only data for field-grown forest trees, those for the Mediterranean oak by

Hättenschwiler *et al.* (1997), indicate beneficial effects of CO_2 in dry years as well.

7.2.4 Community responses to elevated CO_2

While fluxes and pools per unit ground area show rather varied responses to CO_2 enrichment, one set of observations is common to all studies in which more than one species were included in the treatment: no two species were found to respond identically. Hence, no matter whether overall biomass responses or carbon accumulation occurred, responses were species- and even genotype-specific, causing continued CO_2-enrichment to become a biodiversity issue (Bazzaz *et al.*, 1989; Bazzaz 1996; Körner & Bazzaz, 1996; Potvin & Vasseur, 1997).

In the long term, these changes may alter species abundance and community composition, which could have an even larger impact on ecosystem functioning than changes on physiological processes of a given species assemblage (Section 8.1 and 12.2.2). One important lesson from the study of species-specific responses to elevated CO_2 in mixed communities is that the grouping of plants into functional types has had little predictive value for CO_2 responsiveness. C_4 species are not consistently nonresponsive, nor are legumes often more responsive than non-legumes, or fast-growing species more responsive than slow-growing species (Poorter, 1993) when investigated in a competitive setting in their natural environment (Owensby *et al.*, 1993; Table 7.1). For instance, the species that exhibited by far the most pronounced biomass response in a highly diverse calcareous grassland exposed to elevated CO_2 was a slow-growing sedge, whereas the five nitrogen-fixing *Trifolium* spp. in that community did not respond (Leadley & Körner, 1996). In the prairie, C_4 species have been shown to be quite responsive to CO_2-enrichment because of their capacity to take advantage of soil moisture as a result of reduced stomatal conductance (Owensby *et al.*, 1993). Likewise, in the California Mediterranean–climate grasslands, more water remains at the end of the growing season under elevated CO_2. The late-season growth of annuals, especially in dry years, may result in altered community composition in a given and subsequent years and cause seasonal biomass production to increase (Field *et al.*, 1996).

Several studies have also shown strong genotype responses to CO_2, with biomass responses not correlating with photosynthetic responses (Curtis *et al.*, 1996). Selection for responsive genotypes, in particular directions, is one of the most likely responses to elevated CO_2. Since species-specific responses to elevated CO_2 include responses of tissue composition, and thus nutritional quality of tissues, herbivores will be co-affected (Lindroth, 1996; Owensby *et al.*, 1996*b*), and even more subtle responses may be seen with species or genotype-specific mycorrhizal relationships (Sanders, 1996).

In the long term, many effects of elevated CO_2 on ecosystem functioning will be driven by the biogeochemical properties of new species assemblages that will be competitively superior in a high CO_2 world. For instance, grass and dicotyledon communities differ in their rate of root production, which may also affect soil structure and humus production (P.B Tinker, personal communication). Thus, any change in grass/dicotyledon species composition may indirectly affect carbon immobilization. Conversely, any shift from evergreen to deciduous forest or shrubland would accelerate turnover and carbon release from the system.

7.2.5   Modelling ecosystem responses to elevated CO_2

A wide variety of models have been used to investigate and project responses of ecosystems to elevated CO_2. While all of these models share a common objective of simulating responses to CO_2, they differ considerably in the specific mechanisms incorporated. When models are calibrated for specific sites or regions under current climate and CO_2, they often agree, but they may diverge when predicting ecosystem responses under scenarios of future climate change and increasing CO_2. It is essential, therefore, to understand how differences in the formulation of specific mechanisms influence their predictions. Given that results are becoming available from ecosystem experiments on the effects of elevated CO_2, it is appropriate to ask whether patterns of change predicted by models are similar to those observed in experiments. This type of analysis will help identify deficiencies in existing models and needs for further experimental data to improve the models.

How ecosystem/biogeochemistry models implement the elevated CO_2 response

A review of several recent ecosystem model comparison activities (Ryan *et al.*, 1996*a,b*; VEMAP Members, 1995) suggests that there are at least five major ways that direct effects of atmospheric CO_2 have been included in ecosystem models. These include the effects on: (i) stomatal conductance and water use efficiency, (ii) photosynthesis, including acclimation processes and plant respiration, (iii) carbon allocation and growth, (iv) plant structure and phenology, and (v) plant nutrient concentration.

Table 7.3 summarizes the ways in which the direct CO_2 effect is incorporated into eight ecosystem models (Ryan *et al.*, 1996a; VEMAP Members, 1995). Recent versions of many of these models have been improved to include more of the direct effects of CO_2 on ecosystem processes. Table 7.3 shows that most of the models include the direct effects of atmospheric CO_2 on photosynthesis and respiration. Most incorporate the Farquhar photosynthesis model (Farquhar *et*

Table 7.3 *Comparison of the way that direct CO_2 is incorporated into different ecosystem models.*

CO_2 impact	Models							
	BIOME-BGC*	TEM*	CENTURY[R]	GRASS[R]	GEM[R]	HYBRID[R]	PNET-CN[R]	BIOMASS[R]
Stomatal conductance	−20%	−	−20% Tran	Ball *et al.* (1987)	−	Friend (1991)	−?	McMurtrie (1993)
Photosynthesis	F (C_3)	P	−	F ($C_3 + C_4$)	P	F(C_3)	P	F(C_3)
Carbon allocation	−	−	−	I_N	I_N	I_N	−	N_l
Structure & phenology	−	−	−	I_N	−	−	−	−
Leaf C/N	+20%	I_N	+20%	+20%	I_N	−	V	−

F, Farquhar photosynthesis equation (Farquhar *et al.*, (1980); P, Phenological; I_N, indirect effect of model structure; N_l, Function of leaf N content.
*Version of models used by VEMAP (1995).
[R]Version of models used by Ryan *et al.* (1998).

al., 1980), either explicitly or using simplified phenomenological models derived from that model.

In several models canopy photosynthesis is linearly related to absorbed light; this relationship has been derived from simulations of detailed canopy models (e.g. McMurtrie *et al.*, 1992*a*; Kirschbaum *et al.*, 1994), and from assumptions of optimal N distribution within canopies (Haxeltine & Prentice, 1996*a*; Dewar, 1996). Respiration is expressed either using a separation into growth and maintenance components incorporating a Q_{10}-temperature dependence, or by assuming that stand respiration is proportional to canopy photosynthesis (Gifford, 1994*b*; Dewar, 1997).

The impact of CO_2 on stomatal conductance is represented either by published equations relating stomatal conductance to environmental variables or by reducing transpiration water loss by a specified amount. The models that do not explicitly incorporate stomatal conductance all have relatively simple water budget submodels. The increase in the live leaf C : N ratio under elevated CO_2 levels is incorporated into almost all models, with some increasing the C : N ratio by a fixed percentage and others stimulating changes in C : N ratio as a result of internal dynamics.

Table 7.3 shows that few of the models include a direct effect of atmospheric CO_2 on carbon allocation or on plant structure and phenology, reflecting uncertainty as to how CO_2 affects these plant processes. Several models incorporate direct effects of water and nutrient stress on plant carbon allocation. However, the ability of these models to simulate observed changes in carbon allocation from field experiments is uncertain since formal model comparisons with data from elevated CO_2 experiments are limited (Coughenour & Chen, 1997). GRASS is the only model that included the impact of CO_2 on plant structure, associated with the indirect effect of water stress on leaf death rates.

Impacts on Forests and Grasslands

GRASSLANDS   Parton *et al.* (1995) analysed the impact of climate change and increasing CO_2 on seven major grassland ecosystems around the world. The model predicted that doubling current CO_2 would result in increased plant production, soil C levels, and soil decomposition rates. The greatest predicted increases in plant production were in cold desert steppe systems (owing to warming and CO_2 effects), while the greatest increases in soil C were in humid savanna systems. Net N mineralization rates were reduced by increasing CO_2 as a result of the increase in the C : N ratio of leaves. Soil decomposition rates increased because elevated CO_2 reduced transpiration rates and increased the soil water content.

Coughenour & Parton (1996) evaluated the impact of doubled CO_2 levels on a shortgrass steppe and tallgrass prairie using a detailed plant physiology model (GRASS). The results showed that elevated CO_2 increased plant production, root:shoot ratio, live leaf area and N mineralization, while transpiration per unit leaf area was reduced. The total transpirational water loss per unit ground area per year remained constant because of increases in duration and amount of live leaf area that compensated for a reduction in the transpiration per unit leaf area. The results also showed that the average soil water content was increased during the early part (June) of the growing season at the tallgrass site and during the late growing season (July and August) for the shortgrass steppe. Increased soil water content resulted in elevated decomposition and nutrient mineralization rates.

FORESTS Model comparison exercises are an especially useful mechanism for evaluating models (see Chapter 6). Ryan *et al.* (1996b) evaluated the effect of increasing atmospheric CO_2 and climate change on two coniferous forests by comparing results from eight different forest ecosystem models. All of the models predicted that increasing CO_2 would increase aboveground net production and tree biomass, with a smaller effect on soil carbon. Models that had strong links between nutrient cycling and plant production generally predicted smaller increases in plant production because of nutrient constraints. The models disagreed on the response to increased temperature with the differences resulting from the question of whether increased nutrient availability (resulting from increased decomposition rates) would compensate for increased drought stress. In general, this study shows that there are considerable differences in the way forest models respond to changes in climate and atmospheric CO_2.

Multiple ecosystem comparisons and consequences of different formulations
Recently three biogeochemical ecosystem models were compared in their simulations of the CO_2 response of potential natural vegetation for the entire continental United States (VEMAP Members, 1995a). They were BIOME-BGC (Running & Coughlan, 1988; Running & Gower, 1991), CENTURY (Parton *et al.*, 1987, 1993), and TEM (McGuire *et al.*, 1995). The comparison focused on changes in NPP and carbon storage over the entire USA.

Under doubled CO_2, all the models predicted increases in NPP for potential natural vegetation ranging from 5% for CENTURY, to 9% for TEM, and 11% for BIOME-BGC. Changes in carbon storage (vegetation and soils) for the same models were + 2%, + 9%, and + 7% respectively. It is not surprising that there should be differences in the projections since the models differ in how they implement the CO_2 response.

Pan *et al.* (1997) have undertaken a more detailed comparison by evaluating the NPP responses simulated by the three models at specific sites (representing

17 biomes) along temperature and moisture gradients. CENTURY simulated little variation in relative NPP response, with a significant negative relation between precipitation and relative NPP response and no correlation with temperature. For BIOME-BGC and TEM, the ranges in stimulation of NPP were similar, and were far greater than for CENTURY, but the distribution of values differed among biomes. For BIOME-BGC there was a marginally significant negative correlation with precipitation, and a significant negative correlation with temperature. For TEM, there was a negative correlation with precipitation but a positive, highly significant correlation with temperature.

These differences in responses were caused by the way the CO_2 response was implemented (Pan *et al.*, 1997). In BIOME-BGC the effects of elevated CO_2 on NPP resulted from changes in transpiration, soil water and leaf area, with nitrogen feedbacks playing no role. In both CENTURY and TEM, the nitrogen cycle plays a key role in regulating the carbon fluxes. For CENTURY, the decomposition process provides most of the control and the effect of moisture on decomposition is dominant. TEM, however, alters the $C:N$ ratio, which in turn can influence the decomposition rate and the supply of nitrogen; this then feeds back to influence NPP.

An ongoing research effort based on an evaluation of the CO_2 responses of a larger array of ecosystem models (CMEAL Participants, unpublished) aims to improve the representation of the response to CO_2. The models are being run for several sites for which considerable data are available on growth and physiology at ambient and elevated CO_2. In a preliminary model intercomparison four models were run for current and doubled CO_2. Important conclusions from this early experiment include: (i) while the patterns observed among models were often qualitatively similar for particular variables, no two models responded in either a qualitatively or quantitatively similar fashion for the full suite of ecosystem variables; (ii) all of the models predicted a short-term increase in NPP after doubling of CO_2 based principally on increased photosynthesis per unit nitrogen; (iii) models predicted different long-term NPP responses, depending primarily upon assumptions of fixed versus floating $C:N$ ratios for wood or soil and the resultant effect on availability of N for plant uptake; (iv) there were also important differences among models in terms of where N and the increased pool of C are stored in ecosystems at elevated CO_2; (v) all models predicted increased use efficiency of N, water, and light at elevated CO_2, but nitrogen is the only resource for which increased uptake was predicted.

An important lesson from all of the model comparison exercises is the need for good standardisation in protocols to eliminate differences in results that could be due to arbitrary differences in procedures or choices of model parameters. An example of these is seen in patterns of ecosystem nitrogen content and use. Some models have a closed nitrogen cycle, which obviously constrains the ecosystem nitrogen pool and also limits long-term NPP increases; other models have an open nitrogen cycle and predict sustained increases in ecosystem nitrogen and C pools. As noted above, assumptions on wood or soil C : N changes can lead to major differences in predictions (McMurtrie & Comins, 1996). These results highlight the need for better data from experiments on ecosystem pools of N and patterns of distribution within the ecosystem.

Model analyses using new experimental results

Models can be used to evaluate the long-term implications for ecosystems using data derived from shorter term experiments. This section presents results from model simulations undertaken to evaluate the potential consequences of two unanticipated and tentative observations from recent experiments: (i) the reduction in stomatal conductance found in herbaceous systems at elevated CO_2 may not occur for forests, and (ii) the increase in C : N observed in CO_2–fertilized live plant tissue is often not seen in litter. Both observations are contrary to early results and expectations; however, they are the basis for formulations for all current models.

The effects of these two changes were simulated by modifying two existing ecosystem models, CENTURY (Parton *et al.*, 1993 and users manual reference) and G'DAY (Comins & McMurtrie, 1993; McMurtrie & Comins, 1996), applied to a wet Norway spruce boreal forest at Flakaliden, Sweden (Linder, 1995) and a dry *Pinus radiata* forest (Biology of Forest Growth – BFG) near Canberra, Australia (Benson *et al.*, 1992). CENTURY was also applied to the Jasper Ridge annual grassland system in California (Field *et al.*, 1996). These sites were chosen to demonstrate the impact of elevated CO_2 in contrasting ecosystem types. The two models were run to equilibrium conditions using the same observed weather data and then changes in atmospheric CO_2 were imposed (350 to 700 ppm CO_2).

The new understanding was simulated by comparing computer runs where the standard assumptions were used (reduced stomatal conductance and decreased N content of leaf litter) against runs with the new assumptions (no reduction in stomatal conductance for mature trees and no reduction in N concentration in leaf litter). It was assumed that stomatal conductance would be reduced at high CO_2 for grassland, and that the N concentration in leaf litter would be kept approximately constant by reducing internal recycling of N prior to leaf senescence (the fraction of N retranslocated at senescence was reduced from 50% to 40%). In the text these simulations are referred to as STANDARD (old assumptions) and NEW (new assumptions) runs.

Figure 7.4 demonstrates the simulated impact of doubling atmospheric CO_2 on annual net primary production (NPP) at the three sites using the CENTURY

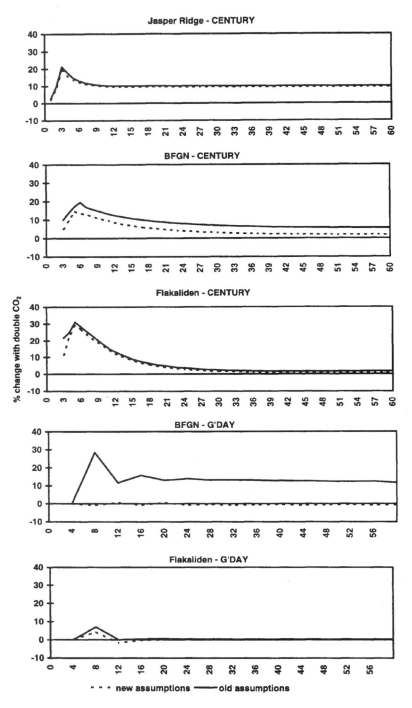

Figure 7.4   The simulated impact of doubling atmospheric CO_2 on annual net primary production (NPP) at Jasper Ridge (USA), Flakaliden (Sweden), and Biology of Forest Growth (BFG – Australia) sites using the CENTURY and G'DAY models (G'DAY was not run for the Jasper Ridge site).

and G'DAY models (G'DAY was not run for the Jasper Ridge site). The results show that following a doubling of CO_2, there is a rapid initial increase in NPP at all sites (+ 5 to 30%) and that this increase is reduced by half or more after 10 years at high CO_2. Comparison of the results for the STANDARD and the NEW CO_2 assumptions shows that the positive response to increased CO_2 is reduced under the NEW CO_2 assumptions and that both models show that this reduction is largest at the BFG site. At the wet Flakaliden site and at the Jasper Ridge site, growth under the NEW assumptions is entirely due to reduced internal N cycling. For both sites there is little difference between simulated NPP under the STANDARD and NEW assumptions, suggesting that altering the N content of dead leaves has little effect on NPP. The difference between the STANDARD and NEW runs would have been more pronounced, however, if we had made the further assumption that changes in litter C : N ratio lead to correlated changes in soil C : N ratio (McMurtrie & Comins, 1996). The need for models to correctly represent C–water interactions is highlighted by the sensitivity of modelled NPP to assumptions about the dependence of stomatal conductance on CO_2.

Figure 7.5 shows the effect of changing the CO_2 assumptions on the average change in annual net N mineralization and soil C levels for the 10 to 20 year period following the doubling of CO_2 levels. The results show that under the STANDARD assumptions there is generally (the only exception is CEN-TURY result for BFG) a decrease in net N mineralization at high CO_2, because the N content of high CO_2 litter is reduced, leading to enhanced N immobiliz-ation during litter decomposition. Under the NEW assumptions litter N con-centrations are higher than under the STANDARD assumptions, so that N immobilization is reduced and N mineralization is enhanced relative to the STANDARD assumptions. Under high CO_2 levels there is generally an in-crease in soil C levels. Using the NEW assumptions results in slightly lower increases in soil C levels, because litter quality is higher, leading to faster turnover of soil C. Thus, both models show that using the NEW direct CO_2 assumptions results in reduced plant responses to doubling atmospheric CO_2 levels with a decrease in the positive effect of CO_2 on plant production and soil C storage.

7.2.6 Progress in predicting elevated CO_2 responses – a lesson from ecosystem physiology

Simple extrapolation from physiological information to ecosystem and global responses would lead to inaccurate predictions about the biotic world under elevated CO_2. This is because such extrapolations do not account for process interactions and feedbacks.

GCTE started with the basic information that there are two primary plant

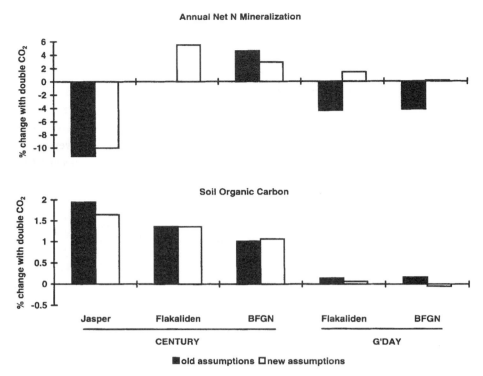

Figure 7.5   Effect of changing the CO_2 assumptions on average change in annual net N mineralization and soil C levels for the 10–20 year period following the doubling of CO_2 levels using the CENTURY and G'DAY models.

responses to elevated CO_2. One type of plant, C_3 plants, would respond positively, and rather strongly, to CO_2, and the other type, C_4 plants, would show little or no stimulation. It was also known that both types of plants would save water under elevated CO_2 and that the tissue produced would be lower in nitrogen and richer in carbon. This information was used to make the following predictions:

- C_4 plants would show limited enhancement under elevated CO_2.
- The enhancement effect would be fairly constant within a photosynthetic type.
- Systems from water-limited environments would show a relatively larger response than systems in other environments.
- Herbivores eating plant tissue from high CO_2 grown plants would have to compensate for the poor quality tissue by eating more (see Chapter 9).
- Decomposition of plant tissue would be slower under high CO_2, resulting in immobilization of nutrients in the litter pool.

What we learned is that:

(i) The projected enhancement of growth by CO_2 predicted from physiological measurements was often not realized. Although increased leaf-level photosynthesis was on average 40% higher under elevated CO_2, aboveground biomass accumulation was only 14% higher (average for nine herbaceous systems). This important finding indicates that whole-plant and ecosystem-level feedbacks were constraining production.

(ii) In systems under high CO_2, C_4 plants actually did comparatively well. This is due in part to the fact that water limitations, which were alleviated under high CO_2, were more important than the direct CO_2 effects in determining competitive outcomes.

(iii) Within C_3 plants there are very different growth responses among species, not necessarily related to conventional functional types, again indicating potential feedbacks on ecosystem processes by changes in community structure. Thus, elevated CO_2 is also a biodiversity issue.

(iv) The predicted increased water savings are observed. In some systems the changing water balance was manifested in a shift in species composition and in others by an increase in canopy development. For water limited systems, the hydrological consequences of elevated CO_2 were often more profound than the direct CO_2-fertilisation effect. Significant increases in soil activity were a consequence of the improved water status.

(v) Most insects appear to show compensatory increased tissue consumption under elevated CO_2. But when given a food choice, net consumption may not change (Arnone *et al.*, 1995). Grazing mammals showed no such compensation, and consequently weight loss occurred (Owensby *et al.*, 1996*b*) (see Chapter 9).

(vi) High CO_2-grown leaf tissues, when naturally senesced, do not often decompose more slowly as hypothesized. The basis for this finding is in part because litter $C:N$ ratio does not always change, as predicted.

7.3 Effects of temperature increase on terrestrial ecosystems

An increased mean global surface air temperature of $1.5 - 4.5\,°C$ is expected to occur within the next century as a consequence of increased atmospheric CO_2 and other greenhouse gases (Houghton *et al.*, 1996). Global warming will affect ecosystem structure and function in many ways, but of special concern are those changes that may result in feedback on the climate system. For example, increased temperature could increase CO_2 efflux from terrestrial ecosystems to the atmosphere, if increased soil organic matter (SOM) decomposition and respiration (efflux that increases exponentially with temperature) is higher than the increased photosynthesis and NPP (a linear to saturating response with temperature; Townsend & Vitousek, 1995; see also Chapter 10). Such an imbalance could further increase atmospheric CO_2 and so act as a positive feedback to global warming.

Much attention has been placed on high latitude ecosystems since General Circulation Models (GCMs) predict the greatest warming in the boreal and tundra regions of the world. Montane systems have also received attention because of their sensitivity to snow cover and soil temperature. Temperature effects in tropical systems could also be important because high rates of respiration result in large quantitative changes in CO_2 efflux in response to small temperatures changes. Therefore, even the small predicted temperature warming in the equatorial regions could release carbon to the atmosphere (Townsend *et al.*, 1992).

7.3.1   Response of high latitude ecosystems to global warming

Although warming will be most pronounced at the poles, Arctic organisms are so well adapted to low temperature and intra-seasonal temperature variations that temperature *per se* does not strongly affect most Arctic processes (Chapin, 1983). For example, photosynthesis, nutrient uptake, and plant growth have broad temperature optima with little difference in rate between 10 and 20 °C (Oberbauer & Oechel, 1989). In laboratory incubations, nitrogen mineralization is unaffected by temperature below a threshold of about 8 °C; above this, mineralization rate increases dramatically (Kielland,1990; Nadelhoffer *et al.*, 1991). Similarly, a 3 °C increase in air temperature in field greenhouses had little effect on plant growth, soil nutrient availability, or species composition over a two- to four year-period in the low and mid Arctic (Chapin & Shaver, 1985, 1996; Havström *et al.*, 1993; Wookey *et al.*, 1993). However, in a high latitude polar desert site, plant growth and reproduction were highly sensitive to small temperature increases above ambient. Of 19 alpine and lowland plant species studied in Switzerland, acclimation to warmer temperatures ranged from full to no acclimation, with complete acclimation occurring in three species (Larigauderie & Körner, 1995).

These adjustments have to be taken into account when building models of responses to climate change. Plants adjust their respiration rates to the prevailing growth temperatures. Thus, information on Q_{10}s derived from short-term measurements of respiration of plants grown at a single temperature may be very misleading when extrapolated to longer time intervals. For example, Gifford (1995) has shown for wheat that a Q_{10} of 1.3 would be more appropriate for longer term extrapolations in models than the commonly used Q_{10} of 2.0. For wild grassland plants Larigauderie & Körner (1995) found a mean 'acclimated' Q_{10} of 1.7 with some species fully acclimated ($Q_{10} = 1$).

Indirect temperature effects associated with changes in thaw depth, nutrient availability, and vegetation will cause substantial changes at high latitudes. Because each Arctic species has a different pattern of response to environment, a 3 °C increase in summer air temperature causes a change in species composition, litter quality, and nutrient availability at a decadal timescale (Chapin & Shaver, 1985; Chapin *et al.*, 1995*b*). Increased nutrient availability increases shrub abundance and decreases the abundance of mosses (an important soil insulator) (Chapin *et al.*, 1995*b*). Some of the vegetation changes observed during the past 15 years are consistent with these experimental manipulations and suggest that regional warming of permafrost, which is more pronounced than can be explained by increases in air temperature (Lachenbruch & Marshall, 1986), could reflect indirect temperature effects on vegetative insulation.

The effects of experimental warming on plant biomass and species composition in subalpine meadows were also primarily indirect (Harte & Shaw, 1995; Harte *et al.*, 1995), with changes reflecting increases in soil temperature and decreases in soil moisture. These changes in soil environment modified the competitive balance between shrubs and forbs, causing a shift toward shrub dominance of the type that normally dominates drier sites. These vegetation changes, in turn, altered shading by the plant canopy and therefore the linkage between energy inputs and soil environment.

Simulation models suggest that a 2 °C temperature increase would cause a 20% increase in thaw depth after 50 years and that warming of 6 °C or 8 °C would cause disappearance of permafrost after 38 or 27 years, respectively (Kane *et al.*, 1992). Permafrost is a critical factor causing a high water table, slow decomposition, and peat accumulation, so temperature increases of these magnitudes, which are within the 10 °C increase predicted at high latitudes with a CO_2 doubling, would cause a qualitative change in the tundra environment.

Decomposition in the Arctic is more sensitive to changes in moisture than temperature (Flanagan & Veum, 1974). Perhaps because of soil drying associated with warmer drier summers during the past decade (Maxwell, 1992), tundra appears to have changed from a net carbon sink to a system that is either a net CO_2 source to the atmosphere (Oechel *et al.*, 1993, Zimov et al., 1996) or is in approximate balance, at least during the growing season (W.C. Oechel & S.A. Zimov, personal communication). Methane flux from tundra also appears much more sensitive to water table depth than to temperature (Whalen & Reeburgh, 1992).

A warming-related increase in depth of the active layer or a change in regional water balance which causes the water table to drop below the soil surface will create a threshold decrease in methane flux to the atmosphere by increasing surface oxidation, leading to a negative feedback to climatic warming. Thus, the net effect of high latitude ecosystems on trace gas feedbacks to climate could depend on the relative magnitude of changes in CO_2 versus CH_4 flux.

Presence of trees at high latitude may be determined more by soil than by air temperature (Murray, 1980), so that expected warming of soils and increase in thaw depth could have a strong effect on treeline advance. A northward advance of treeline could allow greater long-term carbon storage in vegetation. However, the transient response of carbon storage depends on temperature–moisture stimulation of decomposition and carbon loss, which occurs instantaneously, versus a 50–150 year time lag in forest advance and carbon storage (Smith & Shugart, 1993; Starfield & Chapin, 1996; Schulze *et al.*, 1995). Thus, the short-term effects of temperature on whole-system carbon storage may differ from the long-term effects.

Warming is also expected to increase fire frequency in the boreal zone (Flannigan & Van Wagner, 1991), which directly releases CO_2 stored in peat and wood to the atmosphere and increases rates of decomposition (Kasischke *et al.*, 1995). These fire effects on boreal forests could equal or exceed the capacity of individual forest stands to store carbon, further raising questions about the net role of boreal forests as a candidate for the 'missing sink' in the global carbon budget (Ciais *et al.*, 1995).

Northward advance of the treeline in response to climatic warming could act as a positive feedback through reduction in albedo, as snow-covered tundra is replaced by darker forests (Bonan *et al.*, 1992; Thomas & Rowntree, 1992). This albedo feedback by northward-moving forests could have accounted for half of the regional temperature increase observed at high latitudes 6000 yr BP (Foley *et al.*, 1994).

The amplitude of the seasonal change of atmospheric CO_2 concentration has been increasing, with largest increases observed at high latitudes. This could reflect either a longer growing season for plant growth (Keeling *et al.*, 1996*a*), although a 10% change in length of growing season through experimental manipulation of snow cover had no detectable effect on vegetation cover or species composition (Chapin & Shaver, 1996). Alternatively, warming-induced increases in winter respiration could contribute to the increased amplitude of CO_2 at high latitudes (Zimov *et al.*, 1996).

The changes in biotic interactions that will occur with climatic warming are poorly known. A 3 °C experimental warming causes an increase in shrub abundance and a decline in species richness, particularly of insect pollinated forbs, within a decade (Chapin *et al.*, 1995*b*), as in the Colorado subalpine zone (Harte & Shaw, 1995; Harte *et al.*, 1995). These results suggest that migrating caribou, which utilize these forbs during lactation (White & Trudell, 1980), and insect pollinators (Williams & Batzli, 1982) could be adversely affected. Warmer summers also increase insect harassment of caribou with associated declines in feeding and summer energy reserves. By contrast, browsing mammals such as snowshoe hare and moose may benefit from climatic warming because of the proportional increase in shrubs in undisturbed tundra and increased fire frequency with associated increase in the proportion of early successional vegetation.

In summary, direct temperature effects on Arctic processes are subtle, except in polar desert, but indirect temperature effects mediated by changes in nutrient availability, soil moisture, seasonal length, and vegetation have caused, and will cause, changes in energy and carbon balance that are large enough to provide significant positive feedbacks to global warming. Effects of elevated CO_2 are generally small because of prevailing limitation of low temperature, short growing seasons, and nutrients. Two patterns emerge from warming experiments: (i) ecosystem response to warming is greatest in the coldest environments, and (ii) ecosystem responses to warming are less pronounced and have longer time lags than do physiological and growth responses by individual species. Winter warming is likely to be more influencial than summer warming.

7.3.2 Response of mid-latitude and tropical ecosystems to global warming

Less is known about the responses of mid-latitude ecosystems to warming. A network of soil-heating experiments has been established in several ecosystems (Peterjohn et al., 1993). A hardwood forest showed a 1.6-fold increase in CO_2 emissions with soil heating and a 36% decrease in the carbon concentration of the O soil horizon (Melillo et al., 1996a). Warming caused soil respiration to increase 40% in the first year and about 20% the following years; methane oxidation increased 20%.

No warming experiments have been conducted on natural ecosystems in the tropics (see Chapter 9 for a discussion of the effects of increasing temperature on rice productivity). Direct species-level effects in natural ecosystems are likely to be pronounced, but ecosystem-level effects are again most likely to be indirect.

7.4   Effects of changes in water availability on terrestrial ecosystems

Water loss from vegetation is an inevitable consequence of plants exchanging gases with the surrounding atmosphere. It cannot be avoided during carbon assimilation because stomatal opening allows both the inward diffusion of carbon dioxide for photosynthesis as well as the outward diffusion and loss of water from plant cells. Yet maintaining a favourable water balance within the plant is essential for metabolism and survival. The amount of water available to plants is determined by a combination of precipitation, the capacity of soils to store water, and atmospheric evaporative demand. While water is generally a limiting resource in most environments (Kramer & Boyer, 1995), too much standing water may also be damaging to plants because of the soil anoxia that typically results during flooding.

This section discusses recent advancements in ecosystem physiology that have improved understanding of water fluxes, especially in relation to other plant processes. This includes a discussion of how the partitioning of water within ecosystems is regulated by plant cover, and the resulting effects of water availability on plant performance at ecosystem levels. Much of the recent research in water and ecosystems has not dealt directly with the question of elevated CO_2, but rather has focused on (i) understanding dynamics and constraints on fluxes between the vegetation and the atmosphere, and (ii) on establishing those principal aspects of ecosystem physiology that constrain fluxes between the ecosystem and the atmosphere. Recent books and reviews on general aspects of water relations in plants that provide a broad introduction to this field include Jones, 1992; Smith & Griffiths, 1993; Schulze, 1994a; Kramer & Boyer, 1995; and Kozlowski & Pallardy, 1997.

7.4.1   Ecosystem water loss to the atmosphere

Plant transpiration and evaporation from the soil and vegetation surfaces constitute the transfer of water from ecosystems to the atmosphere. Both the structure and physiology of ecosystems combine to regulate this water loss (Jarvis, 1987; Sperry, 1997). Stomatal conductance describes the physiological regulation of water loss at the leaf level, determined by a variety of biological and physical processes, most of which are now well understood (Jones, 1992; Smith & Griffiths, 1993). Unfortunately, leaf-level transpiration does not directly scale up to the whole plant and ecosystem levels because of canopy-level constraints on gas exchange (Jarvis & McNaughton, 1986; Schulze *et al.*, 1996b). These constraints, due largely to ecosystem structure (canopy architecture), are quantified as the aerodynamic conductance, which describes the regulation of water vapour transfer from the leaf surface through the vegetation canopy to the open atmosphere above. If the aerodynamic conductance is large in relation to the potential stomatal conductance, as in meadows, the latter is limiting and transpiration is proportional to the available energy. In contrast, if aerodynamic conductance is small in relation to the potential stomatal conductance, as in forests, the former is limiting and transpiration is proportional to the leaf-to-air evaporative gradient. The magnitude of both stomatal and aerodynamic conductances is not constant, but may change seasonally, or even during the course of a day, with phenology and soil water deficit (Köstner *et al.*, 1992). The Penman–Monteith equation (Penman, 1948; Monteith & Unsworth, 1990) ties together the effects of both the ecosystem structural and leaf physiological components on ecosystem loss of water to the atmosphere, that is, aerodynamic and stomatal conductance, respectively.

While it has been possible to describe the relative effects on stomatal conductance of environmental parameters (Jarvis, 1987), there is no mechanistic model to describe the maximum transpiration performance of different plant functional types with climate. Yet such an understanding is needed to link the impacts of functional types and LAI (Leaf Area Index) with water fluxes to the atmosphere on a regional and global basis. Kelliher et al. (1993, 1995) made progress through analysis of a model which explored relationships between maximum surface conductance (soil surface and stomatal conductance together) and aerodynamic conductance in relation to LAI. Canopy conductance is expected to increase with LAI under conditions of both high and low stomatal conductances, but only at low to moderate LAI values. The relationship is predicted to become less sensitive at LAI values above four and maximum canopy conductance is expected to be achieved by an LAI of six. In contrast, the surface conductance shows a compensatory behaviour associated with soil evaporation (assuming the soil is wet), which is high with sparse vegetation and decreases with increasing LAI values. The Kellier et al. model leads to some interesting predictions that have ramifications for global models. For instance, in ecosystems whose plants have low stomatal conductances, a wet soil without plant cover will have a higher conductance than a vegetated surface. The measured values of maximum canopy conductance from very different ecosystems fall well within the limits of the predictions of this model, indicating that these results provide a framework for integrating ecosystem physiology with global model predictions.

From the Kelliher et al. (1995) analysis, it is clear that there is expected to be a linear relationship between maximum leaf conductance and surface conductance, which would not have been predicted by the Penman–Monteith equation. There is a feedback between stomatal response and plant structure in such a way that if the aerodynamic conductance is high (equilibrium transpiration driven by radiation), plants have high leaf conductance values (e.g. herbaceous species, crop plants). In contrast, if the aerodynamic conductance is low (imposed transpiration driven by the vapour pressure deficit), then the existing plant functional types have low leaf conductances. Both plant functional types may exhibit the same LAI.

Other factors, such as as the nitrogen content of the vegetation, can also affect evapotranspiration. For example, while leaf conductance values are related to leaf structure (Schulze et al., 1994b), there appears to be no functional way of relating leaf structure or leaf conductance to plant nutrition other than via classification of plant functional types. CO_2 assimilation is the common link between stomatal conductance and nutritional aspects of leaf physiology (Marschner, 1995). Schulze *et al.* (1994*b*) showed that for different life forms there are linear relationships between maximum leaf conductance and leaf nitrogen content as well as between leaf surface conductance and the maximum capacity for CO_2 assimilation. A global analysis of these patterns indicates that the greatest potential for CO_2 assimilation is located in industrial regions, where fertilizer loads in combination with elevated dry deposition rates overlap with highly productive agricultural plants.

7.4.2   The regulation of water loss by plants within ecosystems

Actual rates of gas exchange and of leaf conductance are typically lower than the maximal values shown by Schulze *et al.* (1994*b*) because of constraints imposed by light, temperature, humidity, and soil water deficit. Several model approaches have allowed for quantifying the extent to which different parameters limit actual gas exchange rates (Running & Hunt, 1993; Sellers *et al.*, 1997). On average, the ratio of actual to maximum photosynthetic rate is between 40 and 60%.

Recent observations suggest that mechanisms for both feedforward and feedback response exist to limit leaf conductance and that both biophysical and hormonal signals may be involved in this regulation (Schulze, 1994*a*). An important constraint on stomatal regulation of water loss from plants appears to be the avoidance of xylem cavitation, which may irreversibly affect the capacity of roots to supply water to the shoots (Tyree & Sperry, 1989; Meinzer, 1993; Sperry, 1997). Recent analyses indicate that plant species differ in their xylem cavitation characteristics and that variations in this physiological characteristic are closely tied to plant distribution (Langan *et al.*, 1997; Pockman & Sperry, 1997; Sperry, 1997).

7.4.3   The capacity of plants within an ecosystem to exploit soil water

A better appreciation of rooting distributions is critical to an understanding of the carbon and water flux aspects of ecosystem function. Variations in rooting patterns within and among ecosystems determine not only the extent of the exploitable soil profile, but also the extent to which deep soil reserves buffer metabolic activities on both interseasonal and interannual bases. Canadell *et al.* (1996*a*) and Jackson *et al.* (1996*a,b*) have recently synthesized some of the critical global patterns. Jackson *et al.* (1996*a*) found that there is a shift in rooting distribution patterns towards deeper layers in arid systems. At the same time, Canadell *et al.* (1996*a*) demonstrated a remarkable relation between vegetation type and maximum rooting depth which closely paralleled rooting density behaviour. Both of these studies indicate that global models have historically underestimated the exploitation of soils by roots, especially the capacity of roots to exploit deeper soil layers. The incorporation of appropriate rooting-pattern information into the latest generation of Soil–Vegetation–Atmosphere Transfer Models (SVAT) and other biosphere–atmosphere models is now leading to more realistic simulation of the feedbacks from the land surface back to the atmosphere (Kleidon & Heimann, 1998).

Complementing the ecosystem-level studies showing variations in rooting behaviour is the observation of soil-depth partitioning by different functional types within an ecosystem (Sala *et al.*, 1989; Ehleringer *et al.*, 1991; Dawson, 1993*a*). However, it is important to note that seasonal partitioning of root activities in arid land ecosystems can result in functional differences that are not consistent with simple observations of rooting depth patterns. Variations in the amounts and predictability of precipitation inputs as well as in the frequency of extended drought periods during the growing season have led to differences in the extent to which functional types use and rely on surface versus deep roots (Dawson 1993*a*,*b*; Lin *et al.*, 1996). At this point, it is unclear whether classification of functional types with respect to water use overlaps fully with classification of functional types with respect to nutrient extraction from the soil.

7.5 Effects of altered nitrogen deposition on terrestrial ecosystems

No other elemental cycle has been changed by man as much as the global nitrogen cycle (Mohr & Muentz, 1994; Jordan & Weller, 1996: see also Chapter 1) (Fig. 7.6). It has been altered by (i) land-use changes and the introduction of N-demanding species, (ii) fertilizer addition, (iii) cultivation of N_2-fixing species, and (iv) air pollution leading to nitrogen deposition. The total amount of nitrogen being added by man annually already exceeds the amount assimilated naturally. Current values of N deposition are between 0.5 to 2.5 g N m^{-2} year^{-1} in eastern North America and 0.5 to 6.0 g N m^{-2} year^{-1} in northern Europe (Vitousek, 1994).

Nitrogen has a leaky cycle with feedbacks to the soil through nitrate losses to ground water, and to the atmosphere through denitrification. This latter may not only produce N_2 as the ultimate product, but also greenhouse gases such as NO and N_2O depending on soil chemistry. The greatest anthropogenic effect on the N–cycle was originally via harvest, although more recently N deposition has had a greater impact.

The following section focuses on three aspects of the nitrogen cycle: (i) the effect of N deposition on growth, (ii) its interaction with the carbon and other element fluxes, and (iii) its effects on soil properties, especially soil acidification.

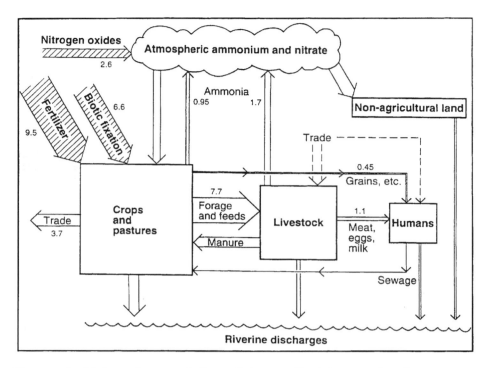

Figure 7.6   Major anthropogenic flows of nitrogen. Shaded arrows show inputs
of newly fixed nitrogen through fertilizer application, biotic nitrogen fixation in
agricultural lands, and production of nitrogen oxides that are converted to nitrate and
deposited from the atmosphere. Unshaded arrows show other anthropogenic fluxes
of nitrogen, including atmospheric deposition of ammonium and nitrate, emission of
ammonia into the atmosphere, flows through the agricultural food chain, return of
nitrogen to crops and pastures in animal wastes, and discharges of nitrogen in rivers.
The widths of the arrows are proportional to the sizes of the flows for the entire
coterminous United States. The numbers near arrows are our calculated estimates
of flows (Tg N y^{-1}). Dashed arrows show potential imports of nitrogen via trade in
agricultural products, which are major sources of nitrogen in many regions, although
the coterminous United States is a net exporter of nitrogen in agricultural products
(from Jordan & Weller, 1996)

The effects of land-use change and of cultivation of legumes on the global
N-cycle are discussed further in Chapter 10.

7.5.1   N availability and N deposition on growth

While effects of fertilizers on crop productivity are well documented
(Marschner, 1995), it is less clear whether anthropogenic atmospheric N deposition enhances growth under otherwise undisturbed conditions. The effect
of N deposition on growth can be demonstrated in forest stands and their tree

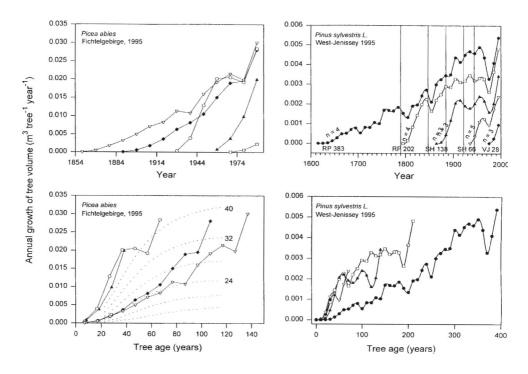

Figure 7.7   Growth of a chronosequence of *Picea abies* stands in Germany and of *Pinus sylvestris* in Siberia. The growth curves are either scaled along real time, or they are compared with respect to age.

ring chronology in Europe (Fig. 7.7; Mund, 1996). The growth rate of young stands was much slower one hundred years ago than in 1996. In addition, the growth rate of old stands has increased at a time when one would expect an age-related decrease in whole stand production; with N deposition the old stand reaches growth rates that are equivalent to those of young stands. A comparison of growth rates with forest inventory yield tables (Assmann & Franz, 1963) indicates that the young stands grow presently at a higher yield class than the old stands did at the time of their establishment, and that the old stands have changed yield classes and presently approach the same yield class as young stands. Similar observations were made with deciduous species (Franz *et al.*, 1993).

This growth response of trees under conditions of N deposition may include some component of response to elevated atmospheric CO_2. Comparable information was therefore collected for a chronosequence of *Pinus sylvestris* stands in Siberia (west Yennesei region), growing far from any source of air pollution (Fig. 7.7). No accelerated growth can be observed in young stands of *Pinus* as compared to old stands at the time of their establishment, and the growth response of old trees reaches a plateau which is interrupted only by N mobiliz-ation due to periodic fires. This suggests that elevated CO_2 had no effect in the absence of N deposition. In contrast, the European stands showed a distinct response to N availability. The possibility cannot be excluded that elevated CO_2 has enhanced this effect, but the change in growth rate is consistent with the history of N deposition rather than with the history of CO_2 rise. It is important to note that the increased NPP in forests under conditions of N deposition does not always result in elevated NEP because of the interaction with disturbance regimes (Schulze & Heimann, 1997).

Melillo *et al.* (1996*b*) argued that the effect of soil warming would result in increased wood growth because the C : N ratio in humus is lower than in wood by a factor of three to five. Therefore each mole of nitrogen from decomposed humus could result in three to five moles of carbon immobilization during wood growth. Bauer *et al.* (1997) demonstrated for European forest trees that wood growth is the dominant sink for N and C in trees. Needle N-concentrations were constant along a European transect although wood growth changed with N-deposition.

7.5.2   N deposition on other element cycles and forest decline

Ingestad & Ågren (1988) demonstrated in laboratory experiments that the growth response to N addition reached a distinct maximum if other elements became limiting. There are numerous cations which can be close to deficiency levels. For instance, magnesium concentrations are often low in silicate soils; iron and potassium are low in calcareous soils; and manganese is low in dolomitic soils. With increasing N deposition, these elements reach a level which results in visible colour changes in foliage, as is shown in European maps of forest decline in which needle yellowing is a predominant feature. The symptom is enhanced by canopy uptake of nitrogenous gases (NH_3 and NO_x), which is not balanced by cation uptake by roots (Schulze, 1995*b*).

Young needles are the dominant sink for magnesium in conifers. During the period of needle growth the old needles of the same branch supply the required magnesium (Lange *et al.*, 1989). Uptake from roots under these conditions is still low due to low soil temperatures. If growth of young needles is inhibited (by pruning of buds on alternative twigs of the same branch), then those old needles that supply new growth turn yellow, while those that do not supply new needles remain green. It has been shown that this response is related to N availability (Oren *et al.*, 1988), because it directly affects the amount of new needle growth (Bauer *et al.*, 1997).

Needle discolouration is only one symptom of forest decline; the symptoms of needle loss and interactions with pests and parasites are geographically more extensive. There is presently no general theory for needle loss, which may be caused by drought on acid soils where root growth is restricted to the topsoil. It may also depend on ozone episodes (Sandermann *et al.*, 1997), or be related to acid mist. It is also possible that old needles in conifers serve as a resource store that is no longer needed under conditions of increased N deposition. The interactions are certainly more complicated than in the case of needle yellowing.

Less clear is the interaction of N deposition with pests and diseases. The forest inventories show an increased abundance of insect outbreaks, mildews and other fungal diseases. It remains unclear if these pests are directly stimulated by N deposition or an indirect consequence of a weakened system.

7.5.3 Nitrogen deposition on soils

Even with maximum growth rates, spruce forests immobilize only about 10% of the total amount of deposited N in wood (Schulze & Ulrich, 1991). The rest is denitrified although N_2 is not always the end product. Sometimes N is incorporated into humus (depending on conditions), or leached to ground water. The latter results in soil acidification, a nonreversible change where the transport of strong acids is balanced by an equimolar transport of cations. Critical loads of acid rain deposition (nitrogen and sulfur compounds) are defined by the capacity of soils to regenerate cations.

The processes of soil acidification are much faster than previously thought, and large areas of forest, agriculture, and natural vegetation are affected (Hildebrandt, 1994). The overall effect of N deposition and soil acidification is an accelerated release of nitrate to ground waters (Durka *et al.*, 1994). About 30% of the atmospheric N deposition may reach ground water in acid soils without interfering with soil biology. Since nitrification seems to be inhibited in acid soils, nitrate is formed from dissolved organic C and dissolved organic N by microbial activity in deep soil layers where base cations are again available but are beyond the reach of roots.

Although soil acidification has been associated primarily with acid rain, this is not the only pathway by which soils may acidify. On poorly buffered soils tree harvesting may have a similar effect as acid rain. Similarly, soil acidification has become a major problem in agriculture. Rice cultivation acidifies soil, while the growth of legumes (e.g. *Lupins* sp.) increases the availability of organic nitrogen in soils but at the same time induces soil acidification due to H^+ excretion. Under natural conditions, nitrogen-fixing species occupy a niche in succession on mineral soils where they increase weathering and soil formation (Marschner, 1995).

The effects of soil acidification on whole watershed function have been reported to last many years after acid inputs stop, and have been suggested to be the cause of decreased biomass accumulation in the Hubbard Brook Experimental Forest in the USA (Likens *et al.*, 1996).

7.5.4   N deposition on species composition

In addition to changes in the physiological performance of ecosystems, N
loading may eventually lead to changes in plant species composition (Jefferies &
Maron, 1997). Ellenberg (1986) scaled the mid-European flora along an indicator value of nitrogen availability. He showed that a large fraction of endangered species grow in habitats which are N-deficient, suggesting that increasing
eutrophication from atmospheric deposition may endanger species biodiversity
in the long term. Results from 12 years of experimental N addition in grasslands
of Minnesota (USA) showed changes in plant species composition, decreased
species diversity, and increased aboveground productivity (Tilman, 1987; Inouye & Tilman, 1988). The major change in species composition was the shift
from C_4 to C_3 grasses, which brought a decrease in the plant tissue $C:N$.

7.6   Effects of UV-B increase on terrestrial ecosystems

The depletion of stratospheric ozone as a consequence of increased an-
thropogenically produced chlorine species has long been recognized (Molina
& Rowland, 1974). Because stratospheric ozone is the main attenuator of solar
UV-B radiation (280–320 nm), UV-B levels are expected to increase at the
Earth's surface. This is of concern because irradiation with sufficiently large
UV-B doses can inhibit the Photosystem II function, damage DNA, produce
free radicals and lead to photomorphogenetic changes (see Caldwell & Flint,
1994 for a review).

In contrast to mid and low latitude regions, plants at high latitudes have
historically been exposed to relatively low UV-B levels. They are therefore
expected to be especially sensitive to increasing UV-B and are subject to the
greatest relative increases. In the past few years ozone reduction at temperate
latitudes has been much higher than previously predicted (Gleason *et al.*, 1993).

Caldwell *et al.* (1989) proposed several mechanisms by which ecosystems
might respond to increasing UV-B (Fig. 7.8). It was considered that many of the
impacts of UV-B would be indirect although there would also be direct effects
on photosynthesis, which would decrease. Some of the indirect effects were
proposed to operate through alteration of biosynthetic pathways that would
alter tissue quality, making it more resistant to herbivores and decomposers.

There are few experimental data to support Caldwell's hypotheses on the
UV-B effects on whole ecosystem processes, and even less data from studies
conducted under realistic field conditions. Nevertheless, some field studies have
shown decreased leaf-level photosynthesis and plant growth at elevated UV-B
(Sullivan & Teramura, 1992; but see Teramura, 1983). For instance, *Pinus taeda*
seedlings exposed to elevated UV-B radiation showed a significant growth

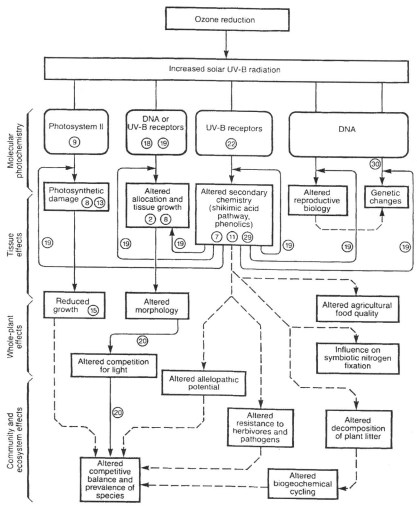

Figure 7.8 Potential consequences of ozone column decrease and corresponding increase in solar UV-B radiation for higher plants at different levels of integration from molecular photochemistry to ecosystems. Putative photoreceptors of solar UV-B that are sufficiently specific to this part of the solar spectrum and that have the appropriate absorption characteristics to result in an appreciable RAF are shown in the round-cornered boxes. These mediate physiological and genetic changes as indicated. The potential implications of these physiological changes for whole-plant function and interactions with other organisms and ecosystem components are also indicated. Solid lines indicate interactions for which there is experimental evidence and the numbers indicate specific references for this information. The dashed lines indicate interactions for which there is as yet no direct experimental evidence. Feedback loops indicate that the flavonoids and phenolics that may be induced by increased UV-B can serve as filtering agents, reducing the flux received by the photoreceptors and, thus, sometimes reducing the response to increased UV-B (from Caldwell *et al.*, 1989).

decrease for only two of seven seed provenances after one year of exposure; however, all groups showed a growth reduction between 12% and 20% after three years of increased UV-B radiation (Sullivan & Teramura, 1992). These results suggest that the effects of increased UV-B are of a cumulative nature and may become very large over the life span of perennial plants such as trees. In addition, the effects of UV-B have been shown to be nonlinear for subarctic grasses; while the equivalent of 15% ozone reduction had a negative effect on growth, 25% ozone reduction had no effect on total plant growth (Gwynn-Jones *et al.*, 1996). It was suggested that UV-B did have a negative effect but increased tillering compensated for the damage within tillering.

Another key process at the ecosystem level is the decomposition of organic matter, which controls nutrient return to soil for microbial and plant growth. Gehrke *et al.* (1995) showed that enhanced UV-B during growth caused a decrease in alpha-cellulose and an increase in tannins in *Vaccinium* sp. leaves after one growing season of exposure. UV-B during litter decomposition decreased the proportion of lignin in the plant residues, decreased colonization by fungal decomposers and decreased total microbial respiration. Since both the *Vaccinium* shrubs under study were deciduous, the effects of UV-B on litter quality and decomposition were obvious within the life span of a leaf.

Species have different responses to elevated UV-B which are not necessarily related to different sensitivity to UV-B radiation but to the growth and allocation response. Some changes in herbivores and detritivores have also been observed as a consequence of changes in the chemical and physical properties of plants (Orth *et al.*, 1990). Caldwell & Flint (1994) therefore suggested that in the long term, elevated UV-B would be most important in altering ecosystem function through changes in species composition.

As the effects of the Montreal Protocol begin to be seen, however, the maximum stratospheric ozone depletion is expected to occur in the next several years, with stratospheric ozone concentrations projected to recover over the next half century (Madronich *et al.*, 1995).

7.7   Effects of tropospheric ozone increase on terrestrial ecosystems

Tropospheric ozone concentrations are currently increasing at a rate of 0.25 ppb yr^{-1} and increasing temperatures, due to climate warming, may exacerbate this trend (Taylor *et al.*, 1994) (Fig. 7.9). Ozone concentrations have tripled in this century and are expected to increase by another 30–40% in the next three decades. Ozone has a direct negative effect on photosynthetic capacity, even at current ambient concentrations (Reich & Amundson, 1985),

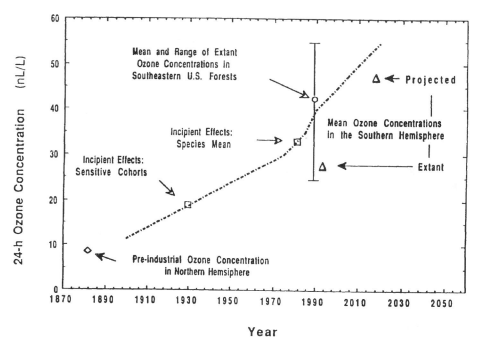

Figure 7.9 Estimate of the long-term changes in tropospheric O_3 in North America, based on extant data for forested landscapes and estimates of the rates of change in tropospheric O_3 in the last century of 0.25 nL L^{-1} yr^{-1} and over the next several decades of 0.5 nL L^{-1} y^{-1} (NAS, 1992). The error bar for the latter part of the 1980 decade indicates the mean and range of 24-h mean concentrations observed in 10 forest landscapes in North America (Taylor *et al.*, 1992) and the corresponding mean (extant and projected in the year 2020) for the Southern Hemisphere ($\triangle$). The inferred preindustrial O_3 concentration ($\diamond$) in the Northern Hemisphere is from Alfonssi *et al.* (1991). The two boxes ($\square$) indicate the O_3 exposure that results in incipient growth effects on seedlings of *Pinus taeda* (loblolly pine); the upper box represents the species' mean, whereas the lower box represents solely the mean of the sensitive cohorts (Taylor, 1994).

the amount of damage being related to the quantity of pollutant that enters a leaf which, in turn, is determined by leaf stomatal conductance. Reduced crop yields due to ozone have been noted throughout the US (Heck *et al.*, 1988). In the longer term, photosynthetic capacity may recover to original levels due to reallocation patterns within the plant.

As is common with other pollutants, ozone has many secondary metabolic effects. Stomatal conductance is reduced in the short term upon exposure to ozone, while leaf duration is reduced in the longer term. This is in contrast to the effects of elevated CO_2, which increases leaf amount (Section 7.2.2). With both CO_2 and ozone, leaf conductance will decrease (Taylor *et al.*, 1994). It has therefore been postulated that increasing CO_2 may, in part, protect plants from ozone damage because of reduced conductance and hence pollutant uptake. Correlations between ozone injury and stomatal conductance have been reported. Volin & Reich (1996) showed that O_3 reduced whole-plant biomass and relative growth rates for a C_3 tree and a C_4 grass, but such effects did not occur when elevated ozone treatments were combined with elevated CO_2. Equally important is the reduction of carbohydrate content induced by ozone exposure, which can be offset by photosynthesis stimulation at elevated CO_2. Thus, ozone and CO_2 responses have the potential to counteract each other.

While much work has been done on ozone effects at the leaf level, little is known about responses at the ecosystem level. A pioneering study in the San Bernadino Mountains of California with a chronic ozone exposure over 30 years showed leaf injury, decreased leaf photosynthesis rates, and premature leaf senescence in five tree species, *Pinus ponderosa* being the most sensitive (Miller *et al.*, 1982). All of the above led to a decreased carbohydrate content with a decrease in radial growth and height of the exposed trees (McBride *et al.*, 1975; Miller & Elderman, 1977). Early senescence and abscission resulted in a thick layer of accumulated pine needles which changed patterns of decomposition. The opening of the overstory and the lowered competitive capacity of the trees led to the development of an understory of shrubs and trees that were ozone tolerant. The weakened *P. ponderosa* became susceptible to root rot, predators (pine beetles) and pathogens (James *et al.*, 1980).

In a controlled field experiment, McLeod & Skeffington (1995) found that ozone had little or no impact on key ecosystem processes, as anticipated from previous growth-chamber and greenhouse experiments. Individual and interactive effects of free-air ozone and SO_2 fumigation on planted tree-stands of *Pinus sylvestris*, *Picea abies* and *Picea sitchensis* were studied in England. Although SO_2 alone, or in combination with ozone, had numerous effects on several ecosystem processes, ozone alone had no effect on growth of any species, foliar leaching or cation concentrations in throughfall, or on mycorrhizae (the latter based on fruit body counts and root harvests for morphotype analyses). In another field study with open top chambers, Boerner & Rebbeck (1995) found that for three deciduous tree species, only *Acer saccharum* showed lower specific leaf area, lower soluble carbohydrate concentrations, and higher lignin content as a result of the elevated ozone treatment; the other two species remained unaffected. However, there was no significant effect on the mass loss from litter over the first year of decay for all species, and N loss was higher only for *A. saccharum*. This species was also reported to reduce fine root growth at elevated ozone, which, in combination with the above changes, has the potential for altering the ecosystem C and N cycling (see also Sandermann *et al.*, 1997).

In summary, little is known about the effects of acute or chronic exposure of ozone at the ecosystem level. However, because species show very different ozone tolerance, it is expected that changes in ecosystem physiology will be ultimately mediated by changes in the species composition.

7.8 Modelling net ecosystem responses to multiple drivers of global change

The previous sections have discussed plant and ecosystem responses to individual global change drivers (CO_2, temperature, N-deposition, UV-B, and ozone). Little is known, however, about interactive effects of those drivers on ecosystem function (Kirschbaum, 1996). Part of the problem stems from the difficulty of carrying out multi-factorial experiments under realistic field conditions at the ecosystem level. The potential complexity of the interactions among the drivers is high, since many show nonlinear responses and nonadditive effects (Hättenschwiler & Körner, 1996; Gwynn-Jones *et al.*, 1996; Wedin & Tilman, 1996). Various modelling approaches have therefore been developed to gain new insights into the ecosystem response to multiple drivers of global change.

At present, however, there are no ecosystem physiology or biogeochemistry models that incorporate the full suite of environmental factors discussed above. One reason for this is the complexity of mechanisms that would need to be included, the tradeoffs between such complexity, and the need to apply and validate such models at the ecosystem or regional level.

The most comprehensive models can simulate many of the interacting effects of CO_2, temperature, precipitation, and nitrogen. Examples of these include the biogeochemistry models described in Table 7.3, and in Chapter 6. Such models include the effects of temperature and water on plant growth and soil properties, including decomposition, and they also incorporate mechanisms by which nitrogen supply controls plant growth both directly and indirectly through soil feedbacks (biogeochemistry models are discussed further in Chapter 10).

The CENTURY ecosystem model (Parton *et al.*, 1993 and users manual reference) and the G'DAY model (Comins & McMurtrie, 1993) were used to simulate the combined impacts of changing atmospheric CO_2 levels, increasing air temperature and N-deposition. Model outputs were compared for the Jasper Ridge annual grassland system in California (Field *et al.*, 1996), a wet Norway spruce forest site at Flakaliden, Sweden (Linder, 1995), and a dry pine forest system (BFG) in Australia (Benson *et al.*, 1992) (see Section 7.2.5 and Fig. 7.4). Ecosystem responses were evaluated for the 10 to 20 year period following the change in environmental conditions (Fig. 7.10). This modelling experiment was devised to simulate effects of doubled atmospheric CO_2 (350 to 700 ppm CO_2), increased air temperature ($+2\,^{\circ}C$), and atmospheric N deposition ($+0.25\ \mathrm{g\ N\ m^{-2}\ year^{-1}}$). These changes in temperature and N deposition

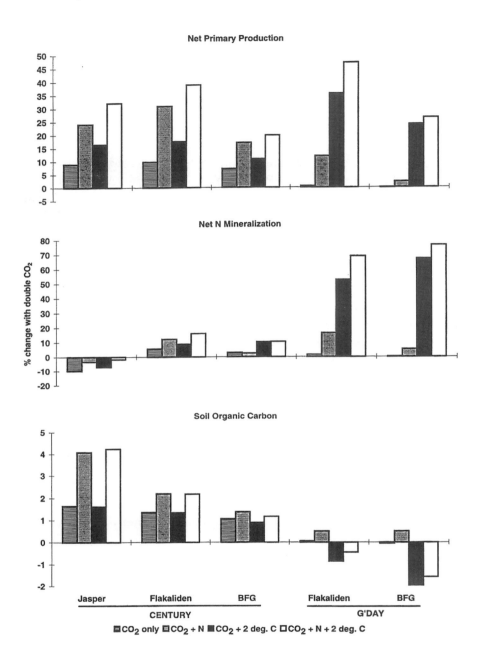

Figure 7.10   The simulated impact of doubled CO_2 (350 to 700 ppm), increased air temperature (+2 °C) and atmospheric N deposition (+0.25 g m^{-2} yr^{-1}) on net primary production, net N mineralization, and soil organic carbon on a wet Norway spruce forest site at Flakaliden, Sweden, and a dry pine forest system (BFG), Australia, using the CENTURY and G'DAY models. The CENTURY model was also run for an annual grassland at Jasper Ridge, California.

are consistent with current projections of GCMs and measured increases in N deposition in highly polluted areas, respectively. The two models were run to equilibrium using the same observed weather data before changes in environmental factors were imposed.

Comparison of the double ambient CO_2 and double CO_2 plus air temperature runs shows that increasing air temperature increases plant production and N mineralization rates for all sites, and decreases soil carbon levels. A similar comparison of the double CO_2 run with the double CO_2 plus N runs shows that adding N results in increased NPP, soil N mineralization rates and soil C storage. Although the results highlight the importance of differences in model formulation, the two models show similar trends, although the CENTURY model tends to have larger NPP responses to increased N deposition while G'DAY has larger responses to increased air temperature. Both models show that the combined effect of increased atmospheric CO_2 with higher air temperature and increased N deposition leads to increased N mineralization and NPP, while soil C storage is decreased by increasing the soil temperature. In summary, (i) increasing N, air temperature and CO_2 together have larger effects on ecosystem properties than increasing CO_2 levels alone, (ii) increasing N inputs results in increases in plant production, N mineralization and soil C storage, while (iii) increasing air temperature decreases soil C levels and increases plant production and N mineralization.

Acknowledgments

We thank Robert Scholes and Sune Linder for reviewing this chapter, Martyn Caldwell for providing information on the UV-B section, and Becky Techau for helping with the CENTURY model runs. We also thank the following people for contributing to Tables 7.1 and 7.2: Jay Arnone, Tim Ball, Peter Curtis, Bert Drake, David Ellsworth, Chris Field, Roger Gifford, Stephan Hättenschwiler, George Hendrey, Jacques Roy, Paul Jarvis, Franco Miglietta, Josef Nösberger, Paul Newton, Richard Norby, Walter Oechel, Clenton Owensby, Catherine Potvin, Jean-Francois Soussana, Boyd Strain, David Tingey, David Whitehead and Jennifer Wolfenden.

8 Ecosystem composition and structure

W. Cramer, H.H. Shugart, I.R. Noble, F.I. Woodward, H. Bugmann,
A. Bondeau*, J.A. Foley, R.H. Gardner, W.K. Lauenroth, L.F. Pitelka
and R.W. Sutherst

*Previous name Alberte Fischer

Box 8.1   Summary

■ The structure of terrestrial ecosystems influences their responsiveness to most
  drivers of global change: for example, growth responses to enhanced CO_2 are
  less at higher levels of organization and over longer periods of observation.

■ The future structure and composition of terrestrial ecosystems will be affected
  by responses at the patch, landscape and global scales. Direct extrapolation from
  the patch to the globe is unlikely to yield realistic projections of ecosystem
  change; landscape-scale processes must be taken into account.

■ A general finding from patch model studies is that many forests appear to be
  sensitive to global change on the time scale of centuries. On shorter time scales,
  e.g. for the next few decades, many forests will show little response due to the
  lag effects in demographic processes. However, in systems where intense
  disturbances are more common, or become more common under global change,
  there will be opportunities for mortality and replacement of existing trees, and
  changes in forest structure and composition may be more rapid.

■ The interaction of global change and landscape phenomena can greatly modify
  both the magnitude and rate of change in community composition and structure. The importance of self-organization in landscape dynamics implies that
  change will not be incremental and smooth, but instead, punctuated and lumpy.

■ Migration plays a critical role in the process of ecosystem adaptation to climate
  change; human modification of landscapes affects the possible velocity of
  migration. Migration rates through the markedly non-random landscapes created by human activities are usually slower than those based on predictions
  derived from theoretical studies based on randomly fragmented landscapes.
  Many species may face a 'double bind' in which they need to migrate in
  response to climate change, but have few places to go and too much hostile
  territory to cross.

Assessments of the response of the terrestrial biosphere to global change are
moving from an equilibrium towards a dynamic representation, triggered by the
recognition that two major problems exist in the equilibrium assessments:

- By definition, equilibrium models simulate no transient changes in vegetation. Therefore, these simulations may at best be used to indicate the direction of possible change but not the time it might take to reach the new conditions.
- Evidence from the past shows that biomes are unlikely to be displaced as homogenous entities. Rather, differences in species' fundamental ecological niches, and their widely varying abilities to migrate, will result in quite different assemblages over a long period of time.

Box 8.2   Future needs

There is a need to:

- gain a better understanding of physiological responses of plants to enhanced CO_2 and changes in temperature and water availability, particularly since vegetation may show different adaptations over various time-scales;
- analyse further the influence of direct or indirect management on landscapes, e.g. through fragmentation or disturbance management; this is a prerequisite for more robust simulations of ecosystem sensitivities at the regional scale;
- improve global assessments of the terrestrial carbon, water and nitrogen cycles, which are necessary for improved predictions of the entire Earth system – such assessments are possible only with further improved Dynamic Global Vegetation Models, which include realistic disturbance regimes and lag times in changing ecosystem structure.

8.1   Introduction: The importance of structure

Most people when told of the structure and composition of a community have a feel for not only its appearance but also numerous other characteristics such as associated animal groups, the climate of the region and its growth patterns. For example, the description 'open Acacia woodland' immediately brings to mind a picture of the savannas of Africa or Australia, alternating greening and drying of the grasses, and groups of large herbivores moving across them. The reason for these immediate and vivid images is that environment, structure, composition and functioning are so intimately interrelated. Without knowledge of each our understanding of an ecosystem is incomplete.

Ecological texts cover the observations and theories as to what determines the structure of the vegetation in a particular climate and what determines the diversity and the composition. This section concentrates on what we know of the effects of changes in driving variables related to global change on the structure, composition and functioning of ecosystems. As ecology has moved from a largely descriptive science to a science with greater emphasis on mechanisms and dynamics, the importance of understanding the structural features of ecosystems has taken something of a lesser position to the importance of understanding the processes or the functioning of ecosystems. It is sometimes easy to forget that the 'structure and function' paradigm (as a classic biologist might use the term) or 'pattern and process' (a phraseology more frequently used by ecologists), are portrayals of two mutually causal agents. Biology has a central tenet – the concept that the form or shape of entities is both modified by and is a creator of function. Processes cause patterns to occur; patterns alter the magnitude and the direction of processes.

For example, Körner (1993) reviewed over 1000 published papers to determine the response of plant systems at several different levels (single plant, cultivated plants, natural vegetation) to elevated CO_2. He found that the higher the level of organization one considers in measuring plant response (leaf photosynthesis, plant growth, ecosystem yield) and the longer period of observation (from hours to years), the smaller were the positive effects of elevated levels of CO_2. The causes of these responses are many, including the tendency for plants to outgrow their pots in longer term greenhouse studies, and thus slow their growth, to biochemical changes leading to the 'down regulation' of photosynthesis in high CO_2 conditions.

There appears to be a hierarchy of structural effects that can alter the response to change in a fundamental process (Table 8.1). The effect of these structurally controlled modifications of system response are nontrivial in many cases – in some of the example cases shown in the table, they can reduce the potential response of the system by 70% or more. Other changes in system structure can amplify potential responses in similar ways.

This chapter deals with one of the major challenges to global change scientists: to predict the effects of changes in driving variables on the structure, composition and functioning of plant communities. Within GCTE there has been an emphasis on developing models of the most important processes to tackle the vexed issue of how to transfer understanding and information at one level (e.g. the individual plant) to other levels in the hierarchical structure of ecosystems. This chapter discusses three levels of organization: patches of vegetation, landscapes, and entire regions or the globe itself.

Table 8.1 *Structure as a mediating factor in the ecosystem response to changes in production-related processes under a doubling of CO_2. There is a wide range of structural responses that can either attenuate or amplify the process response (from Shugart, 1998)*

Level of response	Structural change	Functional implication
Leaf tissue	Change in stomatal index of plants grown in different CO_2 environments. Effect can be seen in plant material collected before the industrial revolution and can be induced under laboratory conditions (Woodward, 1987*b*)	Alteration of the stomatal conduction response of the plant. Implication that plant responses historically may be different from present responses (due to change in stomatal index induced by ambient atmospheric conditions)
Individual plant	Rates of leaf photosynthesis can increase on the order of 50% for C_3 plants in response to a doubling of CO_2 but rates of whole plant growth are often less than 20% of those for control conditions (Körner, 1993)	Structural considerations including photosynthate allocation and internal interactions (e.g. with nutrients such as nitrogen) can moderate the carbon fixation at the leaf level
Plant stand	The increase in stand biomass is less by a factor of about 0.30 (Shugart & Emanuel, 1985) than the increase in growth of the individual plants comprising the stand	Stand interactions (competition, shading, etc.) reduce the stand level biomass increase (or yield) in response to increased growth rates of individual plants
Landscape	Mosaic properties of landscapes alter the stand biomass response to changed conditions (Bormann & Likens, 1979)	Landscapes can be thought of as mosaics in different states of recovery from natural disturbances. Changes in plant and stand process are mediated by the local state of disturbance recovery
Region	The terrestrial surface can alternate between being a source to a sink of carbon in the transient response to environmental change – even in cases in which the long-term response to change is similar to the initial condition (Smith & Shugart, 1993)	Shifts in vegetation in response to change are delayed by large-scale processes involving dispersal, recovery and other inertial effects

8.2   Structural change at the patch scale

8.2.1   Overview

Modelling the effect of changes in driving variables on the structure, composition and functioning of small patches of vegetation is one of the most basic yet challenging tasks facing GCTE. Individual plant models are vital in understanding the response of important crop and forest species to global change. These models, which usually simulate the dynamics of either a single, or very few, species are dealt with mostly in Chapters 7 and 9 of this volume. Patch models are important in understanding how relatively natural vegetation, upon which we depend for so many ecosystem services (provision of clean water, nutrient cycling, carbon sequestration, etc.) respond to global change. Patch models are also the basis for developing the models used at the landscape and global levels of the ecosystem hierarchy.

8.2.2   Application of gap models to forests

Some of the first ecological models dealt with changes in forest composition and structure through time. 'Gap models' were a successful approach (Chapter 6) and they were quickly elaborated to include responses to climatic drivers, as well as to a wide range of disturbances including fire, blow-down, insect defoliation and drought. The rationale of a gap model was to predict the establishment, growth and death of individual trees for all the species potentially able to grow on the site. Simple rules governing reproduction, growth and mortality were sufficient to capture the dynamics of successional processes in most forests. Particular models of disturbances such as fire, blow-down and logging had to be added for particular sites and with these additions the models were useful for exploring a rich set of relationships between climate, management scenarios and the structure and composition of forests.

The apparent success of these models in simulating tree species composition for a wide variety of forests around the world made it tempting to apply them to study the possible impacts of global change, focusing on atmospheric CO_2 concentration and climate (e.g. Shugart & Emanuel, 1985; Solomon, 1986; Pastor & Post, 1988). In the first studies the models were applied as they had been developed for current climate, which certainly was heuristically useful. However, growing concern about their applicability to study global change impacts rose from theoretical considerations as well as from comparative simulation studies (see Bonan, 1993; Bugmann & Martin, 1995; Hänninen, 1995; Bugmann *et al.*, 1996; Loehle & Leblanc, 1996). Consequently, a second generation of gap models was developed that attempted to reflect the climatic controls on plant demography more mechanistically.

In one line of development (e.g. Prentice *et al.*, 1993; Bugmann, 1996*b*), improved formulations were used to describe important processes in the models. These changes increased the robustness of the models while the computational demand remained low. However, the improved formulations cannot take into account more detailed ecological and ecophysiological knowledge. For example, the asymptotic formulation of the height–diameter relationship in FORSKA (Leemans & Prentice, 1989) is undoubtedly more realistic than the parabolic one (Botkin *et al.*, 1972), but it still requires maximum tree height as a model parameter although it is well known that maximum height strongly depends on climatic and soil conditions.

In the other line of development (e.g. Bonan & Van Cleve, 1992; Martin, 1992; Friend *et al.*, 1993, 1997), detailed biophysical and/or physiological submodels for some processes were coupled to gap models to allow for more mechanistic calculations of the environmental effects on tree growth. This enhancement adds many process-based details and provides an explicit way of scaling up from biophysical and physiological knowledge to the ecosystem scale. However, the high requirements for site-specific and species-specific parameters often restricts the application of these models to a few sites. While the parameter estimation problem can be relaxed to a certain extent by basing the models on the concept of plant functional types (e.g. Huston & Smith, 1987; Bugmann, 1996a; Friend *et al.*, 1997), the lumping of many tree species into a few functional groups leads to a loss of detail in the projections of the models (see also the section on plant functional types later in this chapter).

These recent developments have greatly increased our ability to assess the impacts of climatic change on the structure and composition of forests. To date, more than 100 gap model studies have been conducted that dealt with the impacts of global change on forests (for reviews, see Smith *et al.*, 1992c; Dale & Rauscher, 1994; Smith & Shugart, 1996). These studies vary widely with respect to the diversity of models used, the magnitude of the environmental changes and the geographical location and extent they considered. In spite of this heterogeneity, there are a number of general conclusions that can be drawn from these model applications.

1. There is no general response of forests to global change (see Solomon, 1986; Pastor & Post, 1988). Rather, the simulated responses of the systems may range from no change at all to drastic changes in physiognomy, such as the transition from forest to woodland or even grassland, or from tundra to forest. The magnitude of the ecosystem response depends not only on the magnitude of environmental change, but just as much on the location of a particular ecosystem in climate space. Depending on the latter aspect, the sensitivity of forests to global change may vary greatly (see Fig. 8.1, adapted from Bugmann, 1997a).

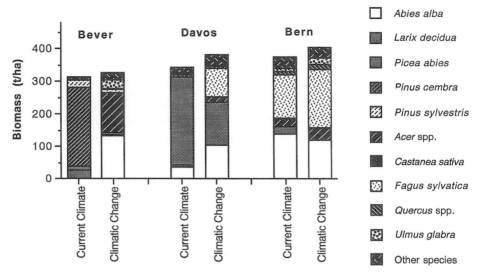

Figure 8.1   Steady-state species composition simulated by the gap model FORCLIM under current climate (left bar of each group) and under a climate change scenario for the year 2100 according to Kienast (1991) (right bar of each group) at three sites along an elevational gradient in the European Alps. Bever: high subalpine site, where a drastic change of the species composition is simulated; Davos: subalpine site, where a shift of species composition is simulated without the currently dominating species going extinct; Bern: low-elevation site, where only minor changes in species composition are simulated. Redrawn from Bugmann (1997*a*).

It can also depend strongly on soil type (e.g. Post & Pastor, 1996).

2. Different environmental scenarios that are constructed to represent, for example, a double-CO_2 world may lead to vastly different projections of the species composition in forest gap models (see Fig. 8.2 from Bugmann & Fischlin, 1996; Smith, 1996). It will probably take considerable time until reliable scenarios of climatic change are going to be available at the regional scale (Kattenberg *et al.*, 1996). Also, models will need to be modified to account for changes in environment factors other than climate (for example, N deposition – see Section 7.5.4).

3. Comparative simulation studies applying several models under the same environmental scenarios showed that seemingly similar models may yield vastly different projections of the future forest composition (e.g. Martin, 1992; Fischlin *et al.*, 1995; Bugmann, 1997*b*). The problem here is that there is considerable uncertainty about the appropriate formulation of environmental influences on demographic processes. For example, Hänninen (1995) showed that two gap models may differ strongly in their response to climatic change depending on the amount of detail incorpor-

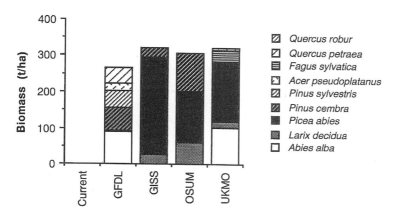

Figure 8.2   Steady-state species composition simulated by the model ForClim for a GCM grid cell in the Central European Alps (46°N, 10.5°W) under current climate (leftmost bar), and for four $2 \times CO_2$ climate change scenarios as described by Lauenroth (1996). Under current climate, the area is simulated to be above the timberline. While the area is becoming forested under all scenarios, widely different species compositions may be obtained depending on the climate scenario that is used. Redrawn from Bugmann & Fischlin (1996).

ated in the simulation of phenology. It is inappropriate, however, to conclude that the more detailed formulation is *a priori* better: in the Hänninen (1995) example, the predictions of the more 'mechanistic' model clearly contradicted experimental results.

4. The sensitivity of some forests to the magnitude of the change signal and the differences in the forest structure that may be simulated by different models under the same change scenario suggests that the best strategy is to set up comparative simulation studies using several scenarios of climatic change and several forest models (e.g. Lauenroth, 1996; Bugmann, 1997*a*). Instead of trying to predict particular future states of ecosystems, this approach aims to assess the sensitivity of forest composition to climatic change, which then may feed into risk assessment procedures that may be important for policy-making.

5. A general finding that emerges from most studies is that many forests appear to be sensitive to global change on the time scale of centuries. On shorter time scales, e.g. for the next few decades, many forests will show little response (Bugmann, 1997*b*) due to the lag effects in demographic processes. However, these projections refer to the long-term consequences of global change, and most studies assume no change in disturbance regime. In systems where intense disturbances are more common, or become more common under global change, there will be opportunities for mortality and replacement of existing trees, and changes in forest structure and composition may be more rapid.

8.2.3   Structural controls on net primary production in grasslands

Only recently has the gap model concept been transferred to vegetation types other than forests. Coffin & Lauenroth (1989, 1990) described a semi-arid grassland based on a gap dynamics approach. The size of the patch simulated by these models is usually < 1 m^2 compared with $c.$ 1000 m^2 for forests but the difference in scale is not the most important factor determining the characteristic dynamics of the grass and forest models (Coffin & Urban, 1993). Perennial grasses with clonal growth were responsive to a broader range of environmental conditions relative to the more restrictive response of individual trees with incremental growth. Asymmetric competition for light by individual trees resulted in different system-level dynamics than for a grassland, where symmetric competition for belowground resources by individual grassland plants is the primary mode of competition.

The grassland model has been used to investigate several scenarios of the effect of global change. Coffin & Lauenroth (1996) compared the response of five functional types (see section later in this chapter) of grassland plants to a climate change scenario. They found that C$_4$ grasses were the most favoured mainly as a result of the temperature increase projected by the GCM. However, these results should be treated with caution because the direct effects of CO$_2$ on water use efficiency of the plants (see Section 7.2.5) were not taken into account.

Other problems arise when trying to develop models based on available data from current climate and atmospheric conditions. The strong positive relationship between annual aboveground net primary production (ANPP) and annual precipitation (as an estimate of water availability) is central to our understanding of productivity patterns and consequently of ecosystem carbon balance in grasslands. The in-depth analysis of this relationship has provided an unusually clear example of the relationship between ecosystem structure and functioning.

Equations relating mean annual ANPP and mean annual precipitation over a series of sites have been found for most of the grassland regions on Earth (Walter, 1971; Lauenroth, 1979; Rutherford, 1980; Sala *et al.*, 1988; Ojima *et al.*, 1996). For example, Sala *et al.* (1988) developed a general relationship between the long-term averages of these variables for a range of sites throughout the central grassland region of the USA (i.e. spatially distributed data). When Lauenroth & Sala (1992) evaluated the relationship between annual ANPP and annual precipitation for a particular semi-arid site in Colorado (USA) using a 52-year time series of production and precipitation data, they found that their relationship (i.e. temporal data) did not fit the spatial model by Sala *et al.* (1988). In dry years, the temporal model overestimated ANPP compared with the spatial model, whereas in wet years it underestimated ANPP.

The difference in behaviour between the spatial and temporal models is that the spatial model is based on the response of a plant community that is in steady state with the precipitation of a particular site, while the temporal model represents the response of a plant community to interannual variations of water availability. These results suggest that in perennial grasslands vegetation structure, defined by the species composition and density of plants, responds slowly to year-to-year fluctuations in water availability. Therefore, in wet years there are too few plants to capitalize on high water availability, whereas in dry years plant density is higher than one would find in an area with low precipitation. This could indicate that most ecological simulation models that contain relationships based upon data from different sites may not provide good estimates of the response of a single site over time to fluctuating weather or to climate change.

Another challenge in modelling the dynamics of grassland communities is to include the full range of responses of plants to herbivory and the response of herbivores to changes in the quantity and quality of forage plants. Some of the issues relating to large mammalian herbivores are raised in the section on pastures and rangelands in Chapter 9.2.3. An even more problematic area is the role of insects and pathogens in the dynamics of managed and natural ecosystems, and how this may change under global change scenarios.

8.2.4 Weeds, pests and diseases

So far there has been limited research on the likely impact of weeds, pests and diseases on the composition and structure of natural ecosystems under global change (e.g. MacDonald, 1992; Sutherst, 1995). In this section, these potential impacts are briefly discussed. The effects of weeds, pests and diseases on the yields of managed production systems are described in more detail in Chapter 9.

Invasive, alien plant species can profoundly affect ecosystem structure, composition and functioning despite the number of species involved being relatively small (MacDonald, 1992). Such weeds are dominating many ecosystems around the world and have been responsible for a large proportion of past extinctions (MacDonald, 1992). Potential invasion of natural areas by exotic weeds is of particular concern (Humphries *et al.*, 1991) when the natural areas adjoin modified habitats that are rich in invasive species. Many ornamental (Lonsdale *et al.*, 1989) and agricultural (Lonsdale *et al.*, 1995) plants with the potential to become weeds have been moved between countries with little consideration of environmental impact. This rate of dispersal of exotic species is expected to accelerate with the volume of international trade, as the World Trade Organization (WTO) protocols come into force (Jenkins, 1996).

With climate change many weed species, such as *Pueraria lobata* (kudzu) and *Lonicera japonica* (Japanese honeysuckle) in North America (Sasek & Strain, 1990), are likely to expand their ranges polewards. Climatic correlations based on the current distribution of two invasive plant species, *Fallopia japonica* (Japanese knotweed) and *Impatiens glandulifera* (Himalayan balsam), in northwest Europe, suggest that under some climate change scenarios they may extend their distribution by as much as 5 ° of latitude northwards (Beerling, 1993).

Interactions at higher trophic levels can also have significant effects on the vegetative component of terrestrial ecosystems. Insect pests and pathogens can play a critical role in the successional dynamics of many terrestrial ecosystems (e.g. the effects of spruce bark beetles in boreal forests), and hence on their structure and composition. Climate change will have variable effects on plant pathogens and insect pests due to their sensitivity to relatively subtle environmental changes. The short generation times and responsiveness to climate and disturbance of many pests and diseases enable them to exploit opportunities created in changing environments. As such, they are likely to be among the first species to respond to environmental change. When introduced into a new or a changed environment, pests and diseases often experience 'ecological release' from their natural enemies, and so are much more competitive than usual (Kingsolver, 1989).

Warm and humid conditions, as are predicted for many parts of the world under future climate scenarios, favour the growth of many plant pathogens. For example, *Phytophthora cinnamoni* is a particularly severe fungal pathogen that poses a major threat to native plants and forests in many parts of the world. As the pathogen is dispersed mainly by human activity, native forests, which are increasingly being subjected to human management, are at risk (Marks & Smith, 1991). In Europe the threat to oak forests from this fungus will likely be exacerbated by climate change (Brasier & Scott, 1994).

Changes in the density or health of grazing herbivores inevitably have consequences for vegetation composition and structure, so there is a need to consider multiple trophic levels when projecting global change impacts on natural ecosystems. For example, when the movement of large ungulates is restricted, as is increasingly the case with rising human population in southern Africa, numbers of the tick *Rhipicephalus appendiculatus* can build up to lethal levels, causing population crashes in the ungulates that, in turn, affect vegetation structure (Lightfoot & Norval, 1981). The impacts of elevated atmospheric CO_2 may lead to plants with poorer plant nutritional value to herbivores (see Chapter 7 and Section 9.2.3). Poor nutrition reduces animal host resistance to parasite infestations (Sutherst, 1987) and has been associated with lethal infestations of the tick *Dermacentor albipictus* on moose in Canada (Samuel & Welch, 1991).

8.3   Structural change at the landscape scale

8.3.1   Spatial dynamics

Chapter 6 describes some of the issues and complexities that arise in modelling the dynamics of landscapes. Modelling the composition and structure of vegetation at landscape scales (100s to 1000s km^2) requires more detail about the spatial heterogeneity of the weather and physical resources (e.g. soils, nutrients) within the landscape as well as models of hard-to-measure phenomena such as long-distance dispersal of plant propagules. Data requirements and computation loads can expand enormously. A question that arose early in studying global change was just how important these fine-grained spatial interactions are in determining broad-scale patterns of ecosystem composition and structure. If their effects are swamped by far more important local processes or by far stronger exogenous processes such as land clearing by humans, then the need to consider fine-grained processes at landscape scales could be avoided.

In many studies landscape processes have been ignored and gap models have been used to assess the impacts of disturbances such as fire, blow-down, land movement and pest outbreaks on communities. However, most disturbances and many other ecosystem processes, such as dispersal and patterns of grazing by herbivores, are landscape phenomena. The disturbances and processes occur in a spatially heterogeneous environment and their effect on structure and composition is dependent on the spatial relationship between elements of the landscape. Many spatial models have shown that strong and relatively stable patterns develop in otherwise homogeneous environments (the phenomenon of self-organization, or self-generating patterns – see review by Perry, 1995). These patterns can change the structure and composition of landscapes. Only a few studies have directly explored whether landscape scale phenomena are crucial for longer-term change of ecosystems.

8.3.2   Disturbance (fire)

Fire is an important issue in global change because it affects a larger area than most other disturbances and the fire regime is affected by several factors that will change in response to global change – for example fuel (litter, dry grass, etc.) production, fuel dryness and ignition frequency. Most climate scenarios predict that warm dry periods will become more common, thus fire fuels are likely to be dry more often (Overpeck *et al.*, 1990). There may also be increases in fuel loads arising from increased rate of litter production, or, possibly, from reduction in litter decomposition rates. The predictions about fuel loads are very uncertain and will vary from community to community and from micro-site to micro-site within communities. This may not matter greatly because it appears that the condition of the fuel and the frequency of ignition events are more important than fuel loads in determining fire regime (Fryer & Johnson, 1988; Beer & Williams, 1995; Bessie & Johnson, 1995).

Some general principles emerged from modelling with simple representations of landscapes and disturbances. For example, Turner *et al.* (1993) showed that two key parameters are sufficient to describe temporal and spatial patterns of disturbed landscapes: (i) a temporal parameter that is the ratio of the average time between successive disturbance events to the time required for a disturbed site to recover to a mature stage, and (ii) a spatial parameter that is the ratio of the size of the disturbance to the size of the landscape. Using these parameters it is possible to identify conditions that lead to qualitatively different landscape dynamics.

Using relatively simple biological representations of landscapes, Bak *et al.* (1988), Green (1989) and Turner *et al.* (1989*b*) found evidence of self-generating patterns arising out of the interplay of succession, fire spread and dispersal, and found that these can be modified greatly by environmental gradients. Holling *et al.* (1996) explored the role of fire as a self-organizing process using four models with increasing degrees of complexity. They found that as soon as the occurrence of past fires began to influence the likelihood of future fires (for example, by the fire spread being a function of forest age) then the system organizes itself into a patchy landscape – even in the absence of any underlying heterogeneity. Work incorporating more detailed models of successional processes and fire spread has confirmed these conclusions (Noble & Gitay, 1996). For example, in simulations of temperate rain forest dynamics in Tasmania, a landscape model developed patterns of frequently burnt grasslands abutting infrequently burned rainforests similar to those observed in southwest Tasmania. In simulations of open woodlands or savannas, trees occurred as scattered clumps.

Noble & Gitay (1996) also compared multiple runs of a point model with landscape versions of the same model, and adjusted fire ignition frequencies to give the same average point fire frequencies. They found that composition and structure differed significantly between the point and landscape versions of the model (Fig. 8.3). Also, the rate of change of the system is often slower in the

Figure 8.3   (a) The distribution of fire frequency across a hypothetical landscape based on a simulation of 1500 years. The landscape is homogeneous in that there are no differences in suitability for the different vegetation types; thus, the patchiness arises from interaction between the vegetation dynamics and different susceptibilities of communities to burning (see also Noble & Gitay, 1996). Darkest shading represents 65 per thousand years and lightest shading no fires. (b) Comparison between point and landscape representations of a temperate rainforest in Tasmania, Australia. In the landscape version, self-generating patterns in the distribution of vegetation types with different burning frequencies delay the dominance of the site by rainforest. From I.R. Noble, unpublished.

(a)

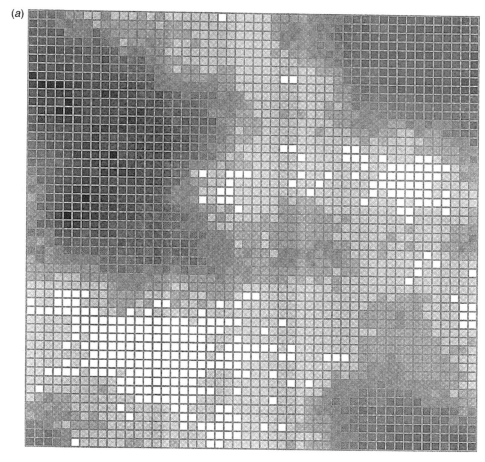

(b)

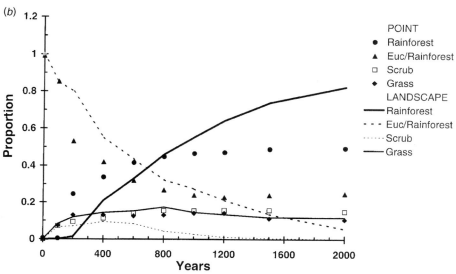

landscape model than in the point model. This arises from self-organizing processes. For example, if a rare sequence of disturbances creates an extensive area of grassland, then subsequent fires spread with higher frequency in that grassland, so delaying successional development to forest. The time taken for dispersal into areas where species have been lost slows the recovery process even more.

These types of studies are only in their infancy. Most are largely theoretical, based on simple process models and on abstract landscapes such as purely homogeneous substrates, simple gradients or some form of structured random landscapes. One of the most comprehensive studies of landscape fires was carried out for Yellowstone National Park (Gardner *et al.*, 1996). In the EMBYR model a geographic information system describing a 30 × 30 km section of the park was used to represent a gridded landscape of 50 × 50 m cells. Detailed weather and fire records were used to drive a fire ignition and spread model. Three climate scenarios were run for 1000 year simulations: 'current climate' and 'wetter' and 'drier' conditions. In the drier scenario fires were more frequent, but with fewer extreme events, leading to a reduction in the amount of forest and an increase in its fragmentation (Swetnam, 1993; Gardner *et al.*, 1996).

There are still many aspects of modelling changes in the structure and composition of communities to be explored. The precise nature of the propagation of disturbances across landscapes and the role of dispersal in determining community change are important questions. There are several theoretical studies of such phenomena (e.g. Lavorel *et al.*, 1993) but the challenge is to translate the theoretical findings to real landscapes on heterogeneous substrates.

Modelling results indicate that some general conclusions about phenomena arising at the landscape scale can be made. It appears that self-organization leads to patchy landscapes in both space and time even with uniform initial conditions on relatively uniform substrates. The relationships between the life history characteristics of the dominant life forms and the disturbance frequency, and between the size of the disturbance and the grain of the substrate, determine the relative stability of any vegetation patterns.

The strongest message for those estimating the effects of global change is that landscape phenomena have the capacity to modify greatly both the magnitude and rate of change in community composition and structure. Self-organization implies that change will not be incremental and smooth, but instead, punctuated and lumpy.

8.3.3 Plant migration and global change

Migration

Many projections of the effects of global change show the reorganization of biomes across the continents under new climatic scenarios. Most of these projections are based on matching existing biomes with existing climates and then mapping these same biomes onto new climate projections. This approach assumes that the Earth's vegetation types and biomes will move as complete units and reach equilibrium with the climate. This is only likely to occur if all the component species of a biome would move at rates sufficient to keep up with climate change. The rates of migration needed to track the changing climate have been estimated to be as much as 1 to 2 km yr^{-1} (Ritchie & MacDonald, 1986). If biomes do not migrate as smoothly as these assumptions suppose, the possible consequences include temporary or long-term disruption of ecosystem processes, reduction of NEP and species extinctions.

Migration is a multi-step process. Plants must successfully disperse propagules to a suitable habitat and these propagules must then successfully germinate, grow and reproduce in order to serve as the source of new propagules for further dispersal. The likelihood of success and the achieved rate of migration depend on the nature of the pre-existing vegetation and the type and frequency of disturbance. Climate change may increases the frequency of disturbance such as fires, blow-downs or droughts, so increasing the opportunities for invasion.

The question 'can species migrate fast enough to keep pace with climate change?' is particularly important in assessing the impact of global change. Its answer requires contributions from palaeo-ecology, biogeography, ecosystem and landscape ecology and mathematical modelling. GCTE sponsored a workshop to bring these viewpoints together and synthesize our current understanding, particularly in relation to migration of plant communities (Pitelka & Plant Migration Workshop Group, 1997). This section summarizes the conclusions.

Plant migrations in the fossil record

Palaeo-ecologists have recorded rapid movement of selected tree species during the early Holocene following glacial retreat. Many of the dominant tree taxa of eastern North America and western Europe spread at rates approaching 1 km yr^{-1} during at least part of the shift from Ice Age to modern range limits. These rapid migration rates were maintained even across broad bodies of water or bands of unsuitable soils (Huntley & Birks, 1983; Woods & Davis, 1989; Kullman, 1996). This observation suggests that migration was driven not by numerous short steps each within a few metres of a parental plant, but by rare long jumps that produce distant outlier populations (Davis *et al.*, 1991). Unfortunately, confirmation is difficult since these individual jumps are difficult to observe in extant communities and the outliers are too sparse to be detected consistently in any pollen record. One rare documented example of such 'jump spread' is the westward migration of *Picea abies* (spruce) across northern Europe during the Holocene. Woody macro-remains of *Picea* have been found in several locations that were hundreds of kilometres in advance of the migrating front (Bradshaw & Zackrisson, 1990; Kullman, 1996).

Theoretical studies support the idea that outliers ahead of the advancing front can increasingly 'pull' the wave forward so that the rate of migration accelerates with time (Kot *et al.*, 1996). In regions of low relief and minimum habitat complexity, migration of plant species is most likely to occur as broad 'waves' of movement such as observed over central North America (Webb, 1987). In topographically complex landscapes, such as the basin and range area of Nevada, migration appears to occur as a series of local expansions centred around 'refugial' microhabitats that harboured protected populations.

The dispersal distances achieved by propagules depend enormously on the type of transport. Air-borne propagules usually fall close to the parent but some well-adapted propagules can drift for huge distances. Extreme wind storms can greatly increase the dispersal of even relatively heavy propagules. Water-borne propagules can be transported for great distances along rivers or over flood plains but there is usually a strong directional characteristic. Propagules transported by animals vary greatly in the distance achieved. The most notable feature of animal mediated dispersal is the degree of selectivity of the eventual position of lodgement of the propagules. This can be in deliberate caches as in food-storing animals or accidentally in preferred resting places. Some species (e.g. blue jays, *Garrulus glandarius*) appear to cache seed species preferentially at the edges of forest, so enhancing forest spread (Johnson & Adkisson, 1985).

A surprising conclusion from the fossil record is that differences in presumed dispersal modes of different species did not seem to be a major factor in the achieved migration rates. This could be due to our inability to detect relatively small differences in migration rate, or it is possible that the rate of climate change may have been the overriding limitation on migration rate. Unfortunately, the fossil record is too sparse to determine whether the rate of spread was limited by dispersal events, landscape features such as the availability of suitable habitat or by the rate of climate change. Thus an understanding of the dispersal process in landscapes and climates of the future will require additional insights from contemporary studies and modelling.

Contemporary invasions

Studies of contemporary plant invasions also provide evidence of rare outlying populations being the source of rapid invasions, thereby reinforcing the importance of 'long jumps' and outlier populations. The invasion of western North

America by the Eurasian grass, *Bromus tectorum* (cheatgrass) is a modern example (Mack, 1986). It appears that the species initially established itself in at least five separate locations in the Intermountain West, all within a decade or two before 1900. Over the next 20 years at least 50 additional foci or outliers were established, but the total area occupied was still relatively small. Then, within only about 10 years (1920–1930), there was a coalescence of these foci such that the grass became prominent throughout 200 000 km² of this arid region.

The study of modern invasions shows that the establishment of new populations is extremely unreliable with many initial colonizations failing. Small inoculations of outlying populations are vulnerable to demographic stochasticity, genetic inbreeding, and breeding problems due to Allee effects (i.e. suppressed recruitment in small populations because the mates are difficult to find or the population is not large enough to exploit resources efficiently, Kot *et al.*, 1996). The net effect of these challenges means that many repeated long jumps, or several replicated outlying populations, are probably necessary to ensure a species range expansion.

Experimental data (Bierne, 1975) show the best predictor of success was simply the number of attempts, with the second best predictor being the size of introduced populations. It can also take a long time before outlying populations build up large enough numbers to send out many propagules, or the large populations required to saturate natural enemies. Lags may be relatively short for widely dispersed, high fecundity species, but far longer for longer-lived, slower-to-reproduce species. Time lags of decades to centuries may be common. Such lags are barely detectable in the palaeo-record, but they are likely to be significant to human affairs during climate change. The lags imply that possibly transient, but nevertheless significant, changes in ecosystem composition, structure and function might be expected as communities migrate. Ecosystems may become much weedier as climate change leads to the decline *in situ* of some species and weedy species take up their resources.

Rules from landscape ecology

A number of modelling studies have allowed landscape ecologists to develop some general rules about the factors that govern the rate of migration. These include:

- The speed of migration is the product of the dispersal rate and rate of population growth at the edge of the invasion.
- There is some threshold of suitable habitat that must be exceeded if the population is to consistently spread across the landscape.
- There is a threshold of suitable habitat and of suitable colonists that must be exceeded for a population to persist after it has spread (Peters, 1991).

■ Modelling (e.g. Dyer, 1995) and empirical (Johnson & Adkisson, 1985) studies suggest corridors are likely to serve only an indirect and limited effect in facilitating movement of animal dispersal vectors.

These rules provide some guidelines as to the important factors to be considered, but one clear message emerging from the large body of theory relating to the effects of landscape fragmentation on animal movements (e.g. Andren, 1994; Bascompte & Sole, 1996; Gustafson & Gardner, 1996; Turner, 1996) is that it is the interplay of species characteristics and landscape pattern that governs the effects of fragmentation. No firm rules can be stated. Also, migration rates through the markedly non-random landscapes created by human activities are usually slower than those based on predictions derived from theoretical studies based on randomly fragmented landscapes.

The effect of human activities

The rates of migration achieved in the future would appear to depend upon a number of factors related to human activity. These include

■ deliberate or accidental cases of long-distance dispersal via human activity;
■ maintenance of outlier populations in gardens and plantations, etc.;
■ changing disturbance regimes;
■ conversion of large areas of habitat to land-uses not conducive to invasion (e.g. to crops, car parks, etc.) and
■ increasing fragmentation of the remaining communities.

Human activities in dispersing propagules great distances either passively on cars, ship or planes are by deliberate introduction and by maintaining populations beyond the normal environmental limits of a species are likely to contribute to dispersal in response to climate change. Botanical and private gardens are large repositories of alien species and have long been sources of plant invasions (Parker, 1977). Global change could trigger a new and largely unforeseen round of invasions as species, currently barely held in check by unsuitable climate, become able to grow as wild populations over newly expanded areas of suitable climate. However, species dispersed this way are only a selection (deliberate or accidental) of the full range of species in a community and thus we cannot expect to see the smooth movement of entire communities across the landscape in response to climate change.

Humans have always manipulated, and sometimes drastically altered, disturbance regimes such as fire and flooding. Disturbances can lead to the mortality of some species while providing opportunities for the establishment of others. It is difficult to predict how disturbance regimes will change under climate change and human response to that change. Some places will be subject to more frequent disturbance, creating opportunities for disturbance-adapted species to spread, while in others previously common disturbance will become less common, leading to the decline in some species. The balance is difficult to predict, but a tenable hypothesis is that disturbances – in the sense of the occurrence of conditions outside the 'normal' range – will become more common under climate change in most regions, and thus species adapted to prospering on disturbed sites will be favoured.

Habitat destruction and habitat fragmentation, two processes that can by themselves drive species to extinction, have the capacity to limit the rate of dispersal of those species not strongly affected by human activities. Many species may face a 'double bind' in which they need to migrate in response to climate change, but have few places to go, and too much hostile territory to cross. For species that have narrow habitat requirements, resulting in only a small proportion of the total area being suitable, fragmentation that involves loss of the preferred habitat will significantly reduce their capability to migrate.

The human modification of land cover associated with habitat fragmentation is likely to influence the behaviour of animals that disperse seeds. Large mammals are often the major dispersal agents of some tree species. Substantial restriction of elephant populations in Africa is likely to restrict the spread of some large-seed tree species (Chapman *et al.*, 1992) while the introduction of large mammals (*Bubalus bubalis* – water buffalo, *Sus scrofa* – wild boar) by humans may contribute to rapid spread, especially of exotic invasive plants (Stone, 1985; Russell-Smith & Bowman, 1992). Sometimes increased fragmentation appears to increase the average movement distances of vectors such as blue jays and thus the dispersal rates of the species they carry (Johnson & Adkisson, 1985). Modelling studies show that if organisms do not suffer enhanced mortality while moving between fragments, and tend to stop moving only when locating an appropriate habitat, then fragmentation can actually increase the speed with which a species adjusts its range to changing conditions (Hengeveld & Van den Bosch, 1997). However, there is a limit to the effectiveness of such phenomena and eventually increased fragmentation must limit dispersal (Malanson & Armstrong, 1996).

Human activity will have several different and opposed effects on dispersal and migration. The most likely outcome is that some species will be advantaged because they are aided in their dispersal by deliberate or accidental human actions; others will be advantaged because they are favoured by a possibly increased disturbance regime. Still other species will be disadvantaged in their dispersal and establishment by the increasing fragmentation of suitable environments. The result is likely to be that some species are well able to keep up with the rate of climate change whereas others will fall behind. The necessary conclusion is that plant – and animal – communities of the future will be

dramatically changed with significant effects on ecosystem processes and biodiversity (see Chapter 11).

8.4   Structural changes of the land biosphere

8.4.1   Past and future changes in vegetation structure

In this section, recent progress in understanding global dynamics of vegetation
structure is reviewed. First, the section describes some of the problems that
arise when modelling concepts that are valid at the local to regional scale are
transferred to continental or global assessments. Next, recent results from
continental and global equilibrium modelling studies are shown. Third, the
developments in the field of dynamic global vegetation models are discussed on
the basis of an intercomparison study made for the present synthesis. Finally,
the importance of these findings for coupled Earth system models is illustrated
using studies with coupled models of biospheric and atmospheric processes.

Projections of the future of the Earth's vegetation as a whole depend on the
development of realistic biosphere models, as well as on appropriate projections
of future climate and land use. Current assessments indicate that broad zones of
terrestrial vegetation are likely to change within the next 50–100 years, due to
either changes in land use or climate, or both (e.g. Kirschbaum *et al.*, 1996*b*).
The regional pattern of these changes, however, is highly uncertain. Structure
and composition are intimately linked with ecophysiological characteristics of
vegetation, but this has only recently been accounted for in terrestrial biosphere
models (see Chapter 6). At the global scale, development of models that
simulate ecophysiology and vegetation dynamics is a prerequisite for the critical
step which is now being made from equilibrium to dynamic models. The
present state of such models, as is reviewed in the remainder of this section,
deals with potentially natural vegetation alone – in the future, such models will
have to be combined with models of changing land use (e.g. IMAGE 2, Alcamo,
1994).

Fossilized remnants from past vegetation, such as pollen and macro-fossils,
have long been known to be suitable for the reconstruction of past climatic
patterns. It is therefore natural to assume that a strong relationship exists
between the broad-scale zonation of vegetation and that of the Earth's climate.
If models can be established that are capable of describing this relationship, then
they provide a predictive tool for the changes that might occur under a future
changed climate (Huntley *et al.*, 1997). This is an important goal since the
productivity of the terrestrial biosphere is the fundamental basis of global food
production (Leemans & Van den Born, 1994). Also, changes in the world's
vegetation, independent of their driving forces, may themselves impact atmos-

pheric circulation. They therefore represent a critical feedback element in global change (Claussen, 1996).

8.4.2   Consequences of scaling-up

General considerations

A frequently heard criticism of global models of the biosphere concerns the limited current knowledge about specific properties of the world's ecosystems. The abundances and ecological characteristics of the world's plant species are poorly known, and knowledge of even the distribution of the more important species is poor. Furthermore, the driving variables of global biosphere dynamics, such as climate and land use, and the characteristics (texture, nutrient status) of the underlying soils are poorly covered by observations. Why, then, is the simulation of these broad-scale processes undertaken, and why should the results of the simulation be considered important contributions to global change science?

First, many studies in global biogeography, both those using numerical models and those using other means of assessment, show clearly that the terrestrial biosphere in fact does have some relatively simple, emerging properties that can be described without too many details. An example is the success of biogeographical models that are able to capture the major structural features of the terrestrial biosphere, such as the distribution of biomes, as a function of climate and soil texture alone (Prentice *et al.*, 1992; VEMAP Members *et al.*, 1995). Second, and as a consequence of this, the fact that a specific model is used in a global framework does not automatically imply that it is expected to be capable of reproducing every possible feature of all ecosystems on Earth. More consistently now than earlier, research is being directed towards specific issues that are to be answered by specific global biosphere models. Among these are the long-term and the interannual changes in total biospheric carbon flux, or the implications that broad-scale changes in land cover might have for the circulation of the atmosphere.

During the development of global models, some abstractions need to be made with respect to the diversity of model representations along four fundamental axes of variation:

- diversity of functional processes (e.g. molecular, plant physiological, competition, dispersal);
- diversity of organisms (species);
- variability through time (e.g. seasonal, interannual, population turnover);
- diversity (and detail) in spatial representation (e.g. lateral fluxes and movements).

For each of these dimensions, recent developments in ecosystem modelling have yielded specific insights that can usually be summarized by asking how much diversity is required to reflect overall system behaviour at a certain scale. Or, in other words, are there levels of detail that can or should be ignored, thereby reducing the overall demands on parameterization or driving data for the resulting model? There are currently insufficient answers to this question to produce direct predictions of the future structure of the biosphere. In the following, examples for these simplifications are discussed with respect to the results from GCTE research.

Species diversity: plant functional types

Plant functional types are now widely considered a necessary and appropriate simplification of species diversity, and they have the advantage that ecosystem types often result more or less naturally from plant functional type assemblages. Also, due to the high diversity of terrestrial ecosystems, an assessment of the impacts of global change on a species-by-species basis is not feasible. Therefore, it is necessary, by some approach, to encapsulate the richness of species' responses to change without knowing the actual range of responses at all relevant scales. Historically this gap of knowledge has been bridged by using functional or structural types of species ('life forms', sensu Raunkiær, 1907; or 'plant types' sensu Box, 1981; for a more in-depth discussion of this history, see Cramer & Leemans, 1993; Smith *et al.*, 1993; Gitay & Noble, 1997). Fundamentally, a functional type (FT) is made up of a number of different species that possess the same or very similar functional responses to the environment; these may or may not share a characteristic physiognomy. For example, species with the C_4 mechanism of photosynthesis may be assigned to a C_4 functional type, which then implies a range of responses to the environment, such as to temperature and CO_2.

Most of the research to date has concentrated on the definition and application of FTs at the level of plants. Plant functional types (PFTs) have been used to describe how current climate controls the distribution and composition of vegetation (e.g. Box, 1981). These early approaches, however, were based on correlations between the present-day climate and the function of a particular type. More recent approaches account for some physiological mechanisms such as cold tolerance and chilling requirements (Woodward, 1987a; Prentice *et al.*, 1992; Sykes *et al.*, 1996; Woodward & Kelly, 1997), but many known functional relationships between plants and their environment are still not covered. The prospects and difficulties concerning the development of an effective approach to predicting functional responses to changing atmospheric chemistry, land use, and climate were addressed at two GCTE workshops*. This section summarizes some of the major findings.

*Held at the University of Virginia (USA), in 1993 (Smith *et al.*, 1997), and at the Potsdam Institute for Climatic Impact Research (Germany), in 1994 (Woodward & Cramer, 1996a)

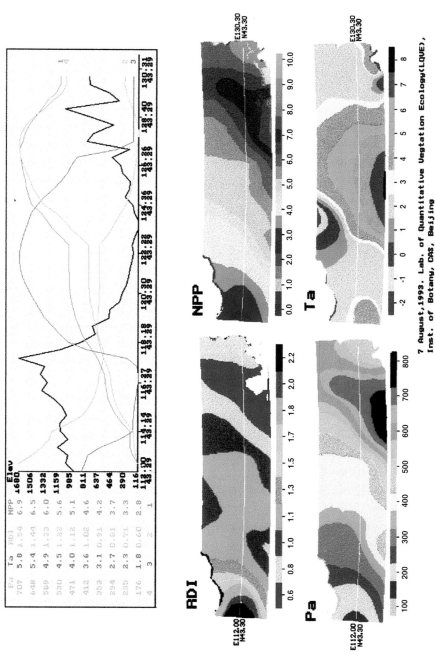

Plate 1 Georeferenced climatic data (radiation, precipitation, temperature) and modelled net primary production (NPP) for the NECT (from X. Zhang, personal communication).

0 269 538 807

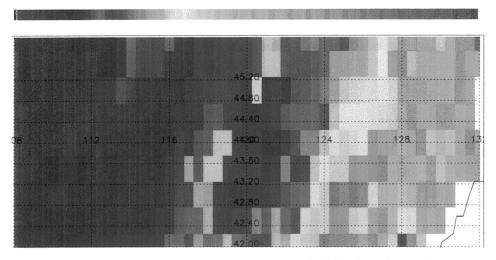

Plate 2 Simulation by the CENTURY model (Parton *et al.*, 1989) along the North East China Transect (NECT) of net primary production (NPP) under current climate (from W.J. Parton, personal communication).

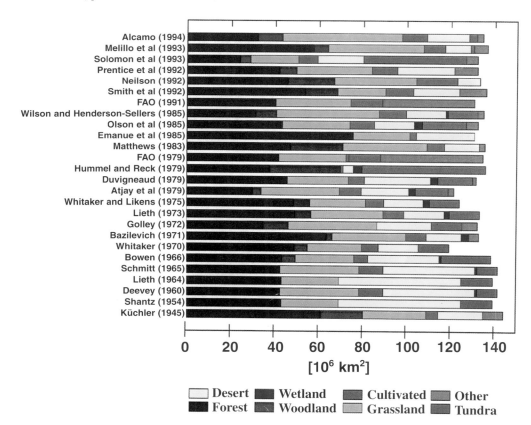

Plate 3 Global assessments for the extent of different land covers. The data are aggregated into coarse classes (Forest, Woodland, Grassland, Tundra, Desert, Wetland, Cultivated and Other) from the original publications or datasets. (Sources: Küchler, 1949; Shantz, 1954; Deevey, 1960; Lieth, 1964; Schmitt, 1965; Bowen, 1966; Bazilevich *et al.*, 1971; Golley, 1972; Lieth, 1973; Whittaker, 1975; Whittaker & Likens, 1975; Ajtay *et al.*, 1979; Bolin *et al.*, 1979; FAO, 1979; Hummel & Reck, 1979; Matthews, 1983; Emanuel *et al.*, 1985a; Olson *et al.*, 1985; FAO, 1991b; Neilson *et al.*, 1992; Prentice *et al.*, 1992; Smith *et al.*, 1992; Melillo *et al.*, 1993; Solomon *et al.*, 1993; Alcamo, 1994).

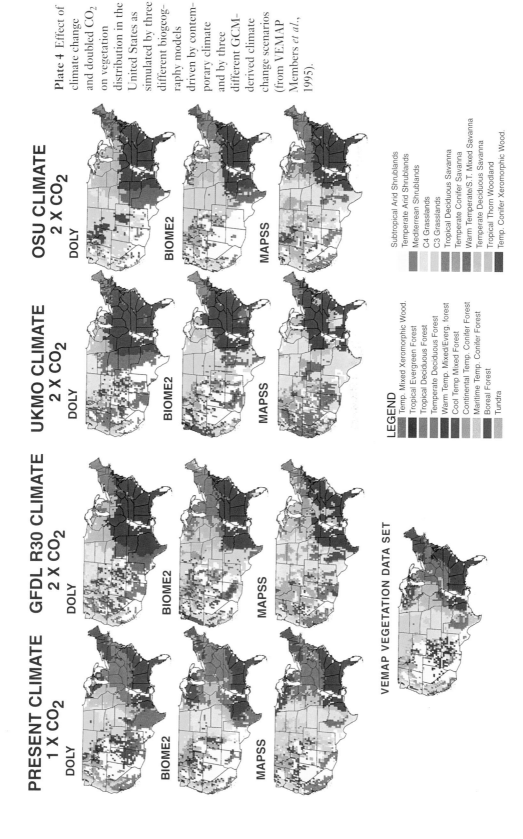

Plate 4 Effect of climate change and doubled CO_2 on vegetation distribution in the United States as simulated by three different biogeography models driven by contemporary climate and by three different GCM-derived climate change scenarios (from VEMAP Members *et al.*, 1995).

PRESENT CLIMATE 1 X CO_2

GFDL R30 CLIMATE 2 X CO_2

UKMO CLIMATE 2 X CO_2

OSU CLIMATE 2 X CO_2

DOLY

BIOME2

MAPSS

VEMAP VEGETATION DATA SET

LEGEND

Temp. Mixed Xeromorphic Wood.
Tropical Evergreen Forest
Tropical Deciduous Forest
Temperate Deciduous Forest
Warm Temp. Mixed/Everg. forest
Cool Temp Mixed Forest
Continental Temp. Conifer Forest
Maritime Temp. Conifer Forest
Boreal Forest
Tundra

Subtropical Arid Shrublands
Temperate Arid Shrublands
Mediterranean Shrublands
C4 Grasslands
C3 Grasslands
Tropical Deciduous Savanna
Temperate Conifer Savanna
Warm Temperate/S.T. Mixed Savanna
Temperate Deciduous Savanna
Tropical Thorn Woodland
Temp. Conifer Xeromorphic Wood.

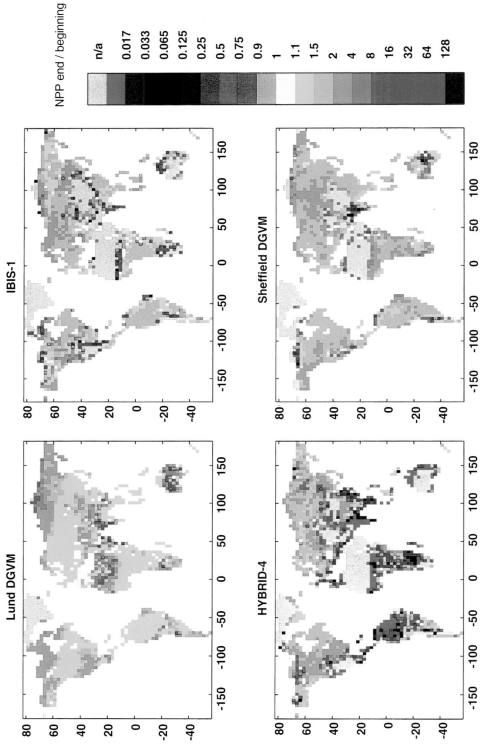

Plate 5 Spatial pattern of changing total net primary productivity (NPP) as a ratio of NPP at the end of the simulation over NPP at the beginning, simulated by the four DGVMs. Yellow to red colours indicate increasing NPP and green to blue colours decreasing NPP.

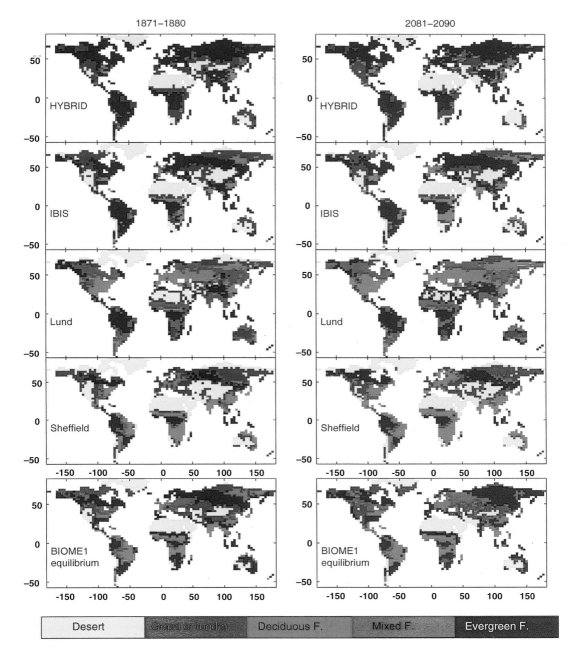

Plate 6 Spatial pattern of changing vegetation distribution simulated by the four DGVMs and by the equilibrium vegetation model BIOME1. Vegetation types have been assigned on the basis of relative LAI values of three major plant functional types (for classification rules see Box 8.3).

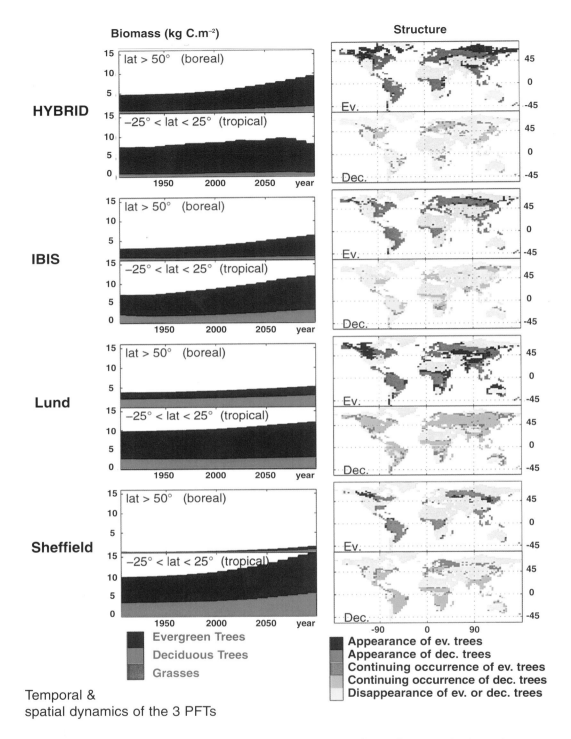

Plate 7 Transient changes in vegetation structure, simulated by the four DGVMs. Left: change in biomass for the major plant functional types from pre-industrial conditions (1860) to 2100 in two latitudinal bands, boreal (north of 50°N, and tropical (between 25°S and 25°N). Right: Changes in distribution of the two tree functional types, evergreen and deciduous, compared between the beginning and the end of the simulation.

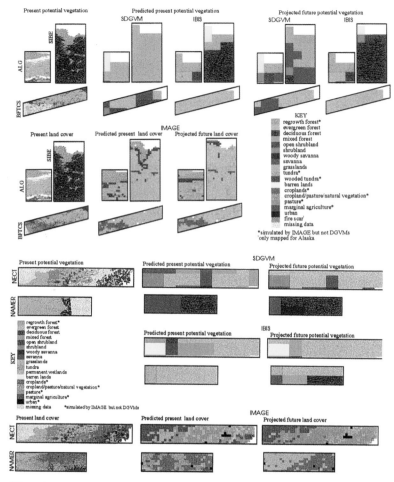

Plate 8 Present and future potential vegetation and land cover for the high latitude (boreal forest–tundra transects): Sources: SIBE (FEST) supplied by IIASA (International Institute for Applied Systems Analysis, Laxenburg, Austria), compiled for the former USSR. Digitized version of the vegetation map of the USSR at scale·1:4M published in 1990, edited by T.I. Isachenko. BFTCS map supplied by the Canadian Forestry Service, Northern Forestry Centre, Edmonton. Alaskan map supplied by the EROS (Earth Resources Observation Systems) Alaska Field Office. Model output for Plates 8 – 10 kindly provided by F.I. Woodward (SDGVM), Jon Foley (IBIS) and Rik Leemans (IMAGE).

KEY
regrowth forest*
evergreen forest
deciduous forest
mixed forest
open shrubland
shrubland
woody savanna
savanna
grasslands
tundra*
wooded tundra*
barren lands
croplands*
cropland/pasture/natural vegetation*
pasture*
marginal agriculture*
urban*
fire scar†
missing data

*simulated by IMAGE but not DGVMs
†only mapped for Alaska

Plate 9 Present and future potential vegetation and land cover for the mid latitude (temperate forest–grassland) transects. Sources: NAMER potential vegetation modified from Küchler (1949) by Coleman as part of the LTER Shortgrass Steppe Project and land use supplied by the Geological Survey's EROS (Earth Resources Observation Systems) Data Center. NECT data supplied by Zhang Xinshi, Institute of Botany, Chinese Academy of Science, Beijing.

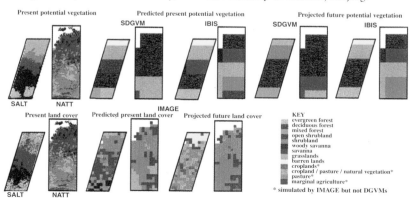

KEY
evergreen forest
deciduous forest
mixed forest
open shrubland
shrubland
woody savanna
savanna
grasslands
barren lands
croplands*
cropland / pasture / natural vegetation*
pasture*
marginal agriculture*

* simulated by IMAGE but not DGVMs

Plate 10 Present and future potential vegetation and land cover for the semi-arid tropical (savanna) transects. Source: NATT extracted and modified from the Vegetation Survey of the Northern Territory, Australia conducted by the Conservation Commission of the Northern Territory. SALT potential vegetation map supplied by the SALT scientific team (Menaut et al., 1995) and land use from the Geological Survey's EROS (Earth Resources Observation Systems) Data Center.

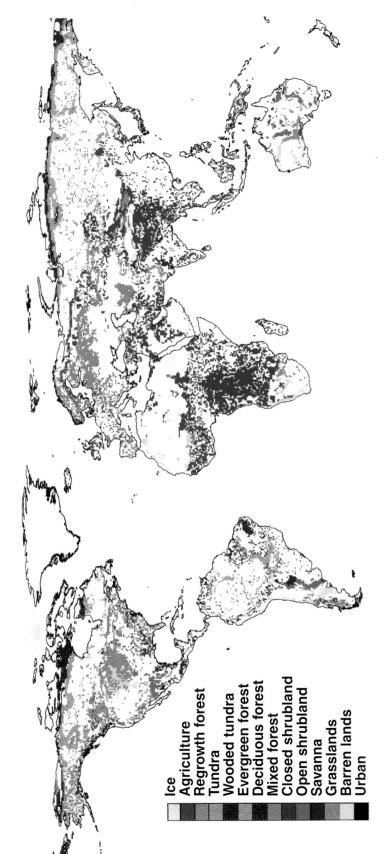

Plate 11 Global land-cover in 2090 relative to 1990 as simulated by IMAGE 2.1 (Alcamo *et al.*, 1996*b* and R. Leemans, personal communication).

Legend:
- Ice
- Agriculture
- Regrowth forest
- Tundra
- Wooded tundra
- Evergreen forest
- Deciduous forest
- Mixed forest
- Closed shrubland
- Open shrubland
- Savanna
- Grasslands
- Barren lands
- Urban

An initial aim was to develop a single, globally applicable 'key' of plant functional types, which could then be used to capture the global range of species and vegetation responses to environmental change. The two workshops have provided the strong message (Bugmann, 1996a; Noble & Gitay, 1996; Woodward & Cramer, 1996b; Bond, 1997) that this global suite is unrealistic due to at least two reasons:

1. Classifications of species into FTs are abstractions of reality – their definition hence depends on the context, scale and question of the specific research context. A biogeochemical model with explicit consideration of photosynthesis, for example, probably needs a distinction between C_3 and C_4 plants (Friend et al., 1997), while other models do not.
2. Functional properties are not always easy to recognize; therefore the derivation of PFTs from observable structural characteristics does not necessarily yield functional classes (Box, 1996; Woodward & Kelly, 1997). An example of this is the current inability to predict a priori the growth responses of a species to CO_2 enrichment. This is a concern, given the high current level of experimentation on this topic (Chapter 7), and at the same time it is a major stumbling block for defining other functional type characterizations of responses to global change.

Nevertheless, there now are many examples of functional type classifications that have a considerable predictive power in global change situations, but which only operate within certain biomes. Such examples include research in semi-arid areas, including desertification problems (Aguiar et al., 1996; Noble & Gitay, 1996; Reynolds et al., 1997; Sala et al., 1997; Walker, 1997), in the Arctic (Chapin et al., 1996a; Shaver et al., 1997) and in temperate vegetation that has been impacted by human activities (Golluscio & Sala, 1993; Thompson et al., 1996; Grime et al., 1997; Sala et al., 1997). Issues of biodiversity have been addressed directly using PFTs in a study of the South African Cape fynbos (Bond, 1997). These successful approaches reflect the dependency of functional types on the scale, context and question that is asked. This is well illustrated by Lauenroth et al. (1997), who indicated that for the prairies of the USA a classification of the vegetation as dominated by one functional type, a C_4 grass type, is inadequate for predicting ecosystem structure and function without the inclusion of details on the life history dynamics.

The use of the concept of plant functional types has tended to confound different purposes and uses. Lavorel et al. (1997) have identified four basic classifications of species to determine different types of groups (Table 8.2). They observed that global change work has concentrated on strategies and functional groups at regional to global geographical scales, whilst general and specific response groups are needed for the scale of human impacts and

Table 8.2 *Types of species classifications into groups (adapted from Lavorel et al., 1997)*

Emergent groups	Inductive classifications based on correlations of traits
Strategies	Species in a group have similar resource use attributes
Functional groups	Species in a group play a similar functional role in ecosystem processes
Specific response groups	Species respond similarly to specific environment factors

management. Most broad classifications converge on major life forms, which is useful for understanding correspondingly broad scale patterns. However, the traits associated with these broad patterns provide little universality in terms of responses to multiple disturbance types at finer resolution of processes – that is, there is a conflict between universality and usefulness, in relation to resolving human impacts. To provide a concerted approach to this problem, Lavorel *et al.* (1997) proposed firstly to accept the fact that different disturbances of environmental gradients may require the use of different suites of functional traits as resolution of processes becomes finer. Secondly, they argue for an hierarchical approach to traits that become systematically less universal as resolution increases. For example, adult and regenerative traits show different patterns of correlation (e.g. Leishman *et al.*, 1992); therefore the regeneration phase traits must be included as well as traits relevant to adult phase responses such as growth form, phenology and grazing responses. The fact that different suites of traits are used for different major groups holds out hope of increased sampling efficiency as well as more appropriate classifications in the future.

Most of the research to date has concentrated on the definition of PFTs at the patch scale; there are just a few publications on the landscape-scale implications of the PFT concept (Noble & Gitay, 1996; Denslow, 1997; Reynolds *et al.*, 1997), and some more studies at the global scale (Box, 1996; Nemani & Running, 1996; Steffen *et al.*, 1996; Cramer, 1997; Leemans, 1997). Many patch-scale studies on PFTs are based on inductive methods ('bottom-up', e.g. Grime *et al.*, 1997), where the attributes of the PFTs are derived from surveys of a large number of species. As the scale increases, this method is gradually replaced by deductive methods ('top-down', e.g. Nemani & Running, 1996). Although the former method appears to be more objective than the latter, in reality both methods are prone to specific biases (Woodward & Cramer, 1996*b*). Therefore, rigorous testing of PFT classifications is required at all scales (Bugmann, 1996*a*).

Global models, such as the Dynamic Global Vegetation Models (DGVMs, Chapter 6 and next section), are generally able to predict the successional dynamics and geographical distributions of vegetation types at the global scale

Table 8.3 *Plant functional types used in four Dynamic Global Vegetation Models*

Type group	HYBRID	IBIS	Lund DGVM	Sheffield DGVM
Trees				
Evergreen	Broadleaf evergreen	Tropical evergreen	Broadleaf evergreen	Broadleaf evergreen
		Warm temperate evergreen		
	Needleleaf evergreen	Cool conifer	Needleleaf evergreen	Needleleaf evergreen
		Boreal conifer		
Deciduous	Broadleaf dry deciduous	Tropical raingreen	Broadleaf raingreen	Broadleaf deciduous
	Needleleaf dry deciduous			
	Broadleaf cold deciduous	Temperate summergreen	Summergreen deciduous	Needleleaf deciduous
	Needleleaf cold deciduous	Boreal summergreen		
Grasses				
	C_3 herbs	C_3 herbs	C_3 perennial	C_3 herbs
			Raingreen C_3 annual	
			Summergreen C_3 annual	
	C_4 herbs	C_4 herbs	C_4 perennial	C_4 herbs
			Raingreen C_4 annual	
			Summergreen C_4 annual	
			Shrubs	

on the basis of a small number of functional/structural types (Cramer, 1997). The maps shown later are based on only three functional types: grasses, evergreen and deciduous trees. Internally, however, most current DGVMs use between seven and ten types, which is usually a compromise between desirable variability and availability of data for parameterization. These small numbers of types appear, at least initially, to be adequate for the aims of DGVMs (Table 8.3, and Chapter 6), i.e. for simulating the feedbacks of vegetation on climate and for projecting the general nature of vegetation change.

8.4.3   Findings from equilibrium simulations of the biosphere

Potential and limitations of equilibrium models

All presently published assessments of biospheric response to climate change are based on equilibrium models of the terrestrial biosphere. When subjected to scenarios of climate change, most of these show strong responses, such as the displacement of biomes or large changes in productivity. However, it is important to note that three major uncertainties exist in these assessments:

- Their results are necessarily strongly dependent on the climate change scenario that is chosen for the assessment. Despite a growing consensus about the likely range of future global average temperatures, there is still large uncertainty with respect to the hydrological cycle and the regional pattern in any climate variables (Kattenberg *et al.*, 1996). Therefore, future assessments of changing vegetation distribution carry the same uncertainty and must be considered scenarios rather than predictions.
- By definition, equilibrium models simulate no transient changes in vegetation. It has been pointed out, however, that the rate of change predicted by current climate models may exceed earlier rates of change (Kirschbaum *et al.*, 1996*b*). Therefore, these simulations may at best be used to indicate the direction of possible change but not the time it might take to reach the new conditions (Cramer & Steffen, 1997).
- There is no basic reason to expect that biomes would be displaced as homogenous entities. Rather, it must be assumed that differences in species' fundamental ecological niches, and their widely varying abilities to migrate, will result in quite different assemblages over a long period of time (Huntley *et al.*, 1997).

Despite these uncertainties, the current suite of global biosphere models has been used intensively for the assessment of possible impacts of global change. Here, a summary of some of these assessments is given.

The VEMAP study

The first phase of the VEMAP exercise (VEMAP Members, 1995) focused on reducing uncertainties in the current understanding of global biospheric functioning by applying three different models of ecosystem structure (MAPSS – Neilson, 1995; DOLY – Woodward *et al.*, 1995; and BIOME2 – Haxeltine *et al.*, 1996), coupled (in factorial combinations) to three models of biogeochemical fluxes (CENTURY – Parton *et al.*, 1993*a*; BIOME-BGC – Running & Hunt, 1993; and TEM – McGuire *et al.*, 1995*a*,*b*), for the coterminous USA. The model combinations were run for a set of environmental scenarios, including current climate and current CO_2 conditions, doubled CO_2 (the direct effect), three GCM-derived scenarios of climate change, and the combination of doubled CO_2 and climate. The coupling between models of ecosystem structure (hereafter called 'biogeography models') and models of biogeochemistry was done without consideration of feedbacks, i.e. the structure was estimated first and then used as input to the biogeochemistry models (results from the biogeochemistry models are discussed in Chapter 7).

The study provided a preliminary indication of the potential sensitivity of natural vegetation in the USA to climate change and changing CO_2 concentrations (Plate 4), as well as a partial measure of the uncertainty associated with the projections. The three biogeography models all incorporate some sort of direct response to changes in CO_2 concentration, but they differ in the specific mechanisms that are considered. In MAPPS, increasing CO_2 causes a reduction in stomatal conductance leading to a reduction in evapotranspiration. The model does not include any effect of CO_2 on photosynthesis and also does not allow for any direct effect on the competitive balance between C_3 and C_4 plants. In BIOME2, in contrast, CO_2 does affect NPP, and this alters the competitive balance between C_3 and C_4 plants, but transpiration is not altered by CO_2. The DOLY model lies between these two extremes in that it incorporates reductions in stomatal conductance and increases in NPP, though there is no effect on the C_3/C_4 competitive balance. It should be noted that none of the three models include the most recent findings of elevated CO_2 research, as described in Chapter 7.

The consequence of these different ways to implement the CO_2 effect is likely to be one reason why the models predict different patterns of biome distribution in the USA in response to doubled CO_2 without climate change (VEMAP Members, 1995). The effect was least for BIOME2 because it does not include a direct effect on transpiration that would become especially important in water-limited systems. On the other hand, since BIOME2 does include an effect on C_3/C_4 competition through NPP, it is not surprising that the model predicted an increase in the relative proportion of C_3 to C_4 grassland. Because of their inclusion of an effect of CO_2 on stomatal conductance, MAPSS and DOLY predicted greater responses to CO_2, including increases in forest cover in the interior of the Western USA and at the transition of forest and prairie. Over all models, the effect of elevated CO_2 in conjunction with climate change generally was to reduce the magnitude of effects seen when climate is changed alone. However, because of its lack of a transpiration effect, BIOME2 tended to predict smaller differences between the simulations run with and without elevated CO_2 than the other two models. Because MAPSS is highly sensitive to both changes in stomatal conductance and LAI, it predicted the most dramatic CO_2 mitigation of climate change-induced changes in moisture availability.

These results highlight the potential importance of direct CO_2 effects for the distribution of ecosystems. Because of the different ways they implement CO_2

effects, the models make different predictions under scenarios of climate change and doubled CO_2. One obvious conclusion is that the models should at least incorporate both effects on photosynthesis or NPP and effects on stomatal conductance.

There are some common features in the responses to climate change for different model combinations, as well as noticeable differences. For instance, all model combinations show a northward migration of vegetation types in the eastern USA as would be expected for a region where currently a marked latitudinal gradient in climate and vegetation exists. The extent of northward movement is greatest for the UKMO climate change scenario (note the distribution of the Warm Temperate Mixed/Evergreen Forest), which predicts the greatest increase in mean temperature. Another region where there are some clearly seen patterns of change in vegetation type is in the transition zone between forest and grassland in the Central USA. However, in this case the three biogeography models often respond differently to specific climate scenarios. Vegetation patterns in the western USA are more complex, probably due to the greater complexity in modelling moisture-related responses as opposed to temperature related ones.

There are important differences among the three biogeography models that help to explain why they sometimes diverge in their predictions and why they can respond differently to the climate change scenarios. In particular, models differ in how they determine potential evapotranspiration, with BIOME2 at one extreme in having evapotranspiration determined only by available energy, and MAPSS using an aerodynamic approach that is sensitive to canopy characteristics; DOLY is intermediate. This results in differential sensitivity to changes in temperature or canopy characteristics. Some of the consequences of these differences can be seen in patterns of forest distribution in the eastern USA and in the balance between conifer forests, broadleaf forests, and savannas.

8.4.4   First results from a comparison of Dynamic Global Vegetation Models

Participating models

The GCTE Synthesis Project provided the first opportunity for comparing four Dynamic Global Vegetation Models (DGVMs, see also Chapter 6): Lund DGVM (I.C. Prentice, Lund University, Lund, Sweden), HYBRID-4 (A. Friend, Institute for Terrestrial Ecology, Edinburgh, UK), IBIS-1 (J. Foley, University of Wisconsin, Madison, USA) and SDGVM (F.I. Woodward, Sheffield University, Sheffield, UK). There are substantial differences between these models. For example, IBIS-1 (Foley *et al.*, 1996) and HYBRID-4 (Friend *et al.*, 1997) are intended to be fully incorporated within atmospheric general circulation models, and therefore include more detailed physical land surface

---

Box 8.3   Procedure for the comparison of Dynamic Global Vegetation Models

Climate drivers and CO_2: All four models were run with a transient climate change scenario produced by the Hadley Centre for Climate Prediction & Research at the UK Meteorological Office[†]. The climate change scenario was derived from a ~240 year simulation of the Hadley Centre's coupled atmosphere-ocean general circulation model (HadCM2) driven by changing amounts of atmospheric CO_2 and sulphate aerosols (Mitchell *et al.*, 1995). This GCM, which has a spatial resolution of 2.5 ° by 3.75 °, explicitly accounts for the combined radiative effect of sulphate aerosols and CO_2 on global climate. The model output roughly corresponds to the calendar years 1860 (preindustrial conditions with CO_2 concentrations of 290 ppmv) to 2099 (with projected CO_2 concentrations at 800 ppmv, see Fig. 8.4), showing an overall warming trend in most areas, particularly in high latitudes. Rather than using the HadCM2 climate model output directly, the DGVMs were driven by the simulated *changes* in climate as compared to their simulated years 1931–60, then superimposed on observed climatic normals for the same period, gridded to the GCMs resolution (Leemans & Cramer, 1991)[‡]. In this way, it was hoped that systematic biases in the spatial pattern of the climate simulated by the GCM would be largely removed, while the trend and interannual variability are retained as much as possible.

DGVM simulations: All DGVMs were first run to equilibrium under pre-industrial climate and CO_2 conditions. During the simulation years from 1860 to 2099, the models were also directly influenced by increasing CO_2 concentrations, identical to those used by the GCM, in order to account for the possible physiological effects on photosynthesis and water balance.

Functional types and output variables: Since the four models used different sets of plant functional types (PFTs), a minimal set of aggregated types was defined: deciduous trees, evergreen trees, and grasses (Table 8.3). For these PFT groups, annual data on net primary productivity (NPP), leaf area index (LAI) and biomass were collected for analysis. Here, the LAI output is used only to generate a simple, rule-based classification of vegetation for mapping purposes. Where the combined LAI of all tree types was less than 1.0, a cut-off value of 0.25 in grass LAI was used to distinguish between desert and grassland. In areas with higher tree LAI than 1.0, the relative proportion of deciduous and evergreen trees was used to distinguish evergreen (evergreen tree LAI > 66% of total tree LAI), mixed (evergreen LAI between 33 and 66% of total tree LAI) and deciduous (evergreen tree LAI < 33% of total tree LAI) forests.

[†] The help of Dr. Richard Betts, UKMO, in the transfer and analysis of these data, is gratefully acknowledged.
[‡] The data set was used in an updated version (v2.1a), which is available upon request from W. Cramer, Potsdam Institute for Climate Impact Research, Potsdam, Germany

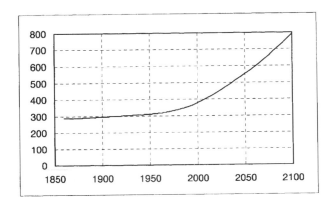

Figure 8.4   Atmospheric CO_2 concentration (in ppmv, historical and future projection), as it is used by the HadCM2 simulation and the Dynamic Global Vegetation Models in this comparison.

parameterizations than Lund DGVM (unpublished) and SDGVM (unpublished). Other differences between the DGVMs arise from their basic assumptions and state descriptions. For example, HYBRID-4 employs an explicit patch-scale description of vegetation dynamics, including a representation of mortality and disturbance, while the other three models scale vegetation dynamics directly to the scale of the entire grid cell. Further differences between the DGVMs arise from the treatment of canopy photosynthesis and conductance, carbon allocation, and disturbance mechanisms (e.g. fire and windthrow). For technical details of the transient simulation, see Box 8.3.

Results

Figure 8.5 shows the development of total global net primary productivity (NPP) through the transient simulation by all four models. Under pre-industrial CO_2 concentrations and climatic conditions, the DGVMs simulate a total global NPP of between approximately 48 and 76 Pg C yr^{-1}. This broad range is similar to the range of estimates of global NPP by equilibrium models (Cramer *et al.*, 1998) – the result thereby illustrates the considerable uncertainty one must accept for this variable. Over the course of the simulation, global NPP also shows significant interannual variability (greatest in HYBRID-4), which is caused by climatic variability or by population processes, or both. Over the course of the simulation, total global NPP increased by approximately 42 to 71%, compared to preindustrial levels. When one of the models (Lund DGVM) was run without changing CO_2 concentrations (results not shown), there was a slight *reduction* in global NPP, indicating that atmospheric CO_2 concentrations (rather than changes in climate) are mainly driving the overall increases in global NPP in this model. For the productivity of the three plant functional type

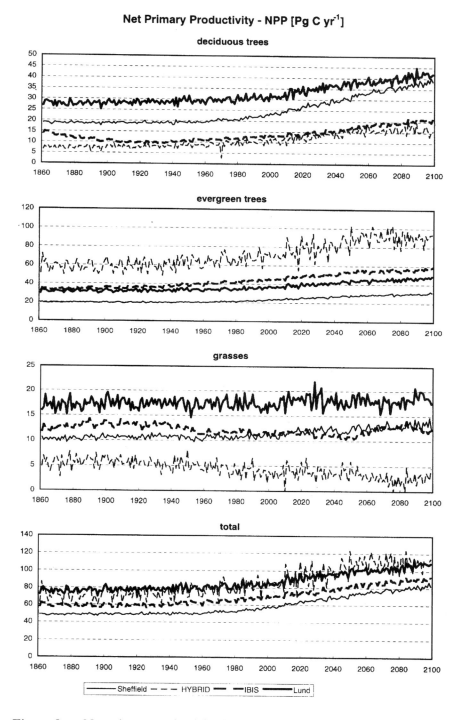

Figure 8.5 Net primary productivity (NPP), in Pg C yr⁻¹, by plant functional types and total, simulated by four Dynamic Global Vegetation Models driven by a transient climate simulation from the HadCM2 general circulation model.

groups (deciduous trees, evergreen trees, grasses), differences between models were greater than for total NPP. No model had a consistent trend for NPP over time of grasses – however, this may mask the possibility of a change in the C_3/C_4 ratio. In all models, tree productivity increased (the slight decrease in deciduous tree productivity shown by IBIS during the first decades appears to indicate that the model was not fully in equilibrium when the transient simulation started).

Total global vegetation biomass increased in all four models (Fig. 8.6), with rates between 1.1 and 2.0 Pg C yr^{-1}. Together, the biomass and productivity values from this comparison indicate that, compared to the equilibrium models (which appear, in one way or another, tuned to present day carbon fluxes, see Ruimy *et al.*, personal communication), the added complexity of transient dynamics also adds to the uncertainty of the flux estimations.

The DGVMs show various geographic patterns in changes of NPP (Plate 5). Generally, the models show similar enhancements of NPP at high latitudes (> *c*. 45 ° north, with the exception of the Canadian Arctic). Model response in arid and semi–arid regions is inconclusive: increases and decreases occur in many areas, except for Pakistan and India, where there is a significant reduction, due to reduced water availability. Three of the four models show general increases in NPP between the tropics and in wet climates, although there is significant variation between model simulations for individual regions. There are also marked differences between the models in the simulations for Australia.

Since these models simulate vegetation structure and biogeochemical fluxes in a coupled mode, it may be more worthwhile to analyse their behaviour with respect to the distribution of major vegetation types (Plate 6). The left half of the diagram clearly exhibits differences regarding the models' ability to simu-late pre-industrial vegetation distribution, but these are partly due to the simplistic classification scheme (see Box 8.1): broadly, the major vegetation zones of the world are recognized by the four models. Comparison with the right-hand part of the diagram (average conditions at the end of the transient simulation) shows that the DGVMs all simulate significant shifts in vegetation distributions through the simulation. In general, these shifts appear to be driven by changes in climate, with rather limited influences of CO_2, except in semi-arid regions. The complex changes in vegetation distribution make it difficult to evaluate the model results on vegetation redistribution by direct comparison with global maps. Inspection of individual grid cells along the time axis (not shown here) reveals that the successional pattern of establishment, growth, co–existence and replacement of population is simulated qualitatively rather well. This means they behave like forest gap models, but with the important added feature of global applicability. More in-depth analyses at a range of different locations will be necessary for assessment of model behaviour in this respect.

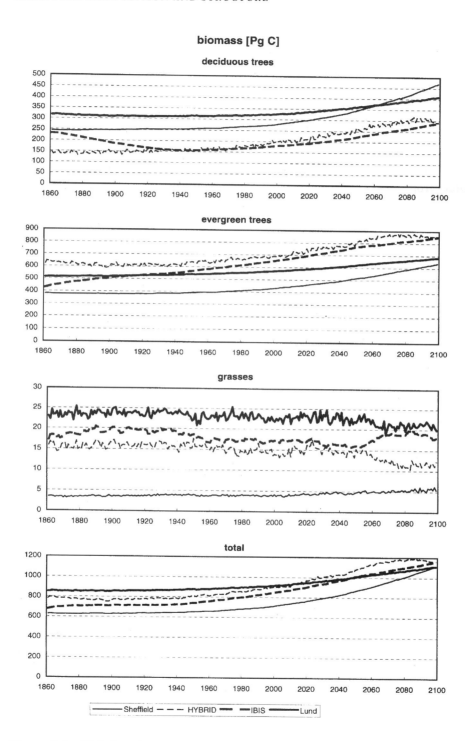

biomass [Pg C]

Figure 8.6 Global total vegetation biomass (Pg C) by plant functional types and total, simulated by the four DGVMs.

Another summary of the dynamics in vegetation structure is given in Plate 7. The left half of the diagram indicates the differential increase in biomass, simulated by all models, for the tropical and boreal zone. The right half shows the component of this change which involves major vegetation redistribution. It must be noted that migration currently is considered to involve no time lags in these models, a feature that will probably change in further developments (Pitelka & Plant Migration Workshop Group, 1997). Due to the warming (particularly at northern high latitudes), trees encroach northwards, but these may be deciduous or evergreen, depending on the models' sensitivity to climatic variables. Two models (IBIS and Lund DGVM) simulate a strong reduction in deciduous trees in the tropics; nevertheless, the overall biomass of these types increases, due to enhanced growth in the remaining areas.

Conclusions

To achieve more plausible and quantitatively reliable simulations, several key issues remain to be resolved. For example, while current DGVMs are able to simulate time-dependent changes in vegetation in relation to climate, such as growth and competitive replacement, there are still other constraints to the transient response of vegetation, such as dispersal of propagules and soil development. Currently there is no accepted paradigm for modelling the spatial dynamics of large-scale vegetation communities mediated by seed dispersal. Initial studies (e.g. King & Herstrom, 1997; Solomon, 1997) have shown that in order to match past plant migration rates as reconstructed from pollen records, it is necessary to take into account (a) the patchiness of species distributions, which often includes small local populations in favourable habitats beyond the main distributional range, and (b) the accumulated effects of low-probability long-distance dispersal (e.g. by migrating animals and birds, tornadoes, etc.) allowing rapid range extension (Pitelka & Plant Migration Workshop Group, 1997). Some parameterizations of these migration processes are now being developed but they have not yet been coupled to DGVMs.

A second key issue is the representation of disturbance in the DGVMs. Most of the DGVMs currently have a disturbance generator but more work is required to improve the realism of this module, especially given the importance of disturbance in the long-term terrestrial carbon budget (see Section 12.2.1). Fire is probably the most important disturbance, and the accurate simulation of change in fire regimes is an early priority. A module is required that estimates (i) the probability of fire outbreak based on changing environment (climate, land use) and the condition of the ecosystem (fuel load, dryness, etc.), and (ii) the areal extent of the fire.

8.4.5 Feedbacks to the atmosphere

The circulation of the atmosphere is strongly affected by the exchange of energy, water vapour, and momentum between the planetary boundary layer and the underlying surfaces. Over ice-free land, these exchange processes are strongly influenced by vegetation and soil properties, including surface albedo, leaf area index, vegetation height, rooting depth, and soil water holding capacity. As a result, increasing attention is being paid to the role of vegetation and soil characteristics as important factors within the climate system, and their importance is likely to increase in a changing global environment. For example, it is likely that future changes in climate will be accompanied by large changes in vegetation cover (Kirschbaum *et al.*, 1996*b*), potentially leading to important changes in land surface characteristics and, ultimately, feeding back onto the atmosphere. In addition, it is likely that the physiological effects of increasing atmospheric CO_2 concentrations could directly alter land surface processes by decreasing stomatal conductance and increasing canopy leaf area. Finally, human land use activities, including deforestation, may significantly alter the climate through significant changes in land surface conditions.

The current generation of climate models represent the biophysical interactions between the terrestrial surfaces and the atmosphere using 'land surface modules' (e.g. BATS – Dickinson *et al.*, 1986; and SiB – Sellers *et al.*, 1986; Xue *et al.*, 1991; Dickinson *et al.*, 1993), which simulate the energy, water, and momentum exchange between land surfaces and the atmosphere. Within global climate models, these modules operate by prescribing the geographic distribution of vegetation and soil types, and their associated biophysical characteristics. A large number of studies have explored the potential sensitivity of the climate system to changes in vegetation cover. For example, numerous studies have used atmospheric general circulation models and land surface modules to examine the atmospheric response to scenarios of large-scale tropical deforestation. These simulations were executed by converting the land cover over a large area of Amazonia from a forest to a grassland. In general, these studies all show large reductions in evapotranspiration, while a majority of them also show significant increases in temperature and reductions in rainfall. It is possible that the changes in climate associated with a hypothetical deforestation of the entire Amazon basin would be of sufficient significance to generate a long-term shift in the expected vegetation. However, the climatic effects of deforestation needs to be examined further in relation to actual patterns of deforestation in contemporary tropical forests. A few preliminary studies have examined the impact of the characteristic scale of deforestation on atmospheric processes, e.g. Eltahir & Bras, (1994), and others.

In another set of sensitivity studies, Bonan *et al.* (1992) investigated the potential effects of the complete deforestation of the circumpolar boreal forest on global climate using the National Center for Atmospheric Research GEN-ESIS climate model (Thompson & Pollard, 1995). The removal of boreal forests increased the albedo of the land surface and resulted in colder air temperatures with respect to the control conditions. The resultant cooling was substantial and affected the temperatures of the entire northern hemisphere. Bonan *et al.* (1992) summarized these results as .'. . the summer cooling caused by deforestation is sufficient to prevent forest regrowth in much of the deforested area. Thus, boreal deforestation may initiate a long-term irreversible feedback in which the forest does not recover and the tree-line moves progressively farther south'.

Just as model simulations in tropical forests and boreal forests have indicated there may be significant feedbacks between the vegetation and climate, vegetation alternation in semi-arid and desert regions has also been shown to have potentially large effects on climate. For example, the conversion of sparsely vegetated semi-arid shrublands to desert represents a substantial change in the features of the terrestrial surface *vis-à-vis* the atmosphere (Charney, 1975). Xue and Shukla (1993) investigated the effects of drought on desertification in the Sahel. They conducted several model experiments that resulted in reduced rainfall to the Sahel when it was converted to desert. The overall implication of the results was to shift the Sahel toward more desert-like conditions.

The broad implications of these results point to a reinforcing feedback between the extant vegetation and climate. Given the degree to which humans are currently modifying terrestrial surfaces, there is a pressing need to further our understanding of the interactions between the atmosphere and the terrestrial biosphere. In particular, direct evidence of climate and vegetation interactions must be sought to evaluate the results of these numerical models. While the results of the above deforestation sensitivity studies cannot yet be directly validated, it is possible to examine climate and vegetation interactions that have happened in the recent geologic past. One particular time period that has been studied extensively is the early-to-middle Holocene (10 000 to 5000 years before present). During this time in the geologic record, differences in the Earth's orbit (mainly in the date of perihelion) amplified the seasonal cycle of solar radiation in the northern hemisphere, thus causing additional heating of northern hemisphere continental interiors in summer and additional cooling in winter. Using climate models, numerous studies have demonstrated how this change in solar radiation would cause a significant warming of the northern high-latitudes and an amplification of subtropical monsoons (e.g. Kutzbach & Guetter, 1986).

The climatic change simulated in paleoclimate simulations is often evaluated using equilibrium vegetation models, and comparing the simulated changes in global vegetation patterns against palaeobotanical data. For example, Foley *et al.* (1994) used the GENESIS climate model (Thompson & Pollard, 1995) and the BIOME1 equilibrium vegetation model (Prentice *et al.*, 1992) to show that, during the period of ~6000 years before present, boreal forests were further poleward than present and tropical grasslands and savannas were present in much of the modern-day Sahara. These two large changes in vegetation cover, which are both strongly supported by the pollen and plant macrofossil data, may have been large enough to influence the climate system. Street-Perrott *et al.* (1990) and Kutzbach *et al.* (1996) have demonstrated the potential importance of changing vegetation cover in the Sahara (i.e. grassland versus desert) on the dynamics of the African monsoon during the early and middle Holocene era. In particular, these modelling results have suggested that the presence of grasslands and savannas in the modern-day Sahara desert amplified the strength of the African monsoon (and the associated precipitation) and thus helped to maintain the Saharan grasslands.

In another set of paleoclimatic studies, Foley *et al.* (1994) demonstrated the potential importance of changes in the boreal forest-tundra ecotone on high-latitude climates during the middle Holocene. In this sensitivity study, two paleoclimate simulations were performed: one with the boreal forest treeline kept at modern conditions, and another with the treeline displaced poleward by approximately 100–300 km. In the case of the modern treeline, the circumpolar region (60° to 90° North) warmed by an average of approximately 1.5°C in response to changes in Earth's orbit. However, there is an additional warming of about 1.5°C induced by albedo differences caused by the poleward displacement of boreal forests in the place of tundra. These results may be relevant to expectations of future climatic change, in which a significant warming of high-latitude regions is expected.

While a number of sensitivity studies have examined the role of changes in vegetation cover on the climate system, only a few papers have discussed the potential role of changing ecosystem function on the climate. For example, the effects of CO_2 fertilization on canopy conductance and transpiration may be large enough to cause substantial shifts in the water balance, and hence alter the atmosphere–biosphere exchange of energy and water vapour (see also Section 7.2.3). A few preliminary studies (Friend & Cox, 1995; Henderson-Sellers *et al.*, 1995; Pollard & Thompson, 1995; Sellers *et al.*, 1996) have examined the potential climatic effects of CO_2-induced changes in canopy conductance. However, possible changes in vegetation structure and composition resulting from CO_2 fertilization have not yet been considered within climate models.

These sensitivity studies have demonstrated that using fixed geographic distributions of vegetation types within climate models severely limits their use in studies of global change. Preliminary attempts at describing the bidirectional interactions between climate and vegetation have been made by incorporating equilibrium vegetation models within atmospheric general circulation models (Henderson-Sellers, 1993; Claussen, 1994; Henderson-Sellers & McGuffie,

1995). For example, Henderson-Sellers (1993) used the simple equilibrium vegetation model of Holdridge (1947) within the CCM1-Oz climate model to predict the global vegetation cover (and the associated land surface characteristics) from the simulated climate. In this work, Henderson-Sellers performed several short model simulations to demonstrate that the coupled dynamics of the climate–vegetation model were stable (with no trends in the simulation) with several regional-scale differences between another simulation made with fixed vegetation patterns.

Claussen (1994) investigated several important issues related to climate and vegetation interactions, including the sensitivity of the coupled system to the specification of initial conditions and the frequency of coupling. Claussen incorporated the BIOME1 equilibrium vegetation within the ECHAM climate model and performed several long simulations, including two that differed in their initial vegetation conditions. In one of his simulations, Claussen initialized the coupled simulation by switching the location of tropical forests and sub-tropical deserts. He found that the climate was affected enough to allow some areas of tropical forest to remain in the south-western Sahara, and desert to remain in India.

A recent study of Betts *et al.* (1997) went one step further in coupling biospheric and atmospheric processes by using the Sheffield University Vegetation Model (the equilibrium predecessor of the Sheffield DGVM) as part of the land surface scheme in a simplified version of the Hadley Centre general circulation model. In this case, not only structural feedbacks (such as those mentioned above) but also physiological factors are considered. In a qualitative sense, this study indicated the potential time lags in structural response of the biosphere, as compared to the direct response in physiology. To add quantitative significance to these findings, however, it is necessary to simulate non-steady state structural and physiology responses (as they are shown by DGVMs), since these differentiate between plant types with respect to their basic life cycle processes, such as establishment, competition and mortality, all of which are likely to influence the length of the delayed response.

There is now increasing activity focused on linking fully dynamic representations of terrestrial ecosystems within climate models. In particular, some of the DGVMs are designed to be fully coupled within climate models. For example, a preliminary attempt at incorporating IBIS-1 within the GENESIS climate model has been completed (Foley *et al.*, 1998). Like the first attempts at coupling atmosphere and ocean models, these coupled simulations have significant needs for future research. These include: (a) the role of initial conditions, and (b) the role of 'flux corrections' in coupled models.

9 Managed production systems

P.J. Gregory, J.S.I. Ingram, B. Campbell, J. Goudriaan, L.A. Hunt,
J.J. Landsberg, S. Linder, M. Stafford Smith, R.W. Sutherst and
C. Valentin

9.1 Introduction

Agriculture and forestry were the first two major industries to be developed.
Over time, other industries arose and the proportion of the population directly
involved in agriculture and forestry has decreased. In some parts of western
Europe only 2% of the population is now directly involved in agriculture
whereas in some African countries it is still over 90%. A similar wide range
exists in intensification and technology with high–input arable agriculture at one
end and shifting cultivation at the other. Forest enterprises are also diverse,
ranging from simple extraction of wood and other products from near-virgin
forest to management and clear-felling of plantations.

Agriculture and forestry depend not only on biological and technological
knowledge and innovation, but also on many social, political and economic
factors including demographic changes, industrialization, structural changes in
the economy, migration and urbanization, and social and institutional arrange-
ments. GCTE has focused predominantly (but not exclusively) on the biological
and environmental constraints to production as outlined by Tinker and Ingram
(1996). This chapter presents some of the major findings to date together with
their implications for future production. It also outlines the research required to
improve assessments of the impact of global change on food and fibre production.

9.2 Factors influencing production systems

9.2.1 Demand for food and forest products

The impacts of global change must be examined in the context of food security
over the next 20 to 25 years and particularly in light of the increasing human
population (see Chapter 1). The global population will increase by about $0.8 - 1$
billion per decade for the next two to three decades (Table 1.1, Chapter 1) with
most of this increase occurring in the less developed nations and almost none in

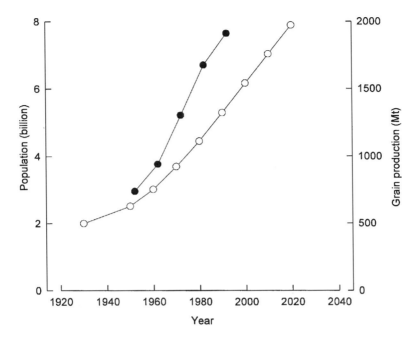

Figure 9.1   Recorded and projected population (○) and grain production (●).
Adapted from Dyson, 1996.

Europe and North America. Together with these changes in total population
there is a global trend towards urbanization. In 1950, 70% of the population
lived in rural areas but it had declined to about 55% in 1995 and is projected
to be about 40% in 2025. This movement of labour away from the land will
increase the pressures towards intensification of crop production. Although
the proportion of the population living in rural areas differs between regions,
the downward trend is evident everywhere. The projections indicate that the
maximum rural population will be achieved between 2010 and 2020 with
decreasing absolute numbers thereafter, except in Africa where the absolute
number of people living rurally will continue to increase for a few more decades.

The growth in human population over the past century has been closely
paralleled by an increase in food production (Fig. 9.1). Generally, production
has increased slightly faster than population so that, for example, 5.8 billion
people today have 15% more food *per capita* than a population of 4 billion had
20 years ago. This increase has not been uniformly distributed and the number
of chronically undernourished has remained relatively constant at about 20% of
the present population of less developed countries. Wherever progress has been
achieved, it has been mainly due to the alleviation of poverty.

Given the close link between global population and food production, and

Table 9.1 *Predicted changes in grain consumption (from Crosson & Anderson, 1994)*

	1988/89 (million t)	2030 (million t)	Annual increase (%)
Less developed countries			
Wheat	266	770	2.3
Rice	309	634	1.3
Coarse grains	300	946	3.2
Total	875	2350	2.3
More developed countries			
Total	803	947	0.4

ignoring the period 1990–1995 when grain stocks were drawn on, it is possible to estimate by extrapolation from Figure 9.1 the required grain production (wheat, rice and maize together supply about 60% of the total carbohydrate). Population growth of 1 billion per decade requires an annual increase in grain production of about 32 million tonnes; more if allowance is made for rising affluence. Assuming no massive change in food distribution, production in the more developed countries is estimated to increase by about 20% between 1988/89 and 2030, whereas increases in the less developed countries will need to average 250% from about 1 billion tonnes per year now to about 2.5 billion tonnes by 2030 (Table 9.1; Yudelman, 1993). The need for increases in production will vary from crop to crop with rice production increasing by two–fold and that of other grains (principally wheat and maize) by nearer to three–fold. Population is not the only driver of production and changes in wealth must also be accounted for. As wealth increases, so the direct consumption of cereals declines and meat consumption increases, requiring more grain per person or an increase in forage production.

Similar analyses for forest production show that the ability to meet the increasing demand during the next few decades varies dramatically between regions. At northern and mid-latitudes the forested area is at present remaining more or less constant (Table 9.2), but at low latitudes the imbalance between reforestation and harvesting is rapidly reducing the area of forested land (Table 9.3). As discussed in Chapter 12, the forests of northern and mid-latitudes currently act as a sink for carbon and though it is proposed by some that they will continue to do so over the next few decades (e.g. Kauppi *et al.*, 1992; Cannell, 1995), this is by no means certain (see Chapter 12). On a global scale there is a net decrease in forested area and standing crop, resulting in the world's forests acting as a carbon source. There is no indication that this situation will change in the near future, so a gradual shift in the balance between

Table 9.2 *Forestry statistics for medium and high latitude countries and Oceania (Japan, Australia, New Zealand). The figures are dominated by softwoods (from UN, 1992)*

Region	Forested area per person (ha capita^{-1})	Standing crop (billion m^3)	Annual gross cut (million m^3)
Europe	0.24	18.5	408
Former USSR	1.43	50.3	518
Canada	4.23	14.9	152
USA	0.78	23.1	620
Oceania*	0.30	5.0	20**

*Australia, Japan, New Zealand
**Australia only

Table 9.3 *Forestry statistics for tropical regions. The data include 90 countries (Africa 40, Southeast Asia 17, and South America 33) and largely refer to natural hardwood-dominated forests (from FAO 1993b)*

Region	Total forested area (million ha)	Forested area per person (ha capita^{-1})	Annual change in forested area (million ha)	Ratio between establishment : deforestation
Africa	527	1.10	4.1	1 : 32
S.E. Asia	311	0.20	3.9	1 : 2
S. America	918	2.30	7.4	1 : 20

regions producing and consuming forest products is expected. This will result in an increasing shortage of wood for construction, pulp and fuel in the developing world.

Forest statistics (UN, 1992; FAO, 1993b) give some indication not only of the trade in wood, the areas of forest and their current rate of deforestation, but also the source/sink situation for carbon (Tables 9.2 and 9.3). These data refer to forests without legal, economic or technical restrictions on commercial forestry (i.e. exploitable forests). Values in Table 9.2 are for temperate zones, but mask some large differences; for example, the average forested area per capita in the Nordic countries is 2.67 ha whereas in the UK it is about 0.04 ha. The values for the tropical regions (Table 9.3) are more uncertain than those in Table 9.2; the data for forested areas are probably reasonable, but estimates given by FAO (1993b) for standing crop biomass can vary by ± 200%. The ratio of wood production (gross cut per annum) to standing crop gives some indication of the overall length of rotation, or return time to cutting, in different regions. The former Soviet Union and Canada have return times of almost 100 years

Table 9.4 *Expected contributions (%) of various techniques to increasing food production between 1988 and 2010 (from FAO, 1993a)*

	Extensification	Intensification Increased number of crops	Intensification Increased yield per crop
Africa (sub-Saharan)	30	17	53
S. Asia	7	13	80
E. Asia	27	9	64
Latin America	28	19	53
Near East/N. Africa	9	20	71
Developing countries (90)	21	13	66

compared to < 50 years in the US and Europe. While this reflects the age structure of the forests in these regions, inaccessibility is also a major factor.

Currently, forested areas per capita are comparable worldwide (Tables 9.2 and 9.3), but the reduction in forested area and the low rates of re-establishment are greater in tropical forests. If these trends continue, and the projected increases in population occur, then the forested area per capita will be markedly lower in tropical regions than in temperate regions.

9.2.2 Changes in land use and intensification of production

There are three major means whereby the projected increases in food and fibre supply in the less developed countries will have to be achieved: first, by expanding the area of cultivated land (extensification); second, intensifying the production system either by increasing the number of crops or cropping cycles sown on a particular area of land or by increasing the yield per unit area of individual crops or both (intensification); and finally, where other economic activities allow, by purchasing food from elsewhere. Globally, no one solution will be appropriate and different regions will cope with the increasing population by different means (Table 9.4). In South and Southeast Asia there is little new land that can be brought into cultivation so that the projected increases in consumption will need to be through increases in yield and intensification (e.g. rice in rotation with an upland crop such as wheat or a legume). In South America, new land is still available for cultivation and a substantial proportion of the projected increases in consumption might be met in this way although the costs of introducing infrastructure, such as roads, is high. In Africa, there is some potential for expansion of the cultivated area but yields will still need to increase substantially because of the paucity of areas with good soils. The principal conclusion of most such analyses is that crop yields per unit area will

Table 9.5 *Required and attainable rice production (t ha⁻¹) in selected countries of South and Southeast Asia (from Hossain, 1997)*

	Current 1992–93	Maximum attainable	Required 2020
Bangladesh	2.7	5.4	5.8
China	6.0	7.6	8.9
India	2.7	5.9	5.4
Myanmar	3.2	5.1	3.7

need to increase to meet demand. For this to occur, the inputs and efficiency of use of water and fertilizer to the existing cultivated land will need to increase.

The requirement for increased rice production in South and Southeast Asia illustrates the differences that will occur even within a single region. Hossain (1997) has calculated the mean yield attainable in different countries of the region allowing for different soil types, availability of water, and proportions of lowland and upland rice production. Some countries will only just achieve the yields required to sustain their consumption by 2020, others will clearly need to import, and others have the potential to export (Table 9.5). Such calculations assume increasing fertilization, maintenance of water supplies for irrigation, and no major outbreaks of disease. These calculations and those of Ramakrishnan *et al.* (1994) and Cao *et al.* (1995) indicate that even if full potential production is achieved, meeting the future food needs of the Indian subcontinent and parts of China will remain a major concern.

9.2.3   Requirements for inputs – nutrients, water and human resources

The increases in production required will inevitably mean that both inputs and offtake of nutrients will increase; about 20 kg N is removed in each tonne of cereal grain harvested and this must be replenished if yields are to be maintained. Additionally, increased urbanization will lead to removal of nutrients from rural areas and their concentration in and around cities. Between 1950 and 1989, fertilizer use increased from 14 million to 146 million tons and was a major factor (together with improved genotypes and irrigation) contributing to the three-fold increase of grain production in that same period.

All crops also require an adequate supply of water. Rainfed cropping is the most widespread system of production and, although more reliable yields can be obtained by irrigation, water for this purpose is becoming increasingly scarce. In some regions, notably the Indian sub-continent and western Asia, water shortage will adversely affect the ability to increase production (Falkenmark, 1997).

An additional, and increasingly serious, factor constraining the production of food will be the availability of labour. Already labour shortages are affecting

Table 9.6 *Actual and potential (in the absence of control measures) losses caused by pests, diseases and weeds to the world's harvests (from Oerke et al., 1994; H. Herren, pers. comm.)*

Crop	Actual losses (%)			Potential losses (%)		
	Pests	Diseases	Weeds	Pests	Diseases	Weeds
Maize	15	11	13	19	12	29
Rice	21	15	16	29	20	34
Wheat	9	12	12	11	17	24
Potatoes	16	16	9	26	24	23
Average	15	14	13	21	18	28
Cassava	13	12	10	50	50	70
Groundnut	13	12	10	30	50	75
Sorghum	13	12	10	30	50	80

production in some regions where, for example, the labour intensive systems of fodder collection, animal rearing and return of nutrients to cropped land are breaking down.

9.2.4   Production losses caused by pests, diseases and weeds

All production systems suffer from losses to pests, diseases and weeds. The losses to the actual harvests of the four major crops range from about 10 to 20% (Table 9.6; Oerke *et al.*, 1994) which, collectively, results in an approximate halving of the global yield attainable under ideal conditions (see Box 9.1). Values for losses to cassava, groundnuts and sorghum crops caused by the yield-reducing agents are more difficult to obtain, but a search of the literature presents a very similar picture. While stem-boring insects appear to cause severe damage to sorghum, mites and mealybugs affect cassava, and several fungal pathogens are particularly severe for groundnuts in India, which grows one-third of the global supply. The data also illustrate the overall importance of weeds as a potential yield reducing factor, while showing that in practice losses are shared almost equally across taxonomic groups.

While agrochemicals have played a major role in reducing losses (Table 9.7), they have a significant effect on production costs, and may adversely affect the environment and biodiversity. Table 9.7 also shows the heavy reliance on agrochemicals for maintaining production, emphasizing the vulnerability of major crops to pests, diseases and weeds. Because of the considerable breeding programmes devoted to wheat and maize, they appear to be less vulnerable than other crops. Reliance on chemicals has also led to the sequential selection of a

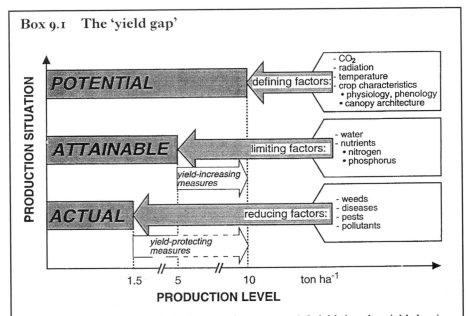

Box 9.1   The 'yield gap'

Many crop simulation models aim to estimate *potential* yield, i.e. the yield that is theoretically possible when there is no edaphic constraint to growth, and based on the fundamental physiological processes of photosynthesis, just limited by the climatic characteristics of radiation and temperature. This theoretical maximum target cannot be reached because environmental factors (e.g. suboptimal nutrients and water) limit productivity, for which farmers try to compensate with nutrient and water additions. Management thus sets the level of *attainable* yield (which can be very close to potential levels where inputs are very high). All crops growing in the field are also exposed to 'yield reducing' factors such as weeds, pests, disease and pollution, which further reduce yield to an *actual* level. Where yield reducing factors are uncontrolled, and hence become severe, actual yield can be a small fraction of potential yield. The difference between potential and actual yield is termed the 'yield gap'. (Figure from Rabbinge & van Ittersum, 1994).

range of different strains of resistant pests, diseases and weeds, leading to concern about the long-term sustainability of reliance on chemicals.

9.2.5   Sensitivity of production systems

Agricultural systems are clearly susceptible to short-term changes in weather (Monteith & Ingram, 1998), and to outbreaks of diseases and pests, often with catastrophic consequences for the local human population. In addition to these often unavoidable, localized problems, political and socio-economic factors interact to influence the level of production at all scales. At a global scale, this complex interaction has contributed to the recent decline in world grain stocks

Table 9.7 *Attainable world production of crops compared with actual yields and the estimated unprotected yields (i.e., if no crop protection measures were implemented) (all data in millions of tonnes) (from Oerke et al., 1994)*

Crop	Attainable	Actual	Unprotected
Maize	729	449	295
Rice	1047	509	184
Wheat	831	548	400
Potatoes	464	273	123
Cassava	623	157	21
Groundnuts	87	23	5
Sorghum	184	58	9

that has resulted from three major factors: first, a gradual decrease in cropped area from 732 million ha in 1981 to 669 million ha in 1995 due, in part, to urban growth, social unrest and government policies to remove land from production; second, little or no growth in water supplies for irrigation since 1990; and finally, a decrease in world fertilizer use since 1989. Together, these factors have contributed to a decline in grain production of 5% between 1990 (when the largest harvest of 1.78 billion tonnes was recorded) and 1995 (1.69 billion tonnes) (Worldwatch Institute, 1996).

9.3 Impact on production: current understanding

The solution to increasing food and fibre supplies lies in developments in both our social, economic and political systems, and in our better use of the resource base and technology (Ingram, 1997). There is often scope for improved management of the resource base but, where this is limited, technological development can make a substantial contribution by improving agroecosystem management to produce crops nearer to their maximum physiological potential, thereby reducing the 'yield gap' (Box 9.1). In many parts of the world, and particularly in those where populations are rising most rapidly, the yield gap is often large; losses to pests, diseases and weeds remain a major problem, and insufficient supplies of water and nutrients are increasingly common.

The general concerns about food security discussed in Section 9.1 are likely to be exacerbated still further by changes in atmospheric composition and associated changes in climate. These additional changes, when added to the projected changes in land use, will affect already stressed and rapidly changing systems in ways we cannot yet predict with any accuracy. GCTE has organized an integrated research effort to determine the relationships between potential yield and yield-reducing factors, and how global change will affect them.

Table 9.8 *Maximum and minimum results from eight models included in the GCTE Wheat Network (see Chapter 3). Model simulations for the growth of hypothetical wheat crops were prepared using common weather datasets (one for Crookston, USA and one for Lelystad, The Netherlands) and forced to use the same time course of leaf area index (LAI) development, as generated by the Cropsim(93) model (from Goudriaan et al., 1994)*

	Date of anthesis (DOY)	Maturity (DOY)	Maximum LAI (m^2/m^2)	Total above-ground dry wt (t/ha)	Grain dry wt (t/ha)
Crookston, Minnesota, USA (spring wheat)					
Prescribed	183	216			
Minimum	182	215	4.5	10.6	3.6
Maximum	187	216	4.5	16.1	8.8
Lelystad, The Netherlands (winter wheat)					
Prescribed	166	207			
Minimum	164	204	7.5	13.5	5.5
Maximum	166	207	7.5	26.4	12.1

9.3.1   Wheat

A large amount of experimental work has examined the impact of environmental variables on the performance of wheat. This work has been reviewed at various times (see Austin & Jones, 1975; Evans *et al.*, 1975; Gate, 1995), and the results used in the construction and calibration of models of wheat growth (van Keulen & Seligman, 1987; Hunt & Pararajasingham, 1995).

Given the large amount of experimentation on wheat, it might be expected that all wheat models would perform in the same general fashion, but perhaps differ in aspects related to the characteristics of specific cultivars and the details of the agronomic practices at specific locations. A series of model intercomparisons has been conducted to evaluate this expectation, and to develop confidence that the outputs are not 'model dependent'. In one comparison (Goudriaan *et al.*, 1994), the performance of models was markedly different even when differences in leaf canopy development were eliminated by using a standard time-course of leaf area as a model input (Table 9.8). Such findings have stimulated work to compare both the performance of different models and to analyse the underlying reasons for differences.

Initial comparative work (J. Goudriaan, unpublished) showed that models did not respond similarly to various environmental factors. For example, the response of daily photosynthetic activity to air temperature was significantly different among three widely used models (Fig. 9.2). In one, the response was

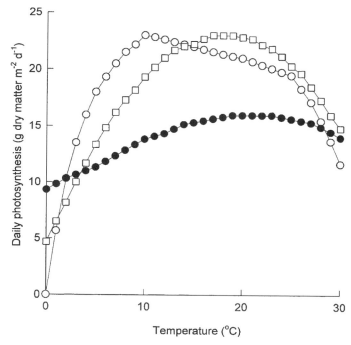

Figure 9.2   The response of daily photosynthetic activity for three models (Sucros, [○] AFRC [●] and CERES-Wheat [□]) plotted against air temperature (from J. Goudriaan, unpublished).

'peaked', whereas in another it was almost flat. Such a contrast possibly reflects differences between cultivars, but it is unlikely that cultivars differ so markedly. It is more likely that the models were developed for application over a limited range of temperatures with little attention paid to their performance outside the range of concern.

Figure 9.2 also indicates that models differ in the values used or simulated for the absolute rate of assimilation at optimum temperature. The magnitude of the differences between models initially suggested gross dissimilarities in crop performance, but because that did not appear to be the case (Fig. 9.3), the more likely explanation was differences in the basis of expression of assimilation rate with, for example, some models including leaf sheaths and others excluding them. This finding emphasizes a common problem when comparing both experimental data and model outputs, namely that of term definition; terms may differ in subtle but significant ways.

Further analyses (e.g. Porter *et al.*, 1995; Jamieson *et al.*, 1998) have indicated that the existing models differ not only in some of the basic parameter values and response surfaces, but in some fundamental aspects of construction. For example, Jamieson *et al.* (1998) showed that for a number of models, a reduction

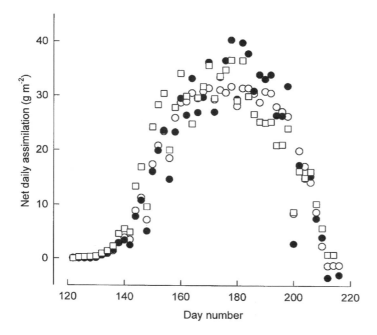

Figure 9.3   Net daily assimilate for three models (Sucros [○], AFRC [●] and CERES-Wheat [□]) for weather data from Cookston, USA (from Jamieson *et al.*, 1998*a*).

in photosynthetic efficiency as moisture stress increased was important in determining the rate of biomass accumulation, whereas experimental data indicated that this factor was a minor contributor to the variation in biomass. Such a finding may have resulted from errors in parameterization but may also reflect a problem in the concepts underlying the models.

Given the reported differences in model performance when using standard input datasets, it might be expected that the current models would also give different predictions when run using changed weather scenarios. This was examined by Toure *et al.* (1995) using four simulation models and weather scenarios for a site in western Canada. For current CO_2 levels and one future climate scenario for temperature and rainfall, two models predicted increases (25% and 3%) in yield for non-irrigated wheat, whereas two other models predicted decreases (36% and 15%) in yield. The authors concluded that 'the predicted impact of climate on spring wheat depends on the model used', an unsatisfactory conclusion for those wishing to use current models for climate change studies.

Besides modelling, recent experimental work has sought to clarify plant responses to several environmental variables. Much of this has concentrated on responses to CO_2 (e.g. Grasshof *et al.*, 1995; Kimball *et al.*, 1995; Wheeler *et al.*,

1996) but there is not yet any clear consensus as to the magnitude of the potential impact of a change in CO_2 concentration on yield. Data from the Arizona Free-Air CO_2 Enrichment (FACE) experiment (arguably the most 'realistic' field-scale experiment yet undertaken) showed that 550 ppm CO_2 increased grain yields by 8 and 12% under well-watered conditions (mean yield 8 t ha^{-1} with ambient CO_2) and by 21 and 25% (mean yield 5.3 t ha^{-1} with ambient CO_2) with crops under moderately severe drought stress (Pinter et al., 1996). More recent studies with improved experimental facilities and with optimal nitrogen and irrigation gave yield increases of 15 and 16% for two growing seasons (B. Kimball and P. Pinter, personal communication). If these latter results are linearly extrapolated to the possible effect of a doubling of CO_2 then, under ideal conditions, yields would be 28% greater; this is comparable with previous chamber-derived data and with the '+30% for C_3 crops' stated by IPCC (1996). However, field crops are frequently subject to nutrient limitations and, under low-N conditions, the FACE experiment only increased yields by 5 and 12% (9 and 22% if linearly extrapolated to doubled CO_2). So while there is still uncertainty about the overall effect on yield, these experiments have emphasized the significance of the interactions between CO_2 and temperature, and the difference in response recorded when crops are grown under different water and temperature regimes.

Other studies on temperature, moisture, and photoperiod, in some cases using different planting dates to achieve contrasting sets of environmental conditions (e.g. L.A. Hunt et al., 1996), have also been reported. These studies have highlighted, first, the differences between soil and air temperature and the need to use whichever is more appropriate in determining tissue temperature (Jamieson et al., 1995); second, the interaction between photoperiod and temperature in determining vegetative growth and reproductive development (e.g. Pararajaysingham & Hunt, 1996); and finally, the subtle differences between genotypes in their response to a change in an environmental factor (e.g. Brooking et al., 1995).

These recent experimental studies, together with the analyses of model performance, give cause for optimism that it will soon be possible to simulate with a reasonable degree of confidence the response of wheat crops to climate change variables. A few cycles of model evaluation followed by analysis of model performance, improvement, and re-examination should lead to the production of models that can be used with reasonable confidence. The availability of datasets that allow for the examination of wheat models both in the major current wheat growing environments, and under combinations of weather variables that would be found under climate change scenarios, will be essential to such a task. Further, it should be noted that the array of cultivars used in recent experimental studies does not appear to be representative of those used in all wheat growing regions of the world, with significant deficiencies for the major wheat producing countries and regions of Russia and the Ukraine, China, India, and the prairies of the USA and Canada.

The deficiencies in the crop models alluded to above along with the shortage of data for many wheat growing regions and cultivars will have to be corrected before predictions can be made with confidence. However, current experimental evidence agrees with suggestions that double pre-industrial levels of CO_2 will be unlikely to increase grain yields by more than about 10% under ideal conditions, therefore probably by only 5–7% under conditions of normal management. The effect on grain quality is uncertain. The increases in production required by an expanding population will thus have to be sought in other ways. This will present challenges because:

1. Wheat is essentially an extensively grown crop. The scope for increasing yields still further in the relatively small areas under intensive management is limited because most of them are already close to potential yield. Those areas where wheat is grown on an extensive basis are generally drier, and are also those likely to be most severely affected by a change in climate.
2. The possibilities for expansion of wheat production into new areas is limited, although there appears to be potential for increasing wheat production in South America and the Ukraine by increasing the area of production, and new genotypes may allow new areas of wheat production in sub-Saharan Africa. In other regions, wheat production can be increased only through intensification.
3. Only a small percentage of the total wheat production is currently traded so that any solution to the problem of grain shortage must essentially be regional or involve a considerable increase in trade.

9.3.2   Rice

Results from the vast amount of research on rice, at the various National Research Institutes in Asia and at the International Rice Research Institute (IRRI), can be found in works such as Yoshida (1981) and in IRRI reports including those of the SARP network (Simulation and Systems Analysis for Rice Production; IRRI, 1995). The Rice Network of GCTE has profited considerably from this work and the SARP network but, being worldwide, it has not been limited to Southeast Asia.

An initial comparison of rice models showed that their predictions for potential production were quite close to the observed values (Peng *et al.*, 1995); the models CERES, ORYZA, SIMRIW and TRYM had an average deviation of 0.8 t ha^{-1} for an actual panicle dry weight of 5.7 and 9.8 t ha^{-1} for the wet season of 1991 and for the dry season of 1992, respectively. A subsequent analysis by the GCTE Rice Network compared the sensitivity of models to temperature and CO_2 (Mitchell, 1996). Results showed that there was a range of predicted decreases in yield of 1.5% to 4.5% per °C increase, with the relative impact being slightly larger in the wet than in the dry season. In another study the effect of temperature was larger: 8% decrease in potential yield per °C (Matthews *et al.*, 1995). For CO_2, the prediction was that the effect of doubling CO_2 from 350 to 700 ppm would increase the yield over a range from 10 to 40%. This range of sensitivities is considerable, but not larger than those found in experiments (Peng *et al.*, 1995).

The effects of temperature are various. Whereas respiration rate increases with temperature, the rate of crop development increases to an optimum temperature of about 30 °C, above which it then declines. Photosynthesis behaves similarly to the rate of development and has an optimum. An important phenomenon is the occurrence of spikelet sterility, outside the range 20–32 °C. As temperature increases from 32 to 40 °C, spikelet fertility steadily decreases to almost zero. Increasing CO_2 does not affect this relationship (Horie, 1993). The effects of temperature are summarised in Box 9.2.

A sensitivity analysis of the models showed that the effect of increasing temperature on a crop grown at IRRI was to be generally negative (Fig. 9.4). However, the predicted effects of temperature increases on yield vary from country to country. For Japan, the effects are generally slightly positive for the northern parts of the country (Horie *et al.*, 1995) but slightly negative in the southern parts. For India, there is a clear disparity between the main (first) growing season in which there is a small negative effect, more than compensated for by the positive CO_2 effect, and the second wet season, in which the temperature effect is clearly strongly negative and cannot be compensated for by the CO_2 effect (Mohandass *et al.*, 1995). Another result was that the increase in temperature tended to have more negative effects in the non-tropical countries, such as Korea and southern China, than in the tropical countries. This is because the temperature during the main growing season is higher in the northern summer than in the tropical, dry monsoon period (Matthews *et al.*, 1995). Because of this, a shift of production towards the dry season may be expected in the tropical monsoon countries. Generally, experimental and modelling efforts indicate that the direct CO_2 fertilization effect will be some-what greater than that estimated for wheat, because rice is generally better supplied with water and nutrients.

Rice cultivation in Southeast Asia is mostly a high input agricultural system, resulting in yields approaching their potential levels to a considerable extent. Clearly, water is amply supplied in lowland rice cultivation (this is in sharp contrast to most wheat production) and the nutrient supply to rice is often greater than that given to wheat. As a result, the remaining yield gap between actual and potential yields is not large in many places and further increases can

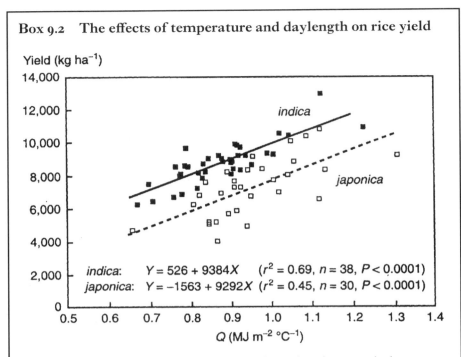

Box 9.2   The effects of temperature and daylength on rice yield

A summarizing characterization of the photo-thermal environment is the concept of the photothermal quotient, which is defined by the total incident global radiation divided by the mean temperature excess above the base temperature (taken as 8 °C), both taken over the period from panicle initiation (PI) to flowering. Because yield is linearly related to the quotient, this ratio can be used to give a quick estimate of the potential yield. (Figure from Matthews *et al.*, 1995).

come only through increased intensification or increases in rainfed rice production. The high actual production will make the yields more sensitive to climatic fluctuations and to pests, weeds and diseases. The traditional technique of transplanting rice seedlings is quite effective in suppressing competition by weeds, but it is labour-intensive and becoming more and more expensive. Direct seeding will increase the risk of yield loss due to weed competition.

Until now, water consumption by evapotranspiration has rarely been an issue, but a shift towards cultivation in the dry season, together with increased intensification, may require this to be reconsidered. Improvement of the water-use efficiency for rice cultivation will be needed as water for irrigation becomes increasingly scarce. Efforts to increase overall production over the next 20–30 years, as with wheat, will be challenging for many reasons because:

1. Rice is near the attainable yield levels in China and parts of India, and shortages of land elsewhere (especially in South and Southeast Asia) will

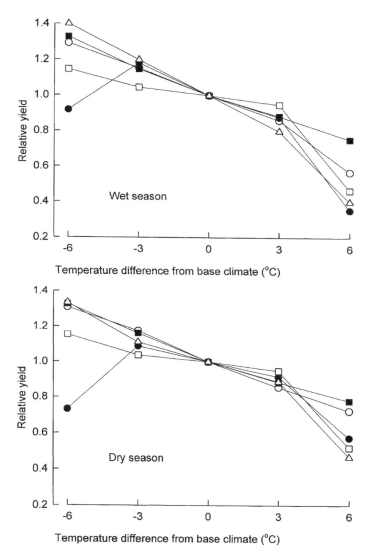

Figure 9.4 The temperature response of five models (Seres-Rice [○], AFRC [●], SIMRIW [□], Trym [■] and VSM [△]), shown as the yield relative to that of the base climate, for wet season (above) and dry season (below) crops grown at IRRI, the Philippines. The model VSM was not run with the base climate alone so the average of yields at −3 and +3 °C was not used for the reference value (Mitchell, 1996).

mean there is limited scope to increase production to that needed to support the projected population by *c.* 2020. There is some potential for expanding the irrigated rice area in the Amazon basin but there are substantial problems with acid soils.

2. There is limited scope for increased international trade as 95% of rice is already consumed by domestic markets.
3. Many rice-growing regions have benefited substantially from the 'green revolution' brought about by genetic and agronomic improvements. There is still scope for further improvements by breeding, but further substantial increases in the irrigated area are not possible because of limitations to water supply.

9.3.3   Pastures and rangelands

A series of papers from the GCTE Pastures and Rangeland workshops (Campbell & Stafford Smith, 1993; Stafford Smith *et al.*,1994, 1995; Odada *et al.*, 1996; Campbell *et al.*, 1997) has reviewed current knowledge about grasslands. To date, global change research on pastures and rangelands has emphasized changes in primary productivity, although it is the resultant effects on livestock that affect food production. The need to handle secondary production, and a focus on coping with climatic variability, sets grazing lands aside from crops and forests. However, within the grazing lands there is a broad continuum of decreasing intensity of management and inputs from intensive, improved pastures through to extensive, native rangeland pastures (Fig. 9.5). Primary production (pasture growth rate) dominates as a topic for global change research in intensive pastures whilst pasture composition, animal responses and social aspects of management become increasingly significant in rangelands. In intensively managed pasture systems, economic considerations are the primary drivers of management decisions, and there is a clear history of adaptive management (such as pasture renewal, agricultural technologies and irrigation) where it is financially worthwhile. This adaptive capacity is important in modifying the potential impact of climate change on pastures (McKeon *et al.*, 1993; Gifford *et al.*, 1996). In contrast, social drivers are especially important in the rangelands and, because of the smaller capacity for adapting management to climatic variability and change, the systems are not well buffered against such changes. The key issues in global change research for pastures and rangelands reflect these differing priorities.

Despite the importance of climatic variability and its interactions with management, the primary emphasis in global change research has to date been on the effects of elevated CO_2. The general conclusion for improved pastures is that, even after acclimation, there will be a CO_2 fertilization effect on potential primary productivity somewhere in the range 0–20% at twice current CO_2 levels (Campbell *et al.*, 1996), but the magnitude of the yield increase will be considerably reduced by water and nutrient limitations where these occur (see Chapter 7). An important recent finding is that responses in a pasture system were nonlinear with respect to CO_2 concentration, with some changes in

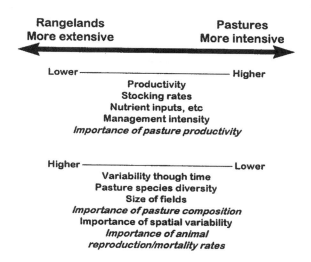

Figure 9.5 Some key axes of variation within the variety of pasture and rangeland agroecosystems that determine research priorities in different systems.

composition and productivity occurring below twice current CO_2 levels (Newton *et al.*, 1995; Campbell *et al.*, 1997). Additionally there will be an increase in leaf $C:N$ ratios (Owensby, 1993; Campbell *et al.*, 1996), with implications for forage quality and animal diets.

The effects of elevated CO_2 will interact with climatic variability; an important effect will be to improve water-use efficiency in dry areas so that, for a given level of climatic variability, interannual fluctuations in productivity are reduced. By simulating one particular sub-tropical grazing system, Campbell *et al.* (1997) found that a 10% increase in pasture transpiration and radiation-use efficiencies resulted in about a 5% increase in mean annual pasture growth and a 3% increase in mean annual liveweight gains in cattle (ignoring any changes in diet quality other than those caused by nitrogen dilution). Most of the increases came about through reducing the variability of productivity between years. Another simulation exercise (G.M. McKeon, unpublished) has shown that a reduction in the restrictions on growth caused by limited soil water could have as much as twice the impact again on long-term mean annual pasture growth, largely achieved through reductions in the variability of pasture production between years.

In general, vegetation composition in higher production, mesic pastures is controlled more by relative growth rates, while in drier rangelands it is affected more by grazing, fire and responses to climate. The evidence on the effects of atmospheric composition on competitive responses between grass species remains equivocal (see Chapter 7).

There is considerable concern that the encroachment of woody plants may increase under global change. Despite suggestions that shrubs in the US may have increased with historical increases in CO_2 (Polley *et al.*, 1994), it is also possible to explain the observed changes as outcomes of changes in grazing and fire regimes (Archer *et al.*,1995). A pot experiment with intact soil cores (J. Brown, unpublished) to assess the effects of elevated CO_2 on the establishment of three shrubs with and without grass competition found that the presence of grass greatly slowed down establishment while the CO_2 treatment had almost no effect on germination, establishment and growth of the three woody species. Thus tree/grass balances are most likely to be determined by changes in the traditional concerns of rangelands ecology (the effects of climate variability and management – especially fire and grazing – on composition, and their interaction with climate change) and it would be inappropriate to further expand research effort on elevated CO_2 studies conducted in isolation from management factors. In this regard, a workshop in Canberra (Stafford Smith *et al.*, 1994) pointed out that increased climatic extremes will increase water redistribution and thus promote perennial woody plants in run-on areas. This hydrologically driven effect is likely to exceed any changes in competitive effects in Australian rangelands and other flat landscapes.

In summary, it is likely that different systems will change composition at different rates. These changes will be driven much more by changes in climate and management than by responses to elevated CO_2. There is a considerable body of work on this issue in the grasslands and rangelands literature, although good models of long-term vegetation change are still uncommon, and implications for grazing livestock are still largely unknown.

The responses of livestock to global change arise both from direct impacts of climate or atmospheric composition, and indirectly through effects on their food supplies and their pests and diseases (see Box 9.3). The direct responses of sheep and cattle reproduction and mortality to climatic effects are reasonably well known (e.g. sheep mortality in relation to exposure – Donnelly, 1984) and global change science raises no special issues here other than creating the need to incorporate the effects of changes in climate and climatic extremes in models of regional productivity. There are no known direct effects of predicted atmospheric composition on animal productivity.

Indirect effects, however, are very likely. For instance, there will be changes in diet quality resulting from the effect of elevated CO_2 on increasing the C:N ratio of most living plant tissues. This may lead to reduced digestibility, but this effect may not be as pronounced in high input pastures because nitrogen fertilizers are applied. In the improved grasslands of northern Europe, Jones *et al.* (1996) found that the increase in C:N in perennial ryegrass had no effect on digestibility, but decreases in nitrogen concentration have been found by Owensby *et al.* (1993, 1994) in tall grass prairies, by Read and Morgan (1996) for a C_3 but not a C_4 grass in short grass prairies, and by Conroy (1992) in an

Australian C_4 grass, among other studies. In short, the implications of increases in CO_2 concentration for diet quality probably depend on the underlying productivity of the system and the nutrient and carbohydrate balances within the forage. Energy-limited livestock grazing productive, protein-rich pastures may be less detrimentally affected by increased C : N ratios than those livestock grazing low protein-containing rangeland forage; intermediate pasture types require further research. Unlike other impacts of CO_2, however, changes in digestibility will be pervasive and additional to any production and compositional changes.

Land use options in rangelands are strongly driven by social issues, especially in less developed nations where a large part of primary production in rangelands is used for subsistence farming. Significant advances have been made in conceptualizing the links between the biophysical system and social institutions in ways that can help to determine how global change may impinge on the interactions. A GCTE-START workshop in southern Africa (Odada *et al.*, 1996) raised the question of whether increasing temporal or spatial variability in resources must be matched by increasing levels of institutional flexibility in time or space in order to meet a given level of ecosystem degradation or productivity. For example, as the spatial scale of resource variability increases, tenure systems may need to incorporate larger areas or may need to be managed at a larger spatial scale; with increasing fluctuations over time, political institutions may need to respond with increasing local flexibility or with increasing resource buffers. Since the major effects of global change in developing country rangelands may be to increase climatic variability and land use fragmentation (e.g. southern Africa – Hulme, 1996; China – Zuozhong, 1996), these conceptual advances in linking human and biophysical dimensions of global change will be vital in achieving sustainable production in these regions.

Because the rangelands are typically located in dry regions, often with xeric soils, they are vulnerable to further degradation and to desertification if overused by people and their livestock. Sustainability collapses when the system is launched into a positive feedback loop with negative impacts. These may occur in both the biophysical and social systems. The consequences of global change in the rangeland areas will differ between regions due to differences in both biophysical and socio-economic attributes; Table 9.9 contrasts, for example, West Africa and Northeast Australia (see also Chapter 12 for a comparison of the IGBP transects in these regions).

The major effects of global change on rangelands over the next 20–30 years are summarized as follows:

1. Any effects of climate or atmospheric change on pasture production in developing countries (increasing CO_2 is estimated to increase pasture

Table 9.9 *The future of rangelands: a comparative example for West Africa and N.E. Australia (numbers in square brackets indicate confidence of prediction: 1 = high, 3 = low.)*

Issue	West Africa	NE Australia
Atmospheric [CO$_2$]	Expect slight net growth enhancement [2] and reduction in forage quality [1]	
Regional Drivers (at 2020)		
Rural population	Net increase despite urbanization [1]	Decreasing [1]
Regional climatic change	Expect slight drying with higher temperatures [2]	Expect <0.5 °C warming, minor changes in rainfall Possible increase in extreme dry events (El Niño) [3]
Linkage to markets	Low but increasing [2]	High [1]
Impacts		
Crop/rangeland boundary	Crops encroach [1]	No significant change/crops recede [2]
Grazing pressure on non-croplands	Increasing [1]	Some decrease in stocking pressures [2] Problems with non-domestic herbivores [1]
Degradation risks		
Soil and water	Increasing soil crusting and rill erosion [2] Increasing local run-off [2] SOM depletion [1] More water points [2]	Increasing risk of soil loss with extreme events [2] Improving condition with LandCare [2] Closure of water points [3]
Vegetation	Decreased soil cover [1] Decreased fallow period [1] Increased fragmentation [1]	Improvement in pasture productivity [2], but invasion by exotic shrubs [1]
Socio-economic factors	Increasing conflicts over land tenure and access (pastoralists/farmers) [1]	Depopulation of rural centres [2] Declining profits for pastoralists [1] Consolidation into large enterprises [2] Use for carbon sequestration [2]

production by less than 7% in this period) will be masked by land-use pressures and short-term climatic variability. This is probably also true for more developed regions of the world, but for the reverse reason; improvements resulting from less intensive land use may exceed the effects due to CO$_2$.

2. The crop/rangeland boundary will encroach on grazing lands in developing countries due primarily to population pressure, with no change to the minor retreat elsewhere.

3. Nutrient 'mining' of lower-fertility soils is likely where cropping continues to expand; there will be no change to the small increase in nutrients in, for example, Australian grazing areas.

4. Encroachment of woody plants into rangelands will increase carbon sequestration, whereas rangelands being converted to cropping land will experience net carbon loss.

9.3.4 Managed forests

A review of short-term studies on single leaves or seedlings of woody plants shows that photosynthetic rate often increases with elevated CO_2 (Ceulemans & Mosseau, 1994). Therefore a common conclusion is that biomass production in forest ecosystems will increase as CO_2 levels rise, and so will the carbon pools. However, up to 60% of the carbon assimilated by photosynthesis in a forest is returned to the atmosphere by plant respiration (see Ryan, 1991). This means that a shift in the balance between photosynthesis and respiration can have a major impact on the net carbon balance. It is predicted that at high latitudes the increase in mean annual temperature will be 4 to 6 °C. This will affect rates of photosynthesis and respiration, but particularly respiration, which could increase by 30–50%. Such an increase would, at least in the short term, counteract the stimulation of carbon assimilation by CO_2. However, both photosynthesis and respiration can acclimate, to some extent, to prevailing climatic conditions so that long-term predictions based on short-term experiments are very uncertain. However, a probable effect is a net loss of carbon from the soil, caused by increased rates of decomposition. Depending on the time-span considered, forest ecosystems may act either as a sink or a source of carbon (see Cannell, 1995). Any increased rate of nitrogen mineralization due to higher temperatures, and, in particular, atmospheric deposition of nitrogen (Thornley & Cannell, 1996) will have a stimulating effect on tree growth on nutrient poor sites.

GCTE has analysed a range of forest experiments using process-based simulation models (e.g. McMurtrie *et al.*, 1990*a,b*; 1994; Ryan *et al.*, 1996*a*; see Chapters 6 and 7). Responses to water stress, temperature and nutrition have been characterized for pines in Australia, Florida, New Zealand, Sweden and Wisconsin (McMurtrie *et al.*, 1994) and for eucalypt plantations in the southwest of Western Australia, a region where General Circulation Model (GCM) simulations indicate a high probability of reduced rainfall as a consequence of climate change (Landsberg & Hingston, 1996). Results revealed similarities in the biology among different species of pine and highlighted the pronounced environmental constraints on their seasonal performance. It is evident, however, that there will be limited progress in predicting tree growth in detail until consistent datasets are available to further develop and validate the models.

It is clear from several long-term forest experiments on a range of species growing in contrasting environments that current rates of biomass production in most forest ecosystems are far below the potential production and that manipulation of nutrient and/or water availability can result in large increases in yield (Linder, 1987; Linder *et al.*, 1996). In the temperate and boreal zones, nutrient availability is the main limiting factor (Fig. 9.6), but in large parts of the Mediterranean and southern hemisphere regions forest growth is primarily limited by water. An experiment carried out on a sandy soil with *Pinus radiata* in the Australian Capital Territory provided clear evidence of the interaction of water and nutrients in terms of their effects on general growth patterns (Linder *et al.*, 1987; Raison *et al.*, 1992). Similar results were obtained in a plantation of *Eucalyptus globulus* in Portugal (Pereira *et al.*, 1989, 1994). This means that it is possible, by improving nutrient and water availability, to increase the production of fibre and wood products by the introduction of more intensive silvicultural methods. These results have considerable bearing on the rapidly growing problem of the shortage of wood/fibre for fuel and local construction in arid and semi-arid tropical and subtropical regions; alleviation can be found, in principle, by the establishment of plantations.

9.3.5   Yield-reducing factors

The consequences of elevated CO_2 and the corresponding changes in climate on communities of crop-related organisms are hard to predict; ecosystem responses to such changes are filtered through complex life systems, involving interactions with host plants, competitors and natural enemies and involving several feedbacks and compensatory links between community members. Furthermore, invertebrate pests and their host plants, and animals and their natural enemies, are affected to different degrees by changes in temperature, rainfall, atmospheric CO_2, and wind speed and direction. Plant pathogens are very responsive to climate, and their distributions and abundances are therefore expected to alter significantly under climate change, as are the distributions and abundances of weeds in croplands and rangelands. The generation of specific predictions requires the linkage of crop growth models and insect or pathogen population models, or crop-weed competition models run under the changed climate scenarios of temperature, rainfall and CO_2.

Despite the obvious complexity of these systems, several researchers have reported progress in the development of modelling approaches to assess likely impacts of global change on pests, diseases and weeds. Models such as CLIMEX (Sutherst *et al.*, 1995) are providing insights into the likely nature and

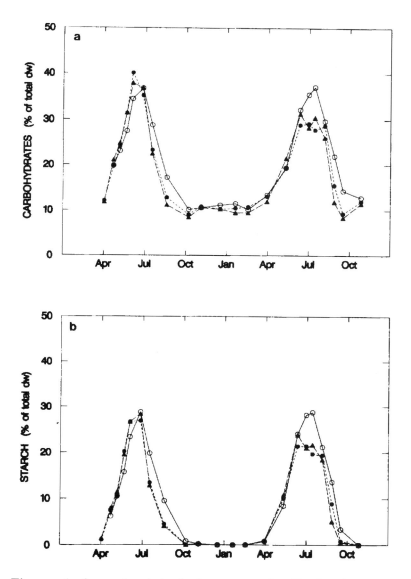

Figure 9.6   Seasonal variation in the concentration (% of total dry weight) of total carbohydrates (a) and starch (b), during two years, in one-year-old needles of Norway spruce trees subjected to different treatments. Note the faster depletion of starch during summer in fertilized trees, indicating that nutrients rather than carbon limit production in this boreal forest. Symbols: Control, open circle; fertilized, filled circle; irrigated–fertilized, filled triangle (from Linder, 1995).

extent of regional impacts. A generic, hierarchical approach to the assessment of impacts of climate change on pests, diseases and weeds and the design of adaptation strategies was described by Sutherst *et al.* (1996). The approach addresses the need for different types of information ranging from rule-based expert opinion, through climate-matching models to linked simulation-GIS results, to answer different policy-related questions.

Teng *et al.* (1996) reviewed progress on a range of modelling approaches with emphasis on linked crop-pest models. Luo *et al.* (1995) used sensitivity analysis of the BLASTSIM model linked to a geographical database of crop growing areas to describe the life cycle of rice leaf blast (*Pyricularia grisea*) under different temperatures and rainfall. Higher temperatures increased the risk of epidemics in cool subtropical zones of Asia, while inhibiting them in Thailand and the Philippines in the warm humid subtropics. Increased rainfall had a negligible effect because the BLASTSIM model simulated a weak relationship between rainfall and dew period.

Sutherst (1996) summarized three levels of impact assessment in Australia. In the first, a rule-based approach was used to relate expert opinion on the climatic preferences of the red-legged earth mite (*Halotydeus destructor*) to a geographical database, in order to estimate the likely change in the distribution under a climate change scenario. The analysis showed that hot wet scenarios led to a substantial reduction in the range of this temperate pasture pest.

In a second approach, the CLIMEX model was used to estimate risks posed by the Queensland fruit fly (*Bactrocera tryoni*) to the horticultural industry in temperate areas. The analysis forewarned of the likely future failure of a proposed sterile insect release approach in an area that is currently marginal for the fly. The projected substantial increase in favourability of the area, and hence excessive population density, could render such an approach ineffective or prohibitively expensive. The example highlighted the value of carrying out such analyses using future scenarios because it had immediate relevance to a major infrastructure decision.

In a third, integrated assessment (see Box 9.3), a population model of the cattle tick (*Boophilus microplus*) was combined with an economic model of the Australian beef industry to estimate the national socio-economic impacts of climate change.

While most of the GCTE effort on the impacts of global change on pests, diseases and weeds has involved the development or use of models (e.g. an assessment of the likely effects of pests, diseases and weeds on agriculture in Australia as a case study – Sutherst, 1996), there has also been significant progress in experimental work.

Several studies with plants (Watt *et al.*, 1995; Roth and Lindroth, 1995) have demonstrated that, in general, insect herbivores feeding on leaves are disadvantaged by high CO_2 levels, when other environmental variables are constant. Insects tend to grow and mature more slowly, consume more plant material and suffer greater mortality at higher concentrations. A decrease in foliar nitrogen is correlated with reduced insect growth but little is known about the effects of elevated CO_2 on plant defensive chemicals. Very little research has been carried out on other insect groups that feed on sap or stems, where plant chemistry may differ from that in leaves. There is also a need to examine the likely interaction of elevated CO_2 with other environmental variables such as moisture and temperature. Harrington and Stork (1995) reported changes in the phenology of aphids in the UK, with the most important responses being earlier emergence in spring with increased reproductive rates of some species in a warmer climate. The variability in responses from different taxonomic groups and the limited range of conditions examined to date indicate that there is much to be done to clarify the likely impacts of pests under climate change.

The impact of global change on plant diseases is similarly complex. Patterns of disease distribution may change and diseases currently rated as of minor importance may become major constraints to crop yields. Impacts on diseases of perennial plants and those caused by soil-borne pathogens may be more serious as chronic stress over several years may lead to progressive deterioration in plant

---

Box 9.3   The effects of cattle tick (*Boophilus microplus*) on yields of beef cattle in Australia.

The cattle tick (*Boophilus microplus*) is a major pest of beef and dairy cattle in tropical and subtropical Australia. It causes losses in productivity through feeding in large numbers and by transmitting malaria-like pathogens, which cause high rates of mortality in non-immune cattle. Tick numbers are highest in warm wet areas so there is concern that global warming will increase the numbers of ticks and allow them to colonize areas that are currently climatically unsuitable. The southern limit of the ticks is currently maintained by a quarantine and eradication program at the New South Wales–Queensland border at a cost of several million dollars per year. In the analysis, potential costs were estimated on the basis that the quarantine line would not be economically sustainable due to the recurrent development of pesticide resistant strains of ticks and the increasing difficulty of eradication under more favourable climatic conditions.

A comparison of the maps overleaf illustrates the estimated value of potential production losses under current climate and a scenario with + 1.5 °C and + 10% rainfall. The number of ticks was estimated to increase in the southern part of their current range where they are limited by the shortness of the warm season, which allows reproduction, and by severe winter mortality. Both these constraints were reduced by global warming and resulted in most of the estimated increases in costs being incurred within the current tick infested area. There was a parallel increase in the potential area affected by ticks in New South Wales, which is currently tick-free. The final estimates of potential increased costs, in terms of nett present value, varied greatly with the scenarios, response strategies and discount rates, but estimates derived under warmer and wetter scenarios indicated substantial extra costs to the Australian cattle industry.

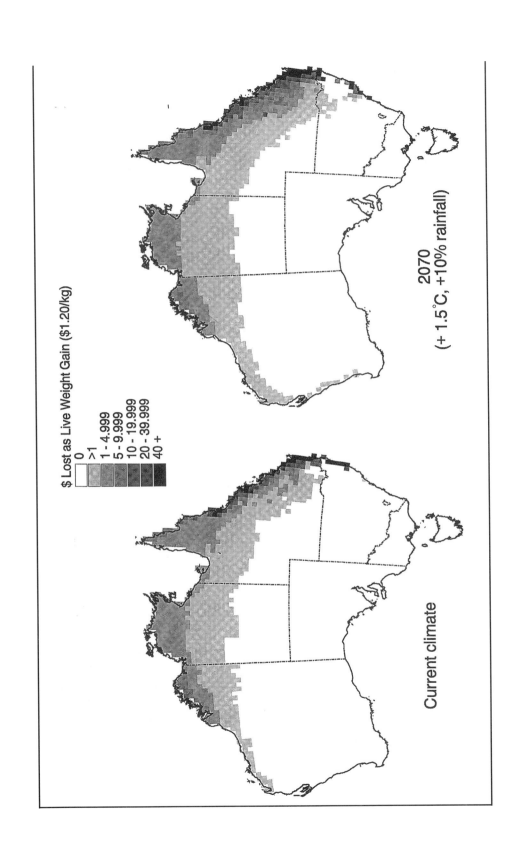

$ Lost as Live Weight Gain ($1.20/kg)

0
>1
1 - 4.999
5 - 9.999
10 - 19.999
20 - 39.999
40 +

2070
(+ 1.5°C, +10% rainfall)

Current climate

health. For example, drought-stressed trees can suffer serious damage from fungi which otherwise exist as latent invaders or endophytes (Lonsdale & Gibbs, 1994). Plant pathogens respond quickly to changes in weather conditions and are resilient under extreme climatic conditions, and climate change may increase their reproduction rate to rapidly overcome host resistance.

The long-term effects of climatic change on diseases have received some attention (Coakley, 1988; Frankland *et al.*, 1994; Sutherst *et al.*, 1996) and the effects of atmospheric pollutants, CO_2 and increased solar ultraviolet-B radiation under climate change have been considered more recently (Coakley, 1995; Manning & Tiedemann, 1995; see Chapter 7). Studies on CO_2–disease relationships within the likely range of atmospheric CO_2 concentrations are, however, rare as many studies in the past have used much higher levels. Variable effects have been observed on both below- and above ground disease. Aerial pathogens are generally more sensitive to high CO_2 concentrations than soil-borne fungi, which are well adapted to very high concentrations of CO_2. Studies of the possible effects of enhanced CO_2 concentration on the efficacy of disease management tactics have only recently begun (Atkinson, 1993).

Effects of elevated CO_2 on diseases result from interactions between hosts, pathogens and the environment. Variable effects have been observed on both above- and belowground diseases. Changes in the physiology and morphology of the host plants are likely to determine the impacts of elevated CO_2 on plant diseases. Incubation periods are reduced, and fecundity is increased in some pathogens under twice-ambient CO_2 (Chakraborty, 1997). Enhanced carbohydrate contents at high CO_2 concentrations encourage the development of some rusts (Manning & Tiedemann, 1995), but inhibit other pathogens such as mildews (Hibbard *et al.*, 1994). Any increase in canopy cover resulting from the fertilization effect of high CO_2 may alter the microclimate and so affect the survival of some pathogens between cropping cycles as well as the multiplication of pathogens during cropping. These studies suggest that there will be little direct effect of elevated CO_2 on pathogens and that impacts of climate change are more likely to occur from altered host plant susceptibility.

Global change will have major effects on the success of weeds in agroecosytems and in natural environments. Disturbance is a primary driver of weed establishment (Grime, 1979) and increased disturbance is an inevitable result of the impending changes in land use, land cover and climate change. The acceleration of international trade and travel is a further driver of change in the status of weeds, with increasing rates of transfer of alien species to new environments. In combination, these drivers represent a powerful force that will favour the spread of alien species around the globe (Mooney, 1996). To date, however, GCTE analyses have only addressed the issue of climate change.

Climate change is likely to have an impact upon the population dynamics and geographical distributions of both annual and perennial weeds (Sutherst, 1996; Sutherst *et al.*, 1996; Teng *et al.*, 1996). Annuals predominate in crop lands, while perennials are mostly associated with rangelands and natural environments. A shift in the competitive balance between weeds and desirable plants is likely because of inherent differences in the ability of plants to respond to changing environmental factors. Changes in climate can stress both natural and agricultural ecosystems, making them more susceptible to displacement by invasive weeds.

Increased temperatures will permit tropical weeds to extend their ranges polewards, but will also limit the ranges of those with vernalization requirements. They will also prevent dormancy of some weeds, thereby reducing the seedbank that presently makes weed control difficult and expensive. C_4 plants out-compete C_3 species at higher temperatures, which will also affect the relative development times of crops and weeds. As some weeds, such as the wild oat (*Avena fatua*), persist partly by shedding seed before the crop is harvested, their development rates relative to those of crops will be important in determining their success. While higher temperatures could favour C_4 weeds, the predicted increase in intensity of rainfall will increase subsurface soil moisture, which is likely to favour the establishment of C_3 woody weeds. Sequences of years with heavy rain will also facilitate the development of thickets by allowing seedlings to grow large enough during periods of abundant pasture growth to escape grazing pressure.

Annual weeds typically have a broad genetic composition enabling them to respond to climatic change before a similar response could be achieved by crops or natural vegetation. Under elevated CO_2, C_3 crops may increase their competitiveness in comparison to C_4 weeds (Patterson & Flint, 1990, but recent ecosystem-level studies do not give clear results, see Chapter 7) whereas the few C_4 crops such as maize are likely to be disadvantaged compared to their C_3 weeds. However, as the higher relative growth of C_3 plants due to the CO_2 fertilization effect may not be fully realized under conditions with limiting nutrients or water, the direct effects of higher temperatures and reduced soil moisture availability are the more likely determinants of the outcome of competition between C_3 and C_4 plants.

The status of any weed under climate change will therefore depend on both the direct effects of environmental variables (increased CO_2 concentration, temperature, and water availability) on plant physiology, and on the weed's competitive interaction with desirable plant species. Interpretation of such interactions will require the further development and use of interplant competition models (Kropff & van Laar, 1993). Application of such models to global change studies has so far been limited to exploratory studies of the likely

Table 9.10 *Some likely impacts of global change on pests, diseases and weeds (after Sutherst, 1996)*

Taxonomic group	Global change driver	Impact
Pests	Temperature and/or moisture	Distribution, abundance and phenology of pests; synchronization of crop and pest phenology; disruption of diapause; the interactions of pests with natural enemies
	Rainfall	Altered incidence of outbreaks of event driven pests, e.g. locusts, armyworm
	Wind	Altered timing, routes and destination of migratory pests using synoptic flows, e.g. budworm, aphids, locust, armyworm.
	Enhanced CO_2	Reduced success of herbivorous insects, except perhaps plant-sucking species.
	Extreme weather events	Changed incidence of many herbivorous pests with increased plant stress
Diseases	Moisture	Altered severity of foliar and root diseases with rainfall
	Temperature	Altered distribution of foliar and especially soilborne diseases which are more difficult to control
	Extreme weather events	Changed incidence of disease with plant stress
	Enhanced CO_2	Increased incidence of foliar diseases with greater LAI
Weeds	Temperature and/or moisture	Changed relative importance of weed species resulting from different physiology, phenology and morphology; C_4 weeds spread with higher temperatures
	Heavier rainfall	Shift in weed seed banks with erosion, increasing weed populations
	Land use changes	Rapid responses of weeds to disturbance from any cause
	Enhanced CO_2	C_3 weeds more successful where temperature and moisture are not limiting

impacts of changes in climate and atmospheric composition on crop–weed competition (Teng *et al.*, 1996).

Table 9.10 summarises some likely impacts of global change on pests, diseases and weeds.

9.3.6   Soils

All aspects of global change will affect soils to some extent, but the nature, severity and consequences of impacts will be highly variable, both spatially and temporally. Change in land use, currently the main manifestation of global change, has far reaching, and often rapid, effects on soil physical structure (e.g.

erosivity) and fertility. This topic is already being researched by many groups and organizations worldwide, but the interaction of land-use change with the other global change drivers of climate and atmospheric composition is less well investigated. The interactive effects of global change on the dynamics of soil organic matter and on soil erosion, both of which are largely controlled by soil biological processes, has been of particular concern to GCTE.

Soil organic matter (SOM) represents a major pool of carbon within the biosphere, estimated at about 1400×10^{15} g globally, roughly twice that of the atmosphere (Powlson *et al.*, 1996; P. Smith *et al.*, 1997), and it will act as both a source and sink of carbon during global environmental change. It is a key element of global change research for several reasons. First, it is important as both a driver of, and response variable to, climate change. Changes in climate are likely to influence the rates of accumulation and decomposition of carbon in SOM both directly through changes in temperature and moisture, and indirectly through changes in plant growth and rhizodeposition. Second, enhanced CO_2 will affect primary production (see Chapter 7), which will potentially increase carbon inputs to the soil; there is evidence that in some systems, enhanced CO_2 increases the allocation of photosynthate to roots (Canadell *et al.*, 1996b; Rogers *et al.*, 1996) and microbial activity (Dhillion *et al.*, 1996). It is, however, no longer thought that elevated CO_2 affects the chemical composition of plant residues and hence their decomposition rate (Ball and Drake, 1997; see also Chapter 7), as distinct from that of living tissue (see Section 9.2.3). Finally, and most significantly, changes in land use can rapidly and strongly alter carbon storage. One of the most striking current examples is the conversion of tropical forest into pasture, but if rising mean temperature permits cropping in cleared boreal forest, SOM oxidation will be induced through tillage. Possible changes in wetlands (reclamation or establishment) may also have strong impacts on sequestration of atmospheric carbon.

Simulation models have incorporated current understanding of the processes controlling SOM dynamics. The GCTE-SOM network has evaluated nine SOM models against seven long-term datasets representing various land uses – forestry, arable crops (many rotations and treatments), and grassland (P. Smith *et al.*, 1997, see Chapter 3). Overall, performance of models fell into two groups, which were significantly different from each other. Group 1 (Daisy, DNDC, Century, RothC, Candy and NCSOIL/NCSWAP) had lower errors than group 2 (ITE Forestry/Hurley Pasture, Verberne and SOMM), possibly because of less calibration of the latter group (main effect), associated with their greater complexity (e.g. water balance, plant growth, etc.). One of the major objectives in the future will be to develop the models so that they can be run in truly 'predictive mode', i.e. without site specific calibration.

Soil degradation may be induced via erosion, chemical depletion, water saturation, and solute accumulation. These are all serious, if site-specific, issues but since it is estimated that nearly one-sixth of the world's land available for agriculture has been degraded already by water or wind erosion (Oldeman *et al.*, 1991), GCTE has concentrated on these aspects. The severity, frequency and extent of erosion will certainly be altered by changes in rainfall amount and intensity, and by changes in wind; such changes are, of course, interlinked and exacerbated by human activity (Valentin, 1996).

Both water and wind erosion are commonly accelerated by land-use change (especially the clearance of vegetation cover), presently the main manifestation of global change. As an example of a regional analysis of potential changes in erosion under global change, the EPIC model was run for two sets of climate conditions for 100 sites in the US corn belt randomly selected from National Resources Inventory sites (Ingram *et al.*, 1996). The first set of conditions was used to investigate the sensitivity of water erosion to precipitation, and included current temperature and CO_2, current wind speeds, and current or changed precipitation (increases and decreases of 10 and 20%). The second set was used to investigate the sensitivity of wind erosion to wind speed and included current temperature and CO_2, current precipitation, and current or changed wind speeds (increases and decreases of 10 and 20%). The results showed that mean water erosion would vary approximately linearly with mean precipitation, with about a 40% change for a 20% change in mean precipitation. In contrast, mean wind erosion would vary nonlinearly with mean wind speed, with an eight-fold increase for a 20% increase in mean wind speed (Fig. 9.7). For a 20% decrease in mean wind speed, wind erosion would decrease four-fold. The dramatic response of wind erosion is attributable to a threshold effect; many of the sites have mean wind speeds above 4 m/s, which is just below the threshold for wind erosion (5.5 m/s; Skidmore, 1965). A 20% change in mean wind speed would have a great influence on the frequency with which the threshold was exceeded, and thus on the frequency of wind erosion events. Two conclusions are readily apparent for this region: first, wind erosion is potentially much more sensitive to climate change than is water erosion; and second, it is critical to understand and predict the wind-speed threshold for wind erosion. Different conclusions might be reached for different regions.

Since erosion processes are both varied and complex, several modelling approaches have been developed for a range of temporal and spatial scales, for erosion by both water and wind. An initial emphasis has been on collating suitable datasets for model comparisons (see Table 3.1, Chapter 3). This is particularly challenging for erosion research because data have to be both spatially and temporally extensive. Also, by its very nature, erosion usually occurs as a result of weather events that may be short-lived and of low

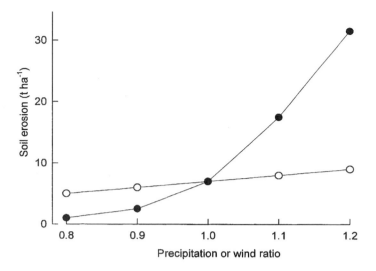

Figure 9.7   Sensitivity of soil erosion in the U.S. corn belt to climate change as estimated using EPIC. Note: For water erosion (○), temperature, CO_2, and wind speed were held at current values, while precipitation volume (expressed as ratio to current volumes) was varied. For wind erosion (●), temperature, CO_2, and precipitation volume were held at current values, while wind speed (expressed as ratio to current speeds) was varied. Each point represents the 100-year average of 100 randomly selected sites.

frequency. Monitoring programmes are therefore a key to providing data for model development and evaluation.

Model comparisons started with an analysis to identify which approaches are most robust for global change studies at the field scale. Five field-scale models were compared using three datasets (Canada, USA, Portugal). The models have significantly different data requirements (relative to the time step, soil properties, etc.). The quality of these data appeared to be a crucial, albeit not easily quantifiable, factor (Boardman & Favis-Mortlock, 1998). Overall, the models performed best on 'home data'. Moreover, better estimates were obtained when long series of events were considered rather than single events. This is due mainly to a compensatory mechanism that consists in overestimating small and frequent events and underestimating rare, major events. Most models failed to simulate soil erosion at the Canadian site because they did not account for the thaw.

It is clearly important to continue evaluating models for SOM and erosion using datasets from a wide range of experimental conditions so that the robustness of the models can be established. This will improve our understanding of each modelling approach, and its limitations and advantages, which will, in turn, help identify where other processes (especially the role of soil biota) need to be accounted for. Such analyses will also increase the confidence with which models can be used in a predictive capacity.

9.4   Impact on production: improving predictive capability

The analyses and syntheses presented in Section 9.3 have allowed more robust models to be developed that will underpin the development of integrated tools for predicting the impact of global change on production. GCTE's initial emphasis was placed at the process level of the many biophysical aspects of agricultural production because it was important to establish clearly the physical limits for particular systems. This section highlights some of the future challenges associated with modelling crops, animal and forest production systems, and the yield-reducing aspects of pests, diseases and weeds.

9.4.1   Field crops

Much crop research is highly site- and season-specific and thus often of limited general applicability. This is because it is often designed to supply information on the performance of a crop on a specific field site, or to help decision-making at a specific field level (e.g. which variety to plant, how much fertiliser to apply, etc.). Variability in the weather between growing seasons has traditionally been accommodated by conducting the same experiment for many years before the data are considered useful for decision-making. However, the pressures of global change stemming from population growth and changes in land use and technology are forcing the adoption of new production systems before there has been time to evaluate their performance over a range of conditions. Further, even when data have been collected over several years, the analytical methods applied to them have often been concentrated on averages rather than on variation with time.

   Agricultural research often concentrates on the components of a system without regard for the interactions among components. This greatly limits the extent to which the results can be used for prediction, analysis or decision-making. These limitations have been recognized by many workers (e.g. Nix, 1984), especially by those in the management and economics disciplines (e.g. Dent, 1993), and steps are being taken to develop new approaches. At one level, this involves extending the number of experimental on-farm and in-production fields. At another level, it involves using simulation models to generate probability information for specific crops on an individual field. Simulation outputs are obtained for many seasons, and for crop rotational sequences in which the management and performance of one crop can affect the options available for the subsequent crop (e.g. the time of harvest of one crop can influence the seeding date of the next crop). Software that facilitates the generation of such probability functions (e.g. DSSAT 3; Tsuji *et al.*, 1994) and the examination of cropping systems (e.g. APSIM; McCown *et al.*, 1996) is now available. The use and improvement of such software, as well as of the crop models themselves, will be essential in ongoing efforts to improve our capability to predict the impact of global change.

Despite many limitations, our understanding of the quality of crop, pasture and forestry models for local sites has improved considerably over the past five years or so, but critical developments remain before we can provide predictions at the agroecosystem level. A general consideration is that nearly all mechanistic crop, pasture and forest models are not spatially explicit; they are point models, only 'valid' either for the plot from which the data were obtained for their parameterization or for a strictly homogenous area. Additionally most production models do not incorporate yield-reducing effects such as pests, diseases and weeds and many do not include carry-over effects of soil nutrients and soil moisture from one year to another so that it is difficult to deal with rotations and cropping systems. Rotations and multiple cropping both involve issues at the enterprise scale that cannot be analysed with separate crop models. Finally, the impacts of changing production can only be assessed by incorporating outputs into an economic model; in commercial agriculture, extra production must be economic.

9.4.2   Grazing systems

Because natural pastures are more complex than crops and the production system includes an animal, problems remain that are different to those of cropping systems. The first problem is that a better appreciation of long-term changes in vegetation composition is required, especially in rangelands where a single pasture may contain over 100 species. Many of the advances relevant to solving these problems will come from the approaches described in Chapter 8 and must be incorporated in models of grazing systems (see Stafford Smith, 1996).

The second problem is that our ability to model animal production is limited. Animal growth (principally sheep and cattle) under current conditions is comparatively well-understood in improved pastures where issues of spatial heterogeneity and diet selection are small (Marshall *et al.*, 1991; Parsons *et al.*, 1994; Freer *et al.*, 1997); it is more poorly handled in rangelands (Baker *et al.*, 1992; McKeon *et al.*, 1993; Ash and Stafford Smith, 1996). Direct effects of climate change on animals are likely to be minimal but indirect effects, acting through changes in pasture quality and productivity, will be more important. In extensive rangelands the effects of productivity on reproduction and mortality are poorly understood even though they are as important as the direct effects on growth (Ash and Stafford Smith, 1996). Additionally, the effects of pests and

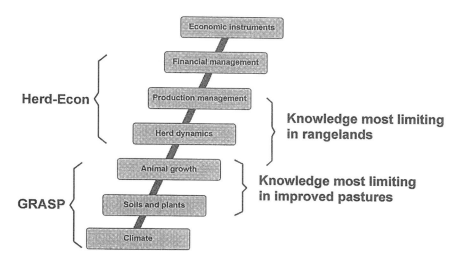

Figure 9.8 Key elements of a system model of a grazing production system, showing where knowledge is probably most limiting in modelling different types of systems – in improved pastures, reproduction and mortality are less important than pasture production and utilization, whereas rangeland herds usually depend critically on births and deaths, as well as changing animal numbers between good and bad years. The simulations used here link the models GRASP and Herd-Econ, which handle the sections shown (from Campbell *et al.*, 1997).

diseases on mortality, and sub-lethal chronic growth deficits, may alter the realization of both potential secondary production and primary production.

The final problem is that of scaling up (Fig. 9.8). At the scale of an individual farm, Campbell *et al.* (1997) have recently used a linked model to test the sensitivity of socio-economic outputs to the potential impacts of CO_2 induced changes in primary productivity as compared to other factors in the system. The results show that the economic effects of elevated CO_2 are unlikely to be greater than those caused by many other biophysical and management factors. This analysis of the whole production system identified those components that were most poorly known, and demonstrated the need to know all processes with a comparable degree of precision.

The ultimate goal of GCTE is to be able to interpret results at regional and global scales. At these scales, many other social, micro- and macro-economic and political issues come into play, and both global adjustments and local management adaptations radically moderate changes in underlying productivity. The traditional approach of economists to deriving elasticity relationships is based on analysing past data; by definition such analyses cannot take account of the ways in which future change may occur. An alternative approach adopted by an Australian project is to use enterprise-scale models to develop regional response surfaces of enterprise profitability to changes in prices (i.e. changes in

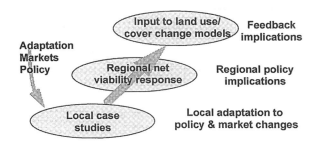

Figure 9.9   Schematic outlining a process of scaling up from local enterprise studies to regional industry viability and hence simplified land use change relationships. Enterprise responses to changes in prices, policy and production may result in management adaptation, which is not captured in a top-down approach to the same problem (After Campbell *et al.*, 1996).

demand), productivity (including changes caused by climatic and atmospheric changes) and policy (e.g. carbon taxes) (Campbell *et al.*, 1996). This project will identify how management is likely to adapt to particular combinations of impacts, and what effect this will have on profitability (Fig. 9.9).

9.4.3   Managed forests

The major component of process-based forest models that needs to be further developed is the carbon partitioning algorithms/sub-models. The first steps in the calculation of forest productivity lead to estimates of the carbon fixed by canopies over selected intervals. The next step requires that this carbon be allocated to components, since the growth response of trees, in terms of carbon allocation to roots and top-growth, determines the time course of leaf area development and the responses of the trees to changes in environmental conditions. Various allocation models are used, ranging from empirical allocation procedures based on observed allometric ratios, to controls on the carbon allocated to foliage and roots determined by plant nitrogen and water status. None of these is free of problems. Landsberg and Waring (1997), in their 3-PG model, used a procedure that begins with carbon allocation to roots, which is influenced by growing conditions, including nutrition, and then uses the ratio of the derivatives of the allometric equations describing foliage and stem mass in terms of stem diameter to obtain carbon partitioning coefficients. This procedure appears to reproduce the foliage and stem development patterns that would be observed in a stand of trees on a particular site growing under specified environmental conditions.

In terms of stand development, the gap-type models provide some basis for analysing establishment success and species distributions, but these have not been used in forestry practice as a basis for the evaluation of management and policy options. Process-based models, which include soil carbon turnover calculations, provide the opportunity to evaluate the sustainability of particular practices and management regimes, but a major limitation of the present generation of these models, in relation to the prediction of forest growth rates and the management of forests for wood production, is the fact that virtually none of them carry the calculations through to stemwood production, which is the thing that interests foresters and forest managers. This will have to be addressed in the near future to establish credibility with the forestry community; the process-based models must be developed to the point where they can compete with conventional mensuration-type models, in areas where these are relatively well-parameterized. They will be then of value not only as scientific tools for analysing ecological systems, but also as tools that can be used for economic analysis.

9.4.4   Pests, diseases and weeds

As with any biological process, pest, disease and weed population dynamics are driven by local environmental influences and can be strongly affected by cultural processes in production systems. These considerations demand that three dimensions be taken into account when conducting and applying research on pest, disease and weed management.

First, pest populations vary dynamically over short periods of time and include feedback loops and time delays so that analyses must be based on a close linkage of biological processes and temporal changes in important driving variables such as temperature, moisture and light. Traditional research using constant laboratory conditions has severe limitations, particularly when deriving cumulative stress functions that relate mortality to daily and seasonal fluctuations of environmental variables. Second, there are compelling reasons for considering pest, disease and weed populations in a spatial dimension, rather than in traditional point experiments. Climatic and other environmental conditions vary spatially resulting in patterning that is a source of valuable information in population ecology (Sutherst et al., 1995). Also, because many pest species are highly mobile, a spatial perspective is desirable to design and implement pest management strategies. Third, effective research must identify points where cultural practices impact on pest populations and decision points in the crop growth cycle where interventions are possible.

In the context of global change, each of these dimensions assumes a much greater importance because the environment will change around the pest and production system. Pests will track movements of production systems in response to global change, so research needs to be integrally linked to the system during the risk assessment process to keep the issues in their proper context.

While pest modelling approaches have often been successful in addressing narrow issues of population dynamics at point sources, it is only recently that such models have been closely linked to models of the crops that they attack (Gutierrez, 1996; Teng *et al.*, 1996). Weed models have of necessity been linked to crop growth models in order to describe competition processes (Kropff & van Laar, 1993), but a wider consideration of whole production systems is required. Pest and disease modellers have also been making progress in incorporating spatial dimensions into modelling analyses, usually by linking their models with GIS (Luo *et al.*, 1995; Williams & Leibhold, 1995). Pest modellers have also been at the forefront of conducting integrated assessments of the impacts of climate change, as illustrated by the example in Box 9.3.

9.5   Agricultural responses to global change

It is clear from the preceding accounts that our ability to predict the effects of global change on global agricultural production is still limited. This is largely due to the fact that global production is the sum of multitudinous individual farmer actions in response to biophysical and socio-economic factors, and we do not, as yet, have an adequate grasp of how global change will affect many of these responses. Different crops respond differently to increases in CO_2 and temperature so that predicting the output of even a single region under climate change is problematic. Such estimates of global production as have been made (e.g. Rosenzweig & Parry, 1994) have usually used a single model of crop growth and applied it universally; an approach which Sections 9.2.1 and 9.2.2 indicate is fraught with difficulties at present because most models have only been locally evaluated. Nevertheless, substantial progress is now being made in refining the biophysical models and evaluating them with more widely distributed data sets. The model testing has been an important initial stage in developing the techniques that will be required for conducting spatial and multi-seasonal analyses with confidence.

9.5.1   Wheat and rice

Global crop production will need to almost double by 2025 to meet the increased demand for food caused by the increase in world population to about 8 billion. In South and Southeast Asia the increase in production will need to occur by intensification on existing land whereas in Africa and South America new land will be needed for cultivation. Land-use change will be the biggest component of global change. Specific points are:

- Under optimum field management, wheat yields may increase by about 15% for 550 ppm CO_2; a 5–12% increase is more likely for average management conditions. This estimate is lower than that suggested by IPCC (1996) as it allows for decreased crop duration (and hence yield) as a consequence of warming and the many factors that reduce yield from potential to actual levels.

■ Major wheat models are being refined but considerable caution is still needed in spatial extrapolation if a single model is used.

■ Rice yields are reduced when temperatures exceed 32 °C at flowering due to spikelet sterility. This finding is unaffected by elevated CO_2 but might differ between genotypes.

■ Major rice models are consistent across a wide range of potential yields and indicate a reduction in yield of about 5% per °C rise above 32 °C. This would largely offset any increase in yield as a consequence of increased CO_2.

■ Elevated CO_2 is likely to enhance yields of rice more than wheat because rice is usually grown intensively with inputs of water and nutrients whereas wheat is grown extensively in drier areas.

9.5.2 Pastures and rangelands

The increasing population in developing countries (particularly in Africa and South America) will mean that cultivated land will encroach further on grazing lands leading to reduced production of livestock on rangelands. This analysis agrees with the IPCC (1996b) conclusion that pastoral systems may be very sensitive to climate change; decreased rainfall would result in decreased pasture production. Specific points are:

■ Doubling current CO_2 will increase NPP in different pastures and range-lands by 0–20% depending on temperature, water and nutrient limitations.

■ A long-term, sensitivity analysis for a subtropical, managed pasture indicates that a 5% increase in pasture growth due to elevated CO_2 would lead to a 3% increase in liveweight gain in cattle because the annual variability of pasture productivity between years would be reduced.

9.5.3 Managed forests

Over the next 25 years, the impacts of global change on managed forests will be mainly attributable to changes in land use, management and atmospheric nitrogen deposition, rather than to changes in climate and atmospheric CO_2. Specific points are:

■ There is considerable scope for improving the productivity of managed forests although, for economic and logistical reasons, this is more likely to be realized in the developed nations.

■ Short-term studies of elevated CO_2 show an increase in plant biomass production. This increase, however, will be reduced in the longer-term as an effect of increased net ecosystem respiration. This reduction may be substantially compensated for by the interactive effects of CO_2 and atmospheric nitrogen deposition; the net effect is uncertain.

■ Forested areas in the humid tropics are decreasing rapidly and the ratio of forested area to population will continue to decrease. This will lead to increasing pressure on hardwood supplies from natural forests which is unlikely to be alleviated by substitution of managed forests/plantations because current establishment rates are inadequate for this purpose.

■ No foreseeable increase in global forested area can make a significant impact on the global carbon balance. However, the mid- and high-latitude forests appear likely to remain a sink for CO_2 while the destruction of forests in the tropics and sub-tropics is a source of atmospheric carbon that will only partially be offset by the establishment of plantations.

9.5.4   Soils

Land management will continue to be the principal determinant of soil organic matter content and susceptibility to erosion during the next few decades, but changes in vegetation cover due to short-term changes in weather and near-term changes in climate will affect SOM and erosion, especially in semi-arid regions. If the projected increase in long-term mean temperature occurs, then rates of SOM oxidation will increase reducing the storage of soil carbon. Specific points are:

■ Plants grown under elevated CO_2 generally increase the allocation of photosynthate to roots which increases the capacity and/or activity of belowground carbon sinks.

■ Models suggest that some of the increased capacity of belowground sinks may lead to increased long-term sequestration of soil carbon, although strong evidence is lacking.

■ Models demonstrate that erosion by water is directly related to the amount of precipitation but that wind erosion increases significantly above a threshold windspeed. The consequences of climate change on the intensity of storms has substantial effects on the amounts of erosion expected from both water and wind.

⑩ Biogeochemistry of terrestrial ecosystems

R.J. Scholes, E.-D. Schulze, L.F. Pitelka and D.O. Hall

Box 10.1   Summary

The sum of known CO_2 sources is greater than the sum of known sinks plus the observed increase in atmospheric CO_2. The difference is known as the 'missing sink', and understanding its origin and reliability is essential if the continued increase of the atmospheric concentration CO_2 is to be halted.

- Although the uncertainty associated with each of the known sources and sinks is sufficiently high that the missing sink may simply be the cumulative result of estimation errors, several lines of evidence suggest that it is real, and located on land.
- The most likely mechanisms for a terrestrial 'missing sink' include the CO_2 fertilization effect, fertilization by nitrogen deposited as a result of human activities, climate trends and changes in land management. The future strength, and even sign, for each of these mechanisms is doubtful. The capacity and stability of the postulated sinks are finite.
- An ongoing, large sink must show up eventually as an increase in a carbon pool. An increase in the size of terrestrial carbon pools over the past century has been demonstrated for a number of locations, but has not been directly proven at a global scale, nor over the full vegetation disturbance cycle. It is likely that the terrestrial pool is thinly distributed over a wide range of ecosystems, and is thus hard to detect.

The carbon cycle is closely and necessarily linked to the cycles of other nutrients (particularly nitrogen, phosphorus and sulphur). Insufficient nutrient supply limits ecosystem-level carbon uptake and storage in many ecosystems. The limiting nutrient in natural terrestrial ecosystems is frequently nitrogen, due to the numerous pathways by which it can leak from the system, and the high energetic cost of extracting it from the atmosphere. Human activities increase both inputs and outputs of nutrients to terrestrial ecosystems. Thus

- Disturbance of global nutrient cycles as a result of human activities is itself a major driver of global environmental change, both directly and indirectly via the carbon cycle.
- Ecosystems downwind of highly industrial areas are being loaded with additional nitrogen and sulphur as a result of the deposition of emissions containing

these elements. This has numerous effects, including possible growth enhancement, water pollution and species composition changes.
- The global phosphorus cycle has been substantially perturbed, particularly
  through agriculture. There is much evidence that phosphorus is an important
  limiting element, particularly in the tropics and in aquatic ecosystems, but the
  consequences for models of the global carbon cycle are unknown.

Box 10.2   Future needs

A key future challenge for biogeochemical research is to establish the nature and
limits of terrestrial carbon sinks, and how they will respond over a period of decades
to centuries to the combined effects of changing atmospheric composition, climate
and land use/cover. More specifically, there is a need to:

- improve our understanding of the effects of elevated atmospheric CO_2 concentration, in combination with changing climate and land use, on NPP in natural
  ecosystems, including the combined effect of warming and rising CO_2 on plant
  respiration and the factors that control the ultimate amount of NPP that finds its
  way into long-lived C pools;
- determine whether the loss of carbon from the soil following warming may be
  partially or completely offset by increases in the carbon input to soils as a result
  of enhanced productivity under elevated CO_2;
- determine the effects of elevated CO_2 on litter chemistry and thus decomposition rate; the litter analyses will need to go beyond simple C : N ratios to include
  structural and non-structural carbon and nitrogen as well as polyphenolic
  compounds;
- gain a better understanding of the role of phosphorus in a high CO_2 world, both
  as a direct limitation to NPP in ecosystems relieved of carbon and nitrogen
  limitation and indirectly as a constraint on biological nitrogen fixation.

10.1   Introduction

A decade ago two basic schools of thought existed regarding the global carbon
cycle response to elevated CO_2: those that believed that terrestrial ecosystem
responses could be extrapolated from laboratory measurements of increases in
photosynthesis; and those that believed that natural ecosystems would show
essentially no response, due to nutrient limitation of growth. Partly as a result of
the work of GCTE-associated scientists, the reality is known to be somewhere
in between. This chapter presents the evolution of our understanding of global
terrestrial biogeochemistry over the past decade.

Biogeochemistry is the chemistry of the surface of the earth (Schlesinger, 1991), involving the study of the cycling of substances between living organisms and the hydrosphere, atmosphere and geosphere. It requires understanding the various forms in which these substances occur, their concentration and spatial distribution in the biosphere, the processes that convert one form into another, the factors which control the rate and path of these transformations, and the consequences of all of the above for life on Earth.

Although 26 elements are known to be essential for life on Earth (Marschner, 1995), and many more are toxic at high concentrations, biogeochemists concerned with global change have focused on carbon, nitrogen, phosphorus and sulphur. Over the period of interest for global change (about two centuries in the past and future), the global cycles of C, N, P and S have altered dramatically, while other elements have been relatively stable. For example, although the increase in atmospheric CO_2 is accompanied by a decline in atmospheric O_2, the decrease is tiny relative to the total atmospheric pool of oxygen (Keeling & Shertz, 1992).

10.2   Key issues

The central theme of this chapter is the global carbon cycle, both because of the role of atmospheric CO_2 in determining the global climate, and because many of the benefits that humans derive from ecosystems – such as food and fibre – have their basis in plant production, which is in turn based on carbon. However, the conceptual model applied in this chapter is that the carbon cycle cannot be viewed outside of the context of the other biogeochemical cycles that are closely coupled to it. A schematic view of the linkages of the carbon, nitrogen, phosphorus and water cycles is presented in Fig. 10.1. Three key issues, of both scientific and policy interest, emerge from this conceptual model.

(i)   Are anthropogenic perturbations of nutrient cycles reducing the capacity of ecosystems to absorb change?

The organisms in ecosystems have evolved over a long period of time to depend on certain sources, levels and processes of nutrient supply. As a result, undisturbed terrestrial ecosystems are very conservative with respect to nutrients; in other words, the losses are small relative to the turnover within the ecosystem. When the nutrient supply is massively disturbed by human action, can the various cycles adjust so that the ecosystem continues to deliver the goods and services on which we depend? This key issue is addressed below first by discussing the mechanisms of ecological buffering (Section 10.3), and then by examining examples of perturbations to the carbon, nitrogen and sulphur cycles (Section 10.4).

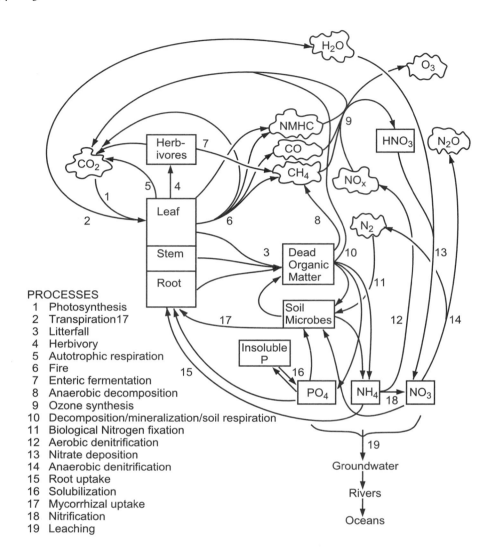

Figure 10.1   Linkages between the carbon, nitrogen, phosphorus and water cycles in a generalized terrestrial ecosystem. The sulphur cycle has been omitted for the sake of clarity, but parallels, in many respects, the nitrogen cycle. The presence of shared pools, processes and controls means that one cycle may be constrained or driven by perturbations in another cycle.

(ii)   To what extent does nutrient limitation constrain terrestrial ecosystem response at the global scale to elevated CO_2 ?

When individual plants are exposed to elevated CO_2, their rate of carbon assimilation increases (see Chapter 7). This effect, which acts to buffer the rise of atmospheric CO_2, is reduced in ecosystems when plant growth is limited by an insufficient nutrient supply. Over what fraction of the earth's surface do such limitations apply, and to what degree? How might cycle interactions reduce or intensify this effect? These questions are addressed in Section 10.5.

(iii)   Is the 'missing sink' terrestrial; and if so, what causes it, and will it last?
A significant part of the global carbon uptake, amounting to about a quarter of the amount of additional CO_2 emitted to the atmosphere by human activities (but only a tiny fraction of the annual exchange of CO_2 between the land, ocean and atmosphere), is unaccounted for. A safe path for future human emissions cannot be plotted until this 'missing sink' is located and understood. The possibility that it is located on land, and if so, its likely future behaviour, are discussed in Sections 10.6 and 12.2.1.

10.3   Ecosystem buffering

Buffering, the capacity of a system to absorb change (i.e. the relative change in the system outputs is smaller than the relative change in system inputs) can have several causes.

10.3.1   Feedback loops

Terrestrial ecosystems are not passive receptors of changes in their environment. Changes in the driving variables lead to alterations in ecosystem structure, composition and functioning, which in turn feedback on the physical and chemical environment. Where the net effect of these responses is to reduce the magnitude of the initial perturbation, the system is said to exhibit negative feedback; where the effect is to amplify the perturbation it has positive feedback. Negative feedbacks can lead to homeostatic (self-stabilizing) conditions, while positive feedbacks are destabilizing. Terrestrial ecosystems show examples of both with respect to global change. The response to a smooth increase in a driving variable (for instance, the steady increase in the atmospheric concentration of CO_2) is therefore unlikely to be smooth, but is likely to exhibit thresholds where the balance of positive and negative feedbacks changes. Such a threshold separates different, finitely bounded stable states of the system. The global ecosystem comprises many hundreds or thousands of subsystems, each with their own thresholds. Prediction of a 'dangerous limit' with respect to the whole globe is therefore a very difficult task.

Negative feedback loops are the source of buffering least likely to lead to unpleasant suprises in the future. An example is the direct CO_2 fertilization effect, which at the leaf level appears to be robust and predictable (Luo, 1996). An increase in the concentration of CO_2 in the atmosphere commonly leads to an increase in the carbon assimilation rate by plants, which helps to reduce the atmospheric CO_2 concentration.

10.3.2   Time lags between uptake and loss

An apparent buffering may also be the result of a time-lag in the system. For example, whether an ecosystem stores or releases carbon is determined by the balance between carbon assimilation (through photosynthesis) and carbon loss (through respiration). Ecosystem carbon assimilation responds positively and almost instantly to increased atmospheric CO_2, whereas the process of decomposition responds only indirectly, through changes in temperature, moisture, and litter quality and quantity – all of which include long delay components. The delays in climate change relative to atmospheric CO_2 change are largely the result of the enormous thermal inertia of the oceans and polar ice caps. The ecosystem delays are a result of the residence times of various biospheric carbon pools, which range from a few weeks for microbial carbon, to thousands of years in the case of the most stable soil organic matter pools. It takes several decades to grow a tree, and then several more for it to decay. The bulk of the soil carbon pool has decadal response times, which would speed up in a warmer, moister world, and slow down if the litter inputs declined in lignin : nitrogen ratio. However, litter quality shows no consistent response to elevated CO_2 (see Chapter 7).

As a result of these time-lags, it would be expected that terrestrial ecosystems would initially absorb CO_2 following an increase in the atmospheric concentration. Once the CO_2 content of the atmosphere is stabilized, as it must be at some stage if dangerous interference with the climate system is to be avoided, the ecosystem respiration component will eventually catch up with photosynthesis, and then exceed it for a century or more. During this period terrestrial ecosystems will be net sources of carbon rather than sinks. Thus, anthropogenic CO_2 emissions must be stabilized well before the atmospheric concentration limit is reached, to allow for the overshoot. When global carbon models are run with a realistically ramped-up atmospheric CO_2 content rather than a step-function to equilibrium doubled CO_2, the effects extend far beyond the time when the atmospheric forcing is ceased (Smith & Shugart, 1993).

10.3.3   Buffering due to system configuration and nonlinearities

A third form of buffering may be inherent in certain system configurations; typically where a large inert pool is dynamically linked to a small active pool. A well-known example is the relationship between the quantity of acid added to a salt-buffered system and the pH of the system. The 'buffer curve' is highly nonlinear. Initially there is a very gradual pH change, but when the buffer capacity has been exhausted, a small acid addition leads to a large response. A similar body of theory can be applied to the capacity of an ecosystem to absorb sulphate and nitrate deposition. The buffer capacity in this case is abiotic, determined by the quantities of various types of minerals present in the soil, but can be amplified by the catastrophic failure of the ecosystem nutrient retention processes, when the biological limits of the participating organisms are exceeded. Thus, a buffered system is advantageous at the beginning of a change process, but disadvantageous as the point of failure is approached. Intrinsic nonlinearities of this sort are also present in biological systems; for example in the concept of 'futile cycles' outlined by Schulze *et al.* (1994*a*).

All of the important interactions between terrestrial ecosystems and the global carbon cycle are nonlinear. Most of these nonlinearities tend to work against long-term buffering. While the carbon assimilation rate increases with increasing atmospheric CO_2, it does so at a diminishing rate. Respiration, in contrast, is an exponentially increasing function of temperature. It is possible to imagine a scenario where warming leads to net CO_2 release from soils, despite the CO_2 fertilization effect, generating further runaway warming. This would be particularly serious where an abrupt threshold exists; for example, the melting of permafrost when the mean annual soil temperature rises above approximately zero degrees or the protective covering of sphagnum is affected (see Section 7.3.1). This would expose large volumes of previously protected soil carbon in the tundra biome to decomposition. The resulting atmospheric forcing could be enhanced if much of the decomposition occurred under anaerobic conditions, resulting in emissions of methane as well.

Paradoxically, soil warming can lead to an increase in ecosystem carbon sequestration under temperate, nitrogen limited circumstances. This is because soil organic matter has a carbon : nitrogen ratio of around 10 : 1, whereas grasses have a C : N of 20 : 1 or more, and trees up to 200 : 1. Provided the nitrogen released by decomposition of the soil organic matter is not lost to the system through leaching or denitrification, it will become incorporated in plant material at a higher C : N ratio, thus sequestering more carbon in the ecosystem (Nadel-hoffer *et al.*, 1995).

10.4 Nutrients as drivers of global change

Human capacity to alter the environment increased dramatically after the Industrial Revolution, approximately two centuries ago. The profound disturbance of global biogeochemistry that followed is one of the major drivers of global change (see Chapter 1). It operates directly on the elemental cycles, as well as indirectly via the global radiative balance and changes in the climate. The anthropogenic effects on elemental cycles can be thought of as two groups of phenomena:

- **Perturbations**, in which the quantity of a given substance entering the ecosystem increases. For example, large inputs of nitrogen, phosphorus and sulphur are entering the global ecosystem as a result of industrial and agricultural activities. They are drivers of global change of comparable impact to the effects of CO_2 on the climate (Vitousek *et al.*, 1997a).
- **Disruptions**, in which the pathways, efficiencies or end products of bi-ogeochemical processes are altered. For example, most agricultural ecosystems are considerably less efficient at recycling and retaining nutrients than the natural ecosystems from which they were derived. The result is a redistribution of nutrients from the soil–plant system into the hydrosphere and atmosphere, sometimes referred to as 'leakiness'.

In reality, the two phenomena are closely linked since the output of one ecosystem is usually the input to another. A large perturbation can cause disruptions by overloading the cycling mechanisms. The disruption results in a perturbation to the ecosystem that receives the leaked products.

10.4.1   Direct perturbations of the carbon cycle

The global carbon cycle has been strongly perturbed during the past two centuries by the injection of large quantities of CO_2 into the atmosphere, as a result of fossil fuel burning and changes in land use. The annual terrestrial photosynthetic flux uses about 15% of the atmospheric carbon pool. Thus, any changes in the land–atmosphere carbon fluxes caused by climate change and land-use factors potentially have large effects on the atmospheric carbon pool.

Carbon emissions associated with land-use change

At present, about three-quarters of the human-induced emissions of CO_2 to the atmosphere result from the burning of fossil fuels and the manufacture of cement, while the remainder result from disturbances to terrestrial ecosystems. When integrated over the past 200 years, the CO_2 emissions associated with land-use change add up to nearly as much (120 Pg C between 1850 and 1990) as the accumulated emissions from the burning of fossil fuels (Houghton, 1994).

When natural ecosystems with a high biomass, such as forests, are felled to make way for low-biomass annual crops or pastures, the difference in the biomass carbon stock is transferred to the atmosphere. Depending on the agricultural practices applied, there may be an accompanying decline in the quantity of carbon stored in the soil. The proportional contributions of soil and biomass vary greatly between ecosystems. Averaging over all global terrestrial ecosystems, the loss of carbon from the soil is thought to be about one-third of the loss from decaying or burned biomass. Much of the information on net carbon emissions resulting from land-cover change comes from forest inventories, which contain uncertainties in both the biomass standing stock and the areal extent of forests. Non-forest ecosystems, such as grasslands, which together comprise 70% of the land surface, are not included in this inventory. Furthermore, forest inventory data rarely includes slash, litter and soil C, and many regions of the world do not have the required inventory data or the means to collect them. As a result, the emissions attributed to land-cover change remain much more uncertain than those resulting from fossil fuel consumption.

The reasons for the decline in soil carbon following the conversion of natural ecosystems to crop agriculture are several (Scholes & Scholes, 1994). Crop plants in general have a lower allocation of carbon to belowground parts than is typical in natural ecosystems. This is a result of years of selection for high aboveground yields, and is the main reason that crop plants must be adequately fertilized, irrigated and protected from competition in order to thrive. Secondly, a large part of the aboveground yield in agroecosystems is removed from the field as harvest rather than entering the soil as litter. Thirdly, the repeated disturbance of the soil through mechanical cultivation speeds up the rate of decomposition of litter within the soil and reduces the degree of physical protection of organic matter. The rate and fraction of carbon loss depends on the size of a poorly understood soil carbon pool with a turnover rate in the order of years to decades. This pool, which appears to consist mainly of leaf and root litter particles small enough to pass through a 2 mm sieve, is a much larger fraction of the total soil organic carbon in tropical than temperate soils (Trumbore, 1993).

Land-use conversions do not necessarily lead to reduced topsoil carbon, even in the tropics. For instance, the soil carbon content in the top metre of pastures derived from forest in the Amazon intially declines, but under good management may eventually increase to a level greater than that of the original forest soil. However, these soils may be very deep, and when the deeper horizons are considered, there may still be a net carbon loss (Trumbore *et al.*, 1995). The top metre of soil in the Amazonian study area of Trumbore *et al.* contains 10–12 kg C m^{-2}, of which 25–40% is susceptable to change over a decadal period, whereas the horizons between one and eight metres contain 17–18 kg C m^{-2}, of which 11–17% turns over in the decadal period. Further uncertainty is provided by the observation that certain African grasses, when planted as pastures in South America, produce large amounts of deep soil carbon (Fischer *et al.*, 1994). All the present estimates of the global soil carbon pool are based on the surface metre or less, which corresponds to the available data, but is an underestimate.

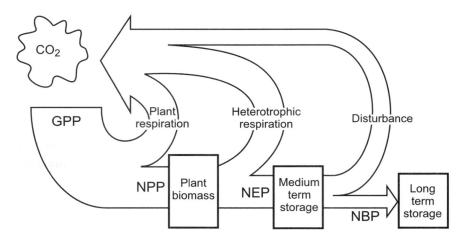

Figure 10.2   The terms used to define various components of ecosystem productivity (Schulze & Heimann, 1998). Note that demonstrating an increase in NPP, or even a positive NEP, is not sufficient evidence for long-term, large-scale carbon sequestration, since this carbon may be lost through infrequent disturbance.

The CO_2 fertilisation effect at ecosystem to global scale

The rate of carbon assimilation by plants through photosynthesis increases with increases in the CO_2 concentration in the leaf atmosphere. The increased photosynthesis may result in increased plant growth (see Chapter 7), which would help to reduce the rate of accumulation of atmospheric CO_2. Such negative feedback loops offer the possibility of creating a self-regulating system, and some authors see the increase in atmospheric CO_2 as a benign phenomenon (e.g. Idso & Kimball, 1993). The relationship between atmospheric CO_2 and carbon assimilation at the scale of an individual leaf is an example of a smooth, nonlinear response. It saturates above about 1000 ppmv CO_2, and will therefore become less and less effective as a negative feedback as the atmospheric concentration rises. Furthermore, there are many steps between photosynthetic carbon assimilation and eventual storage of carbon by the ecosystem, so what is true at the leaf or organism level is not necessarily true at the ecosystem, biome or global scale (Fig. 10.2).

Amthor (1995) reviewed the response of terrestrial plants to increasing CO_2 concentrations in relation to the global carbon cycle, highlighting the uncertainties which hinder our ability to predict global change effects on the carbon cycle. He noted that the effects of elevated atmospheric CO_2 concentration on NPP in natural ecosystems have not been sufficiently studied, particularly with respect to the forest ecosystems, which are thought to dominate global carbon storage and uptake. The combined effect of warming and rising CO_2 on plant respiration remain unclear, and the degree to which carbon assimilation is limited by nutrient availability now and in the future is not resolved.

Most simple global-scale biospheric models use an empirical coefficient β (the 'biotic growth factor') to summarize the effects of atmospheric CO_2 on NPP. It is defined as follows:

$$NPP_e = NPP_0 * (1 + \beta \ln (C_e/C_o))$$

where NPP_e is the NPP at an atmospheric CO_2 concentration of C_e, and NPP_0 and C_0 are the reference level (Bacastrow & Keeling, 1973). This formula is a mathematical simplification that does not capture the actual mechanism of CO_2 response at ecosystem scale, and should therefore not be relied on to be robust when extrapolated into the future.

The magnitude of the global increase in NPP under twice the current atmospheric CO_2 concentration is still unclear. The modal production response of crop species grown under adequate water and nutrient supply is around 30% (equivalent to $\beta = 0.4$), but natural ecosystems subject to nutrient limitation are much less responsive (see Chapter 7). Values of $\beta = 0.2$–0.5 have been used in global C cycle models (Wullschleger et al., 1995). McGuire et al. (1995a) show that for forests the value of β increases with increasing nitrogen status of the plants. A value of 0.40–0.42 mimics the current 'missing carbon sink' (Rotmans & den Elzen, 1993; Goudriaan, 1995; Kheshgi et al., 1996), although it certainly does not represent the complete mechanism, since it conflates the N fertilization effect with the CO_2 fertilization effect.

Extrapolation from the high levels of response obtained from pot and crop experiments predicts that the CO_2 fertilization effect could be responsible for a sink of 2–4 Gt C y^{-1} at present CO_2 levels, more than enough to account for the 'missing' carbon (Gifford, 1994b). In practice, nutrient limitations in natural ecosystems probably restrict it to less than 1 Gt C y^{-1} (Rotmans & den Elzen, 1993; Friedlingstein et al., 1995; Schimel, 1995) – in fact, a recent review by GCTE (see Chapter 7) shows that some systems do not increase biomass accumulation at all under elevated CO_2. The continued existence of this sink assumes that the vegetation of the world does not undergo major structural change as a result of climate change. Although new sinks may arise from biome shifts, the transition period is likely to see existing sinks become sources, as a result of disturbances (see Section 12.2.1).

10.4.2 Nitrogen deposition and fixation as a perturbation of the carbon cycle

The current knowledge of the global nitrogen cycle is summarized in Fig. 10.3. The human impact is substantial – approximately the same amount of nitrogen is introduced into terrestrial ecosystems as a result of human activities as results

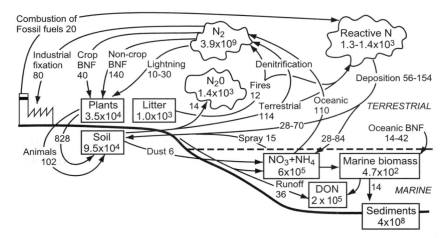

Figure 10.3   The human–perturbed global nitrogen cycle. Note the domination of the pools by atmospheric N_2; as a result the mean residence time of the atmospheric pool is 300 million years. Turnover within the plant–soil loop is approximately decadal, while turnover within plants is approximately annual. From Cornell *et al.* (1995); Nevison *et al.* (1996).

from natural processes (Vitousek *et al.*, 1997b). The flow of nitrogen from the land to the sea is a small proportion of the terrestrial turnover, but makes up a large fraction of the total nitrogen inputs to the sea. About a quarter of this input may be as the result of human activities. The flux from land to sea is assumed to have been historically balanced by losses from the marine environment, including sedimentation and denitrification. It is not known if this is still the case at a global scale.

The global nitrogen cycle has been particularly perturbed by the industrial synthesis of ammonium from N_2, using a process invented in the nineteenth century. Nitrogen is also 'fixed' (i.e. converted from atmospheric N_2 to nitrate or ammonium) indirectly through high-temperature reactions, such as occur in furnaces, internal combustion engines, lightning and vegetation fires, which produce NO_x. The NO_x oxidizes in the atmosphere to NO_3^-, and is deposited in rain or as dust. Finally, biological nitrogen fixing (BNF) crops are now much more prevalent than BNF species were in natural ecosystems, increasing the global total of nitrogen fixed biologically.

At present, anthropogenic nitrogen fixation from all sources (approximately 70 Tg N y^{-1}) equals natural 'preindustrial' fixation in magnitude, but has a rather different spatial distribution. Crop fertilization and BNF crop plantings are concentrated on agricultural land, particularly in the developed countries of Europe and North America. Nitrate deposition occurs within a few hundred kilometres of the source of NO_x emissions, which are also in these regions. As a consequence, nitrogen limitation has been relieved over much of Europe and parts of North America, which is thought to have resulted in increased NPP and possibly increased carbon sequestration. In some cases the ecosystems have become nitrogen saturated, showing no further response to additional nitrogen, and a variety of negative consequences, including a potential decline in the carbon stored by the forest (Schulze & Ulrich, 1991).

Nitrogen fertilizer applied to crops is inefficiently used. The nitrogen cycle in croplands is highly disrupted, due to tillage, the short duration of the crop, harvesting and the use of biocides. Up to 60% of the applied N escapes, either by leaking into the groundwater as nitrate or into the atmosphere as N_2, N_2O, NO_x and NH_3. The leaked nitrate perturbs the nitrogen cycle of downstream freshwater and coastal zone ecosystems. The reactive nitrogenous gases (NO_x and NH_3) that are not immediately taken up by the vegetation canopy are transported away from the site of emission. In the atmosphere they are converted to nitrate, which is deposited back on the land or sea surface.

The form of N escaping to the atmosphere is important because N_2O has about a 120-year lifetime in the atmosphere (it has increased from 275 to $c.$ 310 ppb this century), whereas NO_x has a lifetime of about a day (E. Holland, personal communication).

10.4.3 Sulphur deposition as a disruption to the carbon cycle

The IGBP deliberately excluded the sulphur cycle as a main focus, not because it was considered unimportant or unchanging, but because a substantial body of work had already been done on sulphur as a result of concerns over 'acid rain'. The key role of sulphur aerosols in counteracting the greenhouse effect has brought the sulphur cycle back to centre stage.

The global sulphur cycle has been regionally perturbed by the burning of fossil fuels, which contain 1–2% sulphur. The sulphur is released as SO_2 and a variety of other sulphurous gases, which convert in the atmosphere to H_2SO_4, the main component of 'acid rain'. The sulphur in the acid deposition may initially act as a fertilizer in S-deficient soils, but the long-term cumulative effect is deleterious. The H^+ which accompanies the sulphate anion progressively displaces cations such as Ca^{2+}, Mg^{2+} and Al^{3+} from the soils into the groundwater and rivers. Plants growing in the cation-depleted soils exhibit nutrient deficiency and aluminium toxicity pathologies which stunt their growth (Schulze & Ulrich, 1991; Schulze & Freer-Smith, 1991). Aluminium becomes soluble at low pH, and is toxic to plants and to animals in the freshwater ecosystems into which it leaches. The terminal effect, as yet not generally reached, is tree death (or replacement by herbaceous species), which would result in a release of biomass carbon to the atmosphere.

The global sulphur cycle is frequently, but unpredictably, perturbed by major volcanic eruptions. The 0.5 °C decline in mean annual temperature and the slowdown in the rate of increase of atmospheric CO_2 and CH_4 in 1991 have been suggested to be due to the dust and sulphate aerosols injected into the atmosphere by the eruption of Mt Pinatubo (Schimel *et al.*, 1996). The aerosols reflect solar radiation, reducing the temperature at the ground surface. This reduces photosynthesis, respiration and methanogenesis. The global mean temperature recovered and the atmospheric CO_2 and CH_4 returned to their former trajectories within three years. The incident illustrates the sensitivity of biospheric processes to temperature, and how a localized event involving one elemental cycle can have a profound impact – amounting to several billion tons of carbon – on the global cycle of another element.

Sulphate aerosols and the greenhouse effect

Anthropogenic and natural sulphate aerosols together have an anti-greenhouse warming effect of -0.4 W m^{-2} (Schimel *et al.*, 1996). Anthropogenic sulphate is injected into the lower troposphere, so it settles to earth within weeks. Thus, the anthropogenic sulphate cooling effect is not globally evenly distributed, but is concentrated around the areas of maximum emissions. Major volcanic eruptions, on the other hand, inject sulphate aerosols into the upper troposphere and stratosphere, where they remain aloft for years and have a much wider effect.

10.5   Nutrients as a constraint on the C cycle

The 'biospheric elements' C, N, P and S do not cycle independently of one another, but neither do they necessarily proceed in lock-step, like cogs in a machine. At certain points they interact synergistically. The availability (or unavailability) of one element disproportionately influences the cycling of another, and therefore acts as a threshold or switch. This section discusses the global implications of these linkages.

10.5.1   Ecological stoichiometry

It is useful to consider the elements C, N, P and S together because their cycles are tightly linked by their necessary co-presence in key biological molecules. As a result, the organic portion of their cycles (i.e. the portion in biomass, recently dead material and soil and sediment organic matter) are identical, and their collective release from organic bondage is mediated by the same processes of decomposition.

The proportions of these four elements in broad classes of organic materials is relatively fixed (Table 10.1), leading to the very useful concept of 'ecosystem stoichiometry'. The predictability of these ratios allows the cycles to be linked without much detailed knowledge of the biological species and pathways involved. A key issue when using this principle in forcasting the future state of the

Table 10.1 *The range of concentrations of C, N, P and S in biological materials. Values are means and standard deviations*

Material	Carbon	Nitrogen	Phosphorus	Sulphur
		mg g^{-1} DM		
		molar ratio (C = 100)		
Mature, living leaves				
BNF crops		39.5 ±4.4		
	100	**7.5**		
Cereal crops		34.5±8.8		
	100	**6.6**		
Conifers and sclerophylls		10.7±2.5		
	100	**2.0**		
Temperate deciduous trees		20.8±4.9		
	100	**4.0**		
Temperate evergreen trees		13.6±2.8		
	100	**2.6**		
Tropical deciduous trees		28.6±5.3	0.63±0.06	1.1
	100	**5.4**	**0.05**	**0.09**
Tropical evergreen trees		15.3±4.6		
	100	**2.9**		
Temperate grassland		25.5±8.4		
	100	**4.9**		
Tropical grassland	422±21	10.7±9.8	0.60±0.25	1.0±0.36
	100	**2.0**	**0.06**	**0.09**
Wood	454±10	6.6±3.2	0.8±0.7	0.6±0.4
	100	**1.2**	**0.07**	**0.05**
Soil organic matter				
Temperate	**100**	**9.4**	**0.4**	**0.3**
Tropical	**100**	**4.2**	**0.2**	
Phytoplankton	**100**	**13.6± 1**	**0.85±0.09**	

Sources: Leaves – Schulze *et al.*, (1994*b*); wood– Scholes & Walker (1993); phytoplankton – Anderson & Sarmiento (1994); soil – Sanchez (1976).

biosphere is whether the proportions will remain constant under a different scenario of relative availability.

The narrow ranges of concentration exhibited are thought to be stable, due to the necessity of having given proportions of various classes of molecules in plant, animal and microbial cells in order to satisfy their physiological functions. The elemental proportions in molecules are determined by conventional chemical stoichiometry, which requires fixed molecular composition in order to satisfy molecular structure and function constraints.

The relationships between the proportions of the four elements underpin much of the biogeochemical modelling of recent years (Parton *et al.*, 1988; Melillo *et al.*, 1993). If it is assumed that most terrestrial ecosystems are nitrogen limited, then NPP can be predicted from a knowledge of the nitrogen mineralization rate and the life-form composition of the vegetation. For example, if the nitrogen mineralization is 1 g N m^{-2} y^{-1} in a grassland, the NPP would be about 100 g DM m^{-2} y^{-1}, the carbon assimilation about 50 g C m^{-2} y^{-1}, and the P and S uptake 0.1 and 0.2 g m^{-2} y^{-1} respectively. The nitrogen mineralization can in turn be predicted from the decomposition rate of carbon in litter and soil organic matter.

10.5.2   The importance of 'leaks' in determining nutrient limitation

Most current global biogeochemistry models work on the basis of simple mass balance; nitrogen released from the soil is all available for re-sequestration in biomass. Few handle nitrogen losses adequately. Accurate and complete knowledge of the losses is crucial in determining the long-term carbon sequestration potential of an ecosystem. Semi-process based models of denitrification compatable with the global biogeochemical models are now beginning to appear (Parton *et al.*, 1996), which will help to remedy this problem. Similar attention needs to be given to losses by leaching (especially of dissolved organic nitrogen) and as NO. Biogenic NO losses from the soils of tropical seasonally dry systems are of the same order of magnitude as losses through the burning of vegetation (Poth *et al.*, 1995; Scholes *et al.*, 1997).

10.5.3   The role of water in controlling biogeochemical cycles

The carbon cycle is tied to the water cycle at the shared point of the plant stomata. The C, N, P and S cycles are all linked to the water cycle due to the dependence of soil biological processes, such as decomposition, on the soil moisture content.

The rate-limiting process for carbon assimilation by plants is generally the diffusion of CO_2 through the stomata. This is typically also the rate-limiting step in the evaporation of water from terrestrial ecosystems. As a consequence, there is usually a linear relationship between the NPP of a plant community and the cumulative amount of water it has transpired. The same logic underpins the relationship between production and rainfall in semi-arid regions (Scholes, 1993). The slope of these relationships is often expressed as the Water Use Efficiency (WUE), the grams (or moles) of carbon assimilated per gram or mole of water lost. This term is not used here since it implies direct proportionality, which is not necessarily the case due to potential nonlinearities and a non-zero intercept.

Under elevated atmospheric CO_2 levels the CO_2 diffusion gradient across the stomatal aperture steepens, but unless the atmosphere becomes drier, the H_2O diffusion gradient remains the same. This leads to an increase in carbon assimilation per unit of water transpired at the leaf level. The plant responds by partial stomatal closure in the short term, and reduction in stomatal density and possibly reduction in the leaf area in the medium term. As a result, the ecosystem-level NPP per unit of evaporation increases with increasing CO_2. The increase at the ecosystem scale is smaller than the increase at the leaf scale, since adaptations and other constraints (such as nutrient supply) operate simultaneously. Arid ecosystems (and mesic ecosystems in dry years) benefit most from this effect. It is worth noting, however, that some ecosystems exposed to elevated CO_2 have not shown decreased stomatal conductance (see Chapter 7).

The magnitude of the effect on global transpiration is unknown, since the strength of feedbacks from the atmosphere to the canopy, resulting from the reduced input of water to the atmosphere, is unknown (Amthor, 1995). Reduced transpiration would alter the energy balance of the land surface, leading to warming and an increased vapour pressure deficit in the canopy boundary layer. Modelling results disagree on whether this feedback is sufficiently large to affect the global climate (Henderson-Sellers et al., 1995; Pollard & Thompson, 1995; Sellers et al., 1996). Observational evidence during regional droughts tends to support the view that reduced transpiration can have a regional climate impact, especially in continental regions (Lloyd et al., 1996).

Reduced transpiration is expected to increase soil moisture duration, and in moist systems, groundwater recharge and streamflow. This has been observed in some ecosystem elevated CO_2 experiments (see Chapter 7). The rate of decomposition of litter and soil organic matter in terrestrial ecosystems is under the overriding control of soil moisture and temperature (Couteaux et al., 1995). The N-mineralization rate is therefore linked to the soil moisture content. The strong observed relationships between rainfall and NPP could thus equally be a reflection of moisture limitation of nitrogen mineralization (or of phosphorus uptake), rather than the effects of water stress on stomatal closure, as is generally assumed (Pastor & Post, 1993; Scholes, 1993). If this is the case, higher CO_2 will only indirectly lead to higher NPP, via increased soil moisture and temperature. The effect would be of short duration, until soil C and N levels had equilibrated to the new moisture and temperature regime. The correlations between NPP, nitrogen mineralisation, and evapotranspiration at a biome scale, as they emerge from a biogeochemical model running globally, are shown in Fig. 10.4.

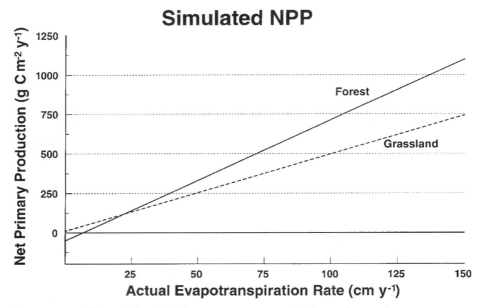

Figure 10.4   The relationships between evapotranspiration and net primary
production as they emerge from the CENTURY model, applied globally. The
structure of this model relates NPP to nitrogen mineralization. The near-linear
relationship between evaporation and NPP, very similar to those observed in
semi-arid lands, is an emergent property. The differences in slope between biomes are
largely due to differences in the C : N ratio of different plant functional types,
indicating that ecosystem composition has direct effects on biogeochemistry. (W.J.
Parton, personal communication).

Water, oxygen and anaerobiosis

The pathways of chemical transformations in the soil, and therefore the products formed, are critically dependent on the oxygen supply, which is in turn
largely controlled by the fraction of the soil pore volume that is occupied by
water (the water-filled pore space, WFPS). If the WFPS exceeds about 60%,
then the diffusion of oxygen from the atmosphere to the sites of decomposition
is restricted, and anaerobic processes begin to dominate. Although the switch
between aerobic and anaerobic states is physiologically quite abrupt at a redox
potential of −150 to −160 mV (Wang *et al.*, 1993), at the scale of the whole soil
profile or landscape the transition is gradual, since the anaerobic microsites,
which form in areas of high oxygen demand even in well-drained soils, expand
gradually until above about 80% WFPS they dominate the profile.

Oxygen supply has a controlling influence on the type and quantity of trace
gases released from the soil (Happel & Chanton, 1993). In aerobic systems, the
main gaseous emissions are carbon dioxide and nitric oxide (NO), and methane

is consumed through oxidation at a low but significant rate. Globally, methane oxidation is estimated to amount to 15–45 Tg y^{-1}, or 3–9% of the methane sink, the same order of magnitude as the annual atmospheric increase (Schimel *et al.*, 1996). Where there is an increased availability of inorganic N, due either to N deposition or deliberate fertilization, the capacity of soil organisms to oxidize methane is reduced. Higher atmospheric methane levels themselves depress the rate of methane oxidation (King & Schnell, 1994). The interactions mentioned in this paragraph are real, but can be responsible for only a small part of the observed global atmospheric methane increase over the past century.

In anaerobic systems, nitrate or sulphate are used as a terminal electron acceptor in place of oxygen. Where nitrate is used, the process results in the emission of N_2 and N_2O, and is known as (anaerobic) denitrification. Denitrification in terrestrial ecosystems is thought to be the major global source of N_2O, but substantial uncertainties remain in the global N_2O budget (Nevison *et al.*, 1996).

Anaerobic decomposition releases carbon in the form of CH_4 and sulphur as H_2S. Isotopic evidence suggests that between 70 and 80% of the global methane source is biogenic (Quay *et al.*, 1991). Anaerobic decomposition is slower than aerobic decomposition, so water-saturated systems (wetlands, including bogs and tundra) are major global carbon stores (455 Pg C, about 30% of the terrestrial total, Oechel *et al.*, 1993), and may be a location of the inferred northern hemispheric terrestrial carbon sink. When wetlands are drained for agricultural or urban development, the soil carbon is released as CO_2 over a period of several decades. The carbon in drained peat may be released catastrophically by wildfire, or may be burned as fuel (Hogg *et al.*, 1992). Wetlands are also important global CH_4 soures (Bartlett & Harriss, 1993), thought to account for 115 (55–150) Tg CH_4 y^{-1}.

The widespread cultivation of paddy rice increases the global extent of anaerobic systems. Rice paddies are thought to contribute 20–100 Tg CH_4 y^{-1} to the atmosphere (6–30% of the total anthropogenic source; Neue & Sass, 1994). Not only is the extent of flooded rice paddies increasing, but as a result of land management practices such as fertilization, there is more carbon and nitrate substrate present to form trace gases. It has been argued that the expansion of paddy rice is largely at the expense of existing wetlands, and thus one anaerobic system replaces another, reducing the net effect.

Anaerobic conditions are also found in the guts of animals. Enteric fermentation in ruminants and, to a lesser extent, termites are sources of methane that together could amount to 100 Tg y^{-1} (Prather *et al.*, 1995). Where the fraction of the biomass processed by these organisms changes as a result of human activities, the result could be a significant change in trace gas emissions. The global domestic livestock population has increased in parallel with the increase in the human population and its rising standard of living. The total number of ruminants, and thus the quantity of NPP consumed by herbivores, is probably currently much higher than at any time in the past, even allowing for steep declines in wild ungulate populations.

10.5.4   Constraint of the carbon cycle by nitrogen

Nitrogen is frequently the most limiting nutrient in terrestrial ecosystems, largely because its terrestrial cycle permits losses either in gaseous or dissolved form. Nitrate is highly soluble, and poorly retained by soils since it bears a negative charge. The many oxidation states of nitrogen have resulted in the evolution of a wide range of bacteria that use nitrogen compounds as an energy source, in the process releasing the gases N_2, N_2O, NO, NO_2 and NH_3. Although the atmosphere is 78% N_2 and certain bacteria and algae are able to convert N_2 to NH_4^+, this is an energetically expensive process. As a result, terrestrial ecosystems are generally short of nitrogen.

Nitrogen limitation is ultimately responsible for the emissions of large quantities of trace gases in tropical savannas. The emission of methane per unit of forage consumed increases as the N content of the forage decreases (Blaxter & Clapperton, 1965). The digestability of plant tissues by bacteria in the rumen is critically dependent on the maintenance of at least 0.8–1% N in the forage. In an elevated CO_2 environment, the nitrogen content in plant tissues decreases (see Chapter 7). If the lower N content is mostly due to increases in leaf starch, as is generally the case in grasses, the forage would remain digestable, but enteric methane emissions would be expected to rise.

Vegetation fires

Nitrogen–limited tropical grasslands and savannas often drop below the 1% N threshold during the dry season, and as a consequence support a low mammalian herbivore biomass in their natural state (see Chapter 9). The grass accumulates, forming a highly flammable, fine, dry fuel, which eventually burns, ignited either by lightning or humans. Pastoralists burn the grasslands to remove the over-burden of dead, unpalatable grass, which shades the new leaves and restricts the access of grazers, and to recycle a portion of the nitrogen immobilized in the soil and dead standing grass. In the process, 40 to 90% of the nitrogen in the dead grass is released to the atmosphere (a process known as 'pyrodenitrification'), further reinforcing the nitrogen limitation of the ecosystem. The other major global biome where large amounts of biomass burns are the boreal coniferous forests, which are also generally N limited.

The emissions from vegetation fires include CO_2, CH_4, smoke particles, CO, NO_x, and a variety of other hydrocarbons and nitrogen and sulphur-containing gases. If the burned ecosystem has a long history of exposure to fire, and the fire frequency and fuel load is not known to have changed in the recent past, it is assumed that the long-term net CO_2 emissions are zero, since the CO_2 emitted by the fire is reassimilated in the period between fires. This is a simplifying assumption, applied for instance in the IPCC methodology (IPCC, 1995), which is untrue at small spatial and temporal scales, and may not be true at regional scales over the past 200 years. It is unclear if the global amount of biomass burned is increasing or decreasing under human influence, since there are local examples of both trends. For example, the savannas of Australia, southern Africa and southern North America have generally increased in biomass, largely as a result of a reduced fire frequency and intensity (Scholes & Archer, 1997). On the other hand, the fire frequency in Canadian boreal forests has increased in the past 30 years (Kurz et al., 1995).

The hydrocarbons CO and NO_x emitted by fires interact to form tropospheric ozone, a powerful a greenhouse gas. It has been estimated that 60% of the global biomass burning occurs in savannas. Biomass burning contributes up to 43 Tg CH_4, 680 Tg CO, 90 Tg particles, 21 Tg NO_x and 420 Tg O_3 to the atmosphere annually (Andreae, 1991). These numbers are probably overestimates, due to overestimation of the area that burns each year and the fuel load involved (Scholes et al., 1996b). Even with the reduced estimates, the trace gas emissions from vegetation fires are globally significant. Wildfires have occurred for millenia and are an integral part of the ecology of fire-prone systems.

Litter quality and the global litter carbon pool

The nitrogen : lignin ratio of plant tissues is a good predictor of decomposition rate in plants that do not have a high polyphenolic content (Aber & Melillo, 1982; Couteaux et al., 1995). The effect of elevated CO_2 on the polyphenolic content, the main quality control on decomposition in tropical litters (Palm & Sanchez, 1990), is unknown. Under elevated CO_2 the N content in live tissues generally falls, suggesting a parallel decreases in litter, leading to a decrease in decay rate (Bottner & Couteaux, 1991; Cotrufo & Inneson, 1995; Gorissen et al.,1995), and thus an accumulation of plant litter. This would act as a negative feedback on further growth in atmospheric CO_2. Experimentation does not always support this hypothesis (see Chapter 7, also Couteaux et al., 1995; Arp et al., 1997). The relative decline in N content in grass leaves is mostly due to the increase in starches and sugars, forms of carbon which are easily decomposed. There are some indications that the decomposition rate of C3 graminoids grown in elevated CO_2 decreases, due to increases in C : N ratio, but there is no effect in C4 species. The lignin content does not seem to be affected.

Higher global temperatures and soil moisture duration act to increase decomposition rates, a positive feedback on atmospheric CO_2. However, in the process they release N from soil organic matter and litter (Van Cleve et al., 1990). In

N-limited ecosystems dominated by woody plants, this would result in an increase in NPP that would exceed the loss of C due to decomposition, and the net transient effect of soil warming would be CO_2 uptake (Bonan & van Cleve, 1992). If the ecosystem is already N saturated, soil warming will lead not only to carbon loss, but to nitrate leakage to the hydrosphere.

Large-scale integrated assessments

The integrated effects of carbon and nitrogen cycling and vegetation change under elevated CO_2 and global warming have been explored at the continental scale by comparing three state-of-the-art biogeochemical models (VEMAP, 1995; see descriptions in Chapters 7 and 8), in factorial combination with three phytogeography models driven by three different GCM-generated climate change scenarios. The three biogeochemistry models were BIOME-BGC, CENTURY, and TEM, and the area of study was the continental United States of America.

Since there currently exists no basis for selecting which of the GCMs, biogeography models, and biogeochemistry models are best or most accurate, several models of each type were employed as a measure of the uncertainty that still exists concerning future patterns of change in climate and ecosystems. The range of results for any particular aspect of biogeochemistry reflects the differences in the assumptions and formulations that are incorporated in the various types of models used in the exercise. All models were run using common input data sets. Consequently, differences in the predictions from similar models can be attributed to dissimilarities among the models in question rather than to different input conditions.

The various model combinations predict widely divergent results, with changes in NPP under doubled atmospheric CO_2 varying from −0.7% to +39.7%, and changes in total carbon storage (plants and soils) varying from −39.4% to +32.3% relative to the present. Thus, on the basis of current knowledge it seems reasonable to predict that the average NPP over all natural terrestrial ecosystems is likely to increase if climate changes, but it is not possible to anticipate even the direction of the response for carbon storage.

Because biogeographic and biogeochemical responses were simulated by different models, it is also possible to separate the responses in NPP and carbon storage into those components that are due to structural (i.e. the geographic extent of different biome types) and functional changes (Fig. 10.5). For some model combinations structural responses dominated, while for others functional responses were most important. In general, structural changes tended to be important when they involved changes in the extent of forest biomes.

It is clear that all three types of models (climate, phytogeography and biogeochemistry) contributed to the wide range in results, but there are certain

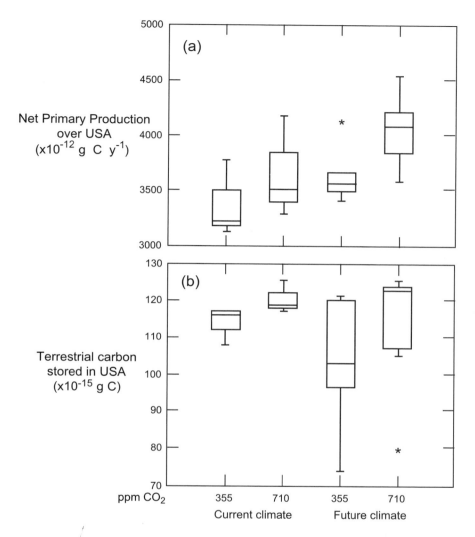

Figure 10.5 Results of the VEMAP experiment, which compared the predictions of various combinations of climate and ecosystem models for (a) NPP, and (b) biospheric carbon storage for the United States of America (excluding Alaska and islands). The box-and-whisker plots show the range of results for the current climate, a change in atmospheric composition only, a change in the climate only and a combined atmospheric and climatic change * = outlier values.

consistent patterns. For instance, TEM tended to predict larger functional responses for both NPP and carbon storage than did either CENTURY or BIOME-BGC. In terms of carbon storage, BIOME-BGC always predicted losses of carbon, CENTURY predicted small gains, and TEM predicted large gains. There are known differences among the biogeochemistry models that

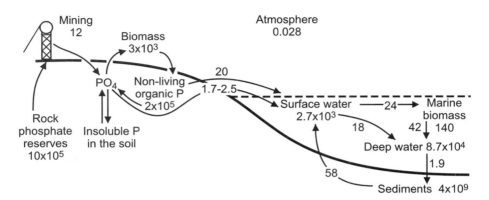

Figure 10.6   The human–perturbed global phosphorus cycle. The key feature is the inexorable flow of P from the land to the oceans, only to be returned in geological time by the uplifting of former ocean sediments to create new areas of land. The biological part of the cycle, while extremely important for life, is a minor ripple on the surface of the huge geospheric and oceanic pools. In contrast to the N cycle, the P cycle is very 'closed', as a result of the absence of significant gaseous forms and the low solubility of phosphate. The key reactions are geochemical rather than biological. From Schlesinger, 1991.

contribute to the differences in predictions (VEMAP, 1995; see also Chapter 7). For instance, in BIOME-BGC, the Q_{10} for decomposition is 2.4, whereas is approximately 2.0 in TEM and CENTURY; this could be a factor that helps to determine the differences in carbon storage. The models also differ in terms of how they treat nitrogen and the constraints it can exert on carbon gain and carbon storage patterns. Note that the variation in model output caused by disagreements between the climate models is as great or greater than the differences between biogeochemical models.

A model experiment covering the Amazon basin, using only one biogeo-chemical model (GEM) and no phytogeographical model, concluded that with a realistic increase in atmospheric CO_2 and rise in temperature, the total carbon stock per hectare in tropical forests could increase by 16% over the next 200 years (McKane *et al.*, 1993).

10.5.5   Constraint of the carbon cycle by phosphorus

Phosphorus is frequently the most limiting nutrient in freshwater ecosystems, due to its low solubility. It is also limiting in some terrestrial ecosystems, particularly those on parent materials inherently low in phosphorus and those that have been exposed to intense weathering. Phosphorus forms insoluble complexes with iron that further reduces its availability, especially in the tropics.

The human influence on the terrestrial P cycle (Fig. 10.6) is large relative to the annual exchanges with the hydrosphere and the atmosphere. The global P cycle has been perturbed by the widespread use of phosphate fertilizer, which represents a transfer from highly localized, geospheric reserves inaccessible to most plants into the biosphere, via agricultural ecosystems. The phosphorus incorporated in agricultural products is transferred through regional and global food trade from rural to urban areas, and from the developing to the developed world. It enters fresh and salt-water ecosystems largely via urban waste streams, causing eutrophication (over-stimulation of production through nutrient additions, leading to oxygen deficiency and loss of diversity) in rivers, lakes, estuaries and coastal zones.

Phosphorus, unlike nitrogen, is probably not directly involved in enhancing the terrestrial carbon sink to any significant degree. This is because its application is highly localized, and it is not readily transported in terrestrial systems. The relocation of phophorus from rangelands and forests to urban areas as a result of harvest could decrease the capacity of these systems for production and carbon sequestration.

10.5.6 Carbon–nitrogen–phosphorus interactions

Although Biological Nitrogen Fixation (BNF) occurs at low rates in free-living organisms, the major global BNF pathway is via bacteria residing symbiotically in the root nodules of certain leguminous plants. The host expends up to 50% of the carbon assimilated through photosynthesis subsidizing the bacteria to perform this function (BNF requires the equivalent of 6.5 g C per gram of N fixed), which may be why even N-limited ecosystems are seldom dominated by BNF plants. In a high-CO_2 world, however, the relative costs of allocating carbon to BNF are decreased and the degree of N limitation increases, suggesting that plants with nitrogen-fixing symbionts should be favoured. Evidence from elevated CO_2 experiments is equivocal (see Chapter 7). The resultant increase in leaf N would increase the photosynthetic capacity of these species, providing a further positive feedback loop. The global consequence could be a widespread change in species composition (including, in agricultural systems, larger areas planted to BNF crops such as soybeans) and a futher increase in nitrogen inputs to terrestrial ecosystems.

The BNF process is critically dependent on the adequate availability of phosphorus (and molybdenum), which may be another reason why BNF is not universal in natural ecosystems (Vitousek & Howarth, 1991), and which may constrain increased BNF under a high CO_2 environment. The availability of P to higher plants growing in low-P soils is typically assured by symbioses between the plant root and vesicular–arbuscular mycorrhizae. As in BNF, the symbionts must be subsidized with carbon from the plant, and may thus become more prevalent under elevated CO_2. The global consequences of increased mycorrhizal infection would be increased NPP, but also increased soil respiration, so the direct impact on carbon sequestration may not be large. The indirect effect, via increased BNF, may permit continued carbon sequestration in biomass.

10.6   The 'missing' carbon sink

The global carbon cycle is currently highly buffered, in that 7.1 ± 1.1 Gt C (the error term represents the 95% confidence interval) is added to the atmosphere annually by human activities, yet only 3.3 ± 0.2 Gt C of this amount appears in the atmosphere (Table 1.2; Schimel *et al.*, 1996).

The quantity taken up by the oceans is limited to 2.0 ± 0.8 Gt by the rate of the physical process of solution of the CO_2 at the ocean–atmosphere boundary. This is confirmed by the isotopic composition of the atmospheric CO_2 (Tans *et al.*, 1990, Siegenthaler & Sarmiento, 1993) and O : N ratios (Keeling & Shertz, 1992). The remaining approximately 1.8 Gt C unaccounted for is assumed to be taken up by terrestrial ecosystems and is referred to as 'missing' since it has not been independently and unequivocally located by direct measurement. Note that the uncertainties associated with the other estimates are sufficient to account for the 'missing' sink if the estimates of the sources were too high and the estimates of known sinks were too low.

The global distribution pattern of atmospheric CO_2 concentrations, their seasonal fluctuations and isotopic composition all suggest a terrestrial ecosystem sink (Enting & Mansbridge, 1991; Ciais *et al.*, 1995); furthermore, they suggest that it is likely to be in the northern hemisphere (Keeling *et al.*, 1996*b*). Changes in the growth rate and extent of temperate forests are thought to be responsible for 0.5 ± 0.5 Gt C of the terrestrial sink.

Models suggest that CO_2 fertilization could be responsible for a large part of the missing C (Harrison & Broekner, 1993; Gifford, 1994*b*; Friedlingstein *et al.*, 1995). Such calculations are controversial. Schimel (1995) has summarized modelling efforts by a number of groups and highlights some problems of limitation to ecosystem C storage: uncertainties regarding the nutrient content of vegetation, the low growth response of native species, and the assumptions about allocation of the extra NPP between short and long-lived tissues. He concludes that CO_2 fertilization could account for 1.0 ± 0.5 Gt C y^{-1}.

Nitrogen fertilization, principally due to deposition of industrial emissions, could contribute to a further uptake of 0.6 ± 0.3 Gt C y^{-1} (Schindler & Bayley, 1993; Hudson *et al.*, 1994). Together, these additional sinks are sufficient to account for the remaining 'missing sink', which is thus no longer causally missing but remains uncertain as to its distribution within the terrestrial biosphere (Schimel, 1995).

Climate change itself could contribute to the missing sink. The interannual variations in the global climate have a measurable effect on the global carbon cycle (Dai & Fung, 1993). On average, the terrestrial carbon fluxes are closely balanced around a net 0 flux (with a slight, overall positive 'sink' – see Table 1.2), but in any one year the flux can be negative or positive. Thus, rising CO_2 could have an indirect effect on carbon uptake, via the climate change it is thought to cause.

A similar analysis can be applied retrospectively. When the estimated year-by-year emissions of CO_2 from fossil fuel combustion since 1850 are compared with the reconstructed atmospheric CO_2 concentrations, and geophysical contraints are applied to calculate the quantity that could have been absorbed by the oceans, it appears that the net flux from terrestrial ecosystems has been 25 Pg C (Sarmiento et al., 1992) to 50 Pg C (Siegenthaler & Oeschger, 1987). Prior to 1940 this flux was mostly positive (i.e. from the biosphere into the atmosphere, as a result of land-cover change), whereas since that date it has been negative (Goudriaan, 1995). Since it is calculated that land-use changes have contributed 120 Pg C to the atmosphere over the period, the remaining areas of untransformed terrestrial ecosystems must be acting as strong sinks (Houghton, 1995).

It seems, therefore, that the terrestrial carbon sink is real, but there is no consensus regarding its location. It has been claimed to have been found in the temperate forests (Kauppi et al., 1992; Sedjo, 1992; Kolchugina & Vinson, 1993). These studies are based on incomplete carbon budgetting. When carbon losses due to harvest and fires are included, the net carbon uptake is close to zero (Houghton, 1993). When forestry statistics report increased growing stocks, they confirm a local terrestrial carbon uptake; but they do not necessarily help explain the missing sink because it neglects respiration. To explain the missing sink, either (i) the observed rates of accumulation must be greater than the rates expected as a result of forest growth following logging or agricultural abandonment, or (ii) the accumulation must be in ecosystems not directly modified by human activity and ignored in analyses of land-use change (Houghton, 1995). In comparing data from deconvolving the historical emissions, land-use change and the atmospheric CO_2 record with data from forest inventories, Houghton (1995) concludes that 'they generally show similar estimates of uptake by vegetation and that the difference is much smaller than the global C imbalance (missing sink) of 1.6 Pg C y^{-1}'. Even though the different approaches are providing similar results, comparative analyses should be pursued because the difference between the two approaches offer one of the only means for direct assessment of the missing sink.

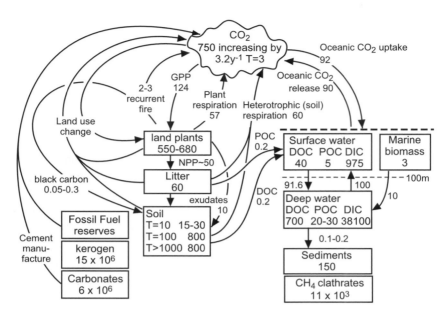

Figure 10.7   The global carbon cycle. The fluxes in this figure are average values in Pg C y^{-1}; the pods are in Pg C. T refers to the mean retention time in the pool, in years. In reality the fluxes vary greatly within and between years, in addition to the overall increasing trend due to rising atmospheric carbon. All the estimates are subject to measurement uncertainty of up to 20%. Note that although the largest pools are in the form of carbonate rocks, fossil fuels and deep ocean water, these pools have a very slow turnover rate. In contrast, the atmospheric, vegetation and soil organic matter pools have turnover rates of decades to centuries, making them the focus of global change research. From Schimel (1995); Kirschbaum *et al.*, (1996*a*), Amthor (1995); Klein Goldewijk & Leemans (1995).

Terrestrial carbon pools

Under conditions of nutrient limitation, which are thought to be widespread in undisturbed terrestrial ecosystems, the rate at which the ecosystems can sequester carbon is determined by the rate of input of the limiting nutrient, up to the point where it is no longer limiting. The total carbon storage capacity, on the other hand, is largely determined by the climate and soil. The two main carbon pools in terrestrial ecosystems are in plant biomass and in the soil (Fig. 10.7). The upper limit of the accumulation of plant biomass is largely controlled by the climatic factors that determine both what functional types of plants can exist in a given environment and the prevailing disturbance regime. There is, for instance, an approximately hundred–fold difference in the biomass per unit area of forested and grassland ecosystems.

How much of the predicted increase in NPP of terrestrial ecosystems (as a result of elevated CO_2, climate change and nitrogen deposition) finds its way into long-lived C pools is unknown, but crucial for understanding whether the current apparent terrestrial ecosystem C sinks will continue unchanged in the future, or increase or decrease.

Sequestration as black carbon

There are few biogeochemical processes that remove carbon semi-permanently from the biospheric cycle at rates meaningful to the study of global change. 'Black carbon', which is carbon in the form of highly condensed compounds (H : C molar ratio < 0.2), is virtually immune to microbial or chemical decomposition and is therefore effectively inert. A fraction of the fuel consumed by vegetation fires is converted to black carbon. The size of the fraction depends on the oxidation state of the fire, ranging from about 12% in flaming combustion to close to zero in smouldering combustion (Kuhlbusch et al., 1996). Four-fifths of the black carbon formed remains on the ground as charcoal, while the rest is given off in the smoke as soot. The soot can remain airborne for days or weeks, contributing to the aerosols that counteract the greenhouse effect and acting as a nucleus for a variety of atmospheric reactions and the condensation of clouds, which further counteracts global warming. The global estimates for black carbon formation through vegetation fires are 50–270 Tg, with most of the uncertainty associated with the amount of biomass burned. Thus vegetation fires present a paradox – in the medium term they release greenhouse gases such as CH_4 into the atmosphere, while in the long term they sequester C in a very secure form.

Black carbon can also be produced by microbes as the endpoint of the humification process. The precise mechanism of this process, and the chemical structure of the compounds produced, is poorly known (Haumaier & Zech, 1995), but the product is essentially indistinguishable from the black carbon produced by fires. This raises the possibility that part of the highly chemically resistant carbon in the soil has its origins in fire, rather than, as previously though, in microbial processes.

If black carbon cannot be decomposed, why does it not completely dominate the soil carbon? It is suggested that the main mechanism of its removal from the soil profile is by erosion, either by water or wind. The ultimate repository is thus the ocean (Verardo & Ruddiman, 1996).

The soil carbon pool

The size of the soil carbon store depends on the depth of the soil profile, its water regime (wet soils store more carbon than dry soils), temperature (cold soils store more than warm soils) and the type and quantity of clays present (more clay means more carbon; high-activity clays are more effective than low-activity clays). Long-term carbon storage in terrestrial ecosystems requires that the carbon be converted to a form that is partly protected against decomposition. This is typically a chemically resistant or physically occluded form of soil organic matter. Physical protection is afforded by complexation with clays, or in soil aggregates. It is disrupted by soil disturbance, for instance by ploughing. The processes leading to chemically resistant soil carbon are poorly understood (see the discussion above) but whatever their origin, there is no indication of enhanced production of such compounds under elevated CO_2.

Warmer soils contain less carbon than cooler soils, all else being equal. The globally integrated modelled sensitivity of the soil carbon store to increases in global temperature is in the range -11 to -39 Pg C $°C^{-1}$ (Schimel *et al.*, 1994, Post *et al.*, 1994). Correlative studies based on current climate-soil carbon relationships suggest -14 Pg C $°C^{-1}$ (Buol *et al.*, 1990). Ecosystem-scale soil warming experiments show an increase in soil respiration. The loss of carbon from the soil following warming may be partially or completely offset by increases in the carbon input to soils as a result of enhanced productivity under elevated CO_2 (subject to the uncertainties discussed above), but little is known definitively about the effect of high CO_2 on litter quality and how this might work through to soil carbon pools.

Due to these uncertainties, changes in the global litter and soil carbon stores cannot be accurately predicted at present. Current elevated CO_2 experiments have not been maintained for long enough to detect small changes in a large soil carbon pool with statistical certainty.

The plant biomass carbon store

The second major biospheric carbon store is in plant biomass, primarily in living wood. The global wood biomass can increase in two ways – either by increased growth of existing woody plants or by increased dominance of woody plants in areas that were previously dominated by grasses or herbs. Accelerated tree growth in the past century relative to previous periods has been detected in several places (Kauppi *et al.*, 1992; Sedjo, 1992), but appears to be more associated with nitrogen deposition and temperature changes than with elevated CO_2.

The survival and dominance of trees in a given environment requires air and soil temperatures above a certain minimum (trees are absent at high latitudes and altitudes), sufficient soil moisture (trees are absent from deserts, except in drainage lines) and infrequent fire. Warming and increased rainfall could therefore lead to more wooded landscapes. The rate and extent of future biome migrations is not known. Since such movements generally require disturbances and mortalities to open the space for invasion, during the transient phase to a new equilibrium they could well be associated with loss, rather than gain, of biomass (see Chapters 8 and 12; Smith & Shugart, 1993).

In tropical savannas, the carbon storage is determined by the proportions of C_3 trees to C_4 grasses. Theory predicts C_3 species to be advantaged by elevated CO_2, but C_4 species by higher temperatures. Elevated CO_2 experiments give mixed results on this point (see Chapter 7). The role of tropical grasslands and woodlands – their soil C sinks and their relative abundance of C3 and C4 species – in the carbon cycle has implications for the interpretations of atmospheric ^{13}C data related to the location of terrestrial C sinks (Ciais et al., 1995). Phillips & Gentry (1994) present data similar to that presented for temperate forests that suggest that tropical forests may be accumulating biomass where they are not subject to land-use change.

In the past the possibility of tropical soils accumulating C on any scale in response to climate change has been generally dismissed as being unlikely. However, recent analyses of the NPP of tropical grasslands (Long et al., 1992), organic C in tropical savannas and woodlands (Scholes & Hall, 1996), introduced and native deep rooted grasses in South America (Fischer et al., 1994; Nepstad et al., 1994) and maximum rooting depth and root distributions (Canadell et al., 1996a; Jackson et al., 1996a) has resulted in a re-evaluation of the role of tropical soils in present and future C storage. Model results suggest that tropical grasslands and savannas might be soil C sinks (Parton et al., 1995).

It seems likely that the 'missing' terrestrial carbon sink is not in one place, but is distributed over the globe, and probably over the error terms in the various estimates as well. It will be very hard to detect, given that when distributed it represents a small increment on top of large existing pools, which are themselves difficult to estimate accurately.

Ecosystem and global C cycle models are inadequately tested against measured terrestrial C fluxes. One of the difficulties in doing so is the mismatch between the temporal and spatial scales at which primary production data have traditionally been collected (annually, on a patch of one hectare or less) and the scales at which they are modelled (hourly to monthly, on grid cell of several hundred km^2). Technological developments in the field of micrometeorology will help to address this problem by making it possible to measure CO_2 fluxes hourly over a large area. Such flux measurements include heterotrophic respiration, and therefore, if continued for a sufficient time, represent NEP. NPP must be inferred by estimating the respiration component independently or from nightime measurements, and adding it to the NEP. However, even if ecosystem-scale flux measurements are continued in wooded ecosystems for many years, it is unlikely that they will cover the full life cycle of the trees, and will therefore fail to detect the episodic, catastrophic release of carbon associated with major disturbances such as harvesting, fire and windthrow. They are also blind to carbon losses due to the removal of harvested material, the leaching of dissolved organic carbon and other landscape-scale transport processes. It is therefore not surprising that the flux measurements to date have all reported positive NEP at an annual scale (i.e. a net carbon sink). This is not proof of long-term terrestrial biospheric uptake (see Section 12.2.2).

These long-term measurements confirm that net ecosystem production is strongly influenced by climate. The Harvard Forest, a temperate mixed forest in northeastern USA, showed an annual variation in CO_2 uptake of 1.4 to 2.8 t C ha^{-1} y^{-1} over the period 1991–5 (Goulden *et al.*, 1996). As a thought experiment, if the Harvard Forest responses were representative of the whole world, the effect of global interannual climate variability would be to modify the C exchange in the northern hemisphere by 1 Pg C y^{-1} or more. Thus, it is feasible that part of the 'missing sink' could be the result of the climate change which has already occurred.

The reliability and persistence of the terrestrial carbon sink

A crucial question is whether the ocean and terrestrial ecosystems will continue to act as a large buffer in the global carbon cycle, or whether their buffer capacity will be exhausted before the human population switches to non-carbon emitting forms of energy. If the buffer capacity is finite and reachable, this will set an upper 'safe limit' to the cumulative carbon emissions, and international agreements will be needed to allocate individual national contributions to this amount.

As has been discussed above, many of the processes that could be contributing to the terrestrial carbon sink are temporary or decline in effectiveness as the atmospheric carbon dioxide concentration increases. The pools into which the carbon may be flowing are finite, and some, like biomass and litter, are relatively insecure, since they have a high turnover rate. There is thus no reason to believe that the carbon sink capacity of the biosphere is effectively infinite, and some reasons to believe that it could reverse and become a source (see Chapter 12), possibly with little warning and no chance of avoidance.

10.7   Remaining uncertainties

A key question remains unanswered: How will terrestrial ecosystems respond, at the global scale and over a period of decades to centuries, to the combined effects of changing atmospheric composition, climate and land use/cover (see Section 12.2.1)? Such responses will be linked by positive and negative feedbacks with accompanying changes in temperature, water and nutrient availability and pollutants. Endeavours to disaggregate the effects on NPP, plant functional type distribution, ecosystem respiration, net biospheric carbon uptake, nutrient availability and ecosystem water use have not yet yielded clear answers, primarily because of the lack of sufficient long-term experimentation with natural vegetation. The objective of this research would ultimately be to quantify the sink capacity of terrestrial ecosystems, and identify threshold rates of perturbation of the carbon cycle that should not be exceeded.

The effects of elevated CO_2 on litter chemistry and thus decomposition rate require further research. The litter analyses will need to go beyond simple $C:N$ ratios to include structural and non-structural carbon and nitrogen as well as polyphenolic compounds. The current soil carbon models will need to represent this complexity.

The role of phosphorus in a high CO_2 world, both as a direct limitation to NPP in ecosystems relieved of carbon and nitrogen limitation and indirectly as a constraint on biological nitrogen fixation, is an important area for future research. Such research would need to focus on the effects of increased carbon substrate availability on mycorrhizae and nitrogen-fixing organisms.

⑪ Global change, biodiversity and ecological complexity

O.E. Sala, F.S. Chapin III, R.H. Gardner, W.K. Lauenroth,
H.A Mooney and P.S. Ramakrishnan

Box 11.1 Summary

Causes of change in biodiversity:

- Changes in land use and introduction of alien species are the most important causes of rapid shifts in biological diversity. Over the coming decades, the effects of land-use change may be altered by climatically induced changes in frequency and severity of disturbance. The direct effects of changes in atmospheric composition and climate will increase with time.
- Changes in land use often lead to increased habitat fragmentation, affecting species persistence, abundance and diversity at the landscape scale. Models and empirical studies confirm the existence of species-specific thresholds. When these thresholds are passed, species decline in abundance, break up into isolated populations, and ultimately may go extinct.
- The invasion of exotic species into natural systems either through accidental or purposeful introductions (e.g. in biotic control efforts) is a powerful driver of global change with increasing threats to the viability of indigenous species. Habitat fragmentation and altered disturbance regimes enhance the invasibility of alien species.
- Altered land use is affecting diversity most strongly in humid tropical and temperate regions. Changes in climate and atmospheric composition may be more important in arid and arctic/boreal ecosystems.

Consequences of changes in biodiversity:

- Species diversity affects ecosystem processes through both species traits and species numbers. Species that have largest effects on ecosystem processes include those that alter:
    - (i) frequency and intensity of disturbance;
    - (ii) water and nutrient supply;
    - (iii) trophic structure.
- There is currently no clear understanding of the relationship between species number and ecosystem processes. However, maintaining a high species number increases the probability of retaining species that:

(i) effectively acquire resources under differing conditions;

(ii) tolerate extreme events.

Therefore, maintaining high species number in ecosystems could result in higher productivity, nutrient retention, and buffering of ecosystems against unanticipated effects of global change.

■ Genetic diversity in crops has become increasingly important to enhance resistance against new varieties of pests and diseases. High diversity of crops and associated species will increase the probability of dependable yield of low-input agriculture in unpredictable environments.

■ Changes in landscape diversity may affect the spread of fire, pests, and waterborne materials.

Box 11.2   Future needs

There is a need to:

■ examine diversity at scales other than species, i.e. genetic diversity, functional type diversity, and, at larger scales, landscape diversity to understand the interactions within and between these levels of organization, the effects on trophic pathways, and the consequences for ecosystem functioning;

■ understand the resource requirements and population dynamics of species most responsible for ecosystem functioning, i.e. keystone and dominant species, especially in fragmented landscapes;

■ intensify efforts to understand the extent, frequency, or severity of disturbance that can be tolerated by ecosystems without permanent loss of biodiversity and disruption of system-specific processes;

■ undertake more observational studies of current invasion events of exotic species in natural communities and analyse past invasion events to determine (i) the effects of biological invasions on biodiversity and ecosystem functioning, (ii) the characteristics that make an ecosystem vulnerable to invasions, and (iii) the characteristics that make an organism a successful invader;

■ extend local and short-term studies on biodiversity/ functioning relationships to different regions and longer time periods;

■ understand the significance of species diversity within and among functional types (i.e. group of species with similar effects on ecosystem processes) with respect to the stability of ecosystem processes under global change.

11.1   Introduction

Large changes in biological diversity are occurring at a global scale simultaneously with changes in land use, composition of the atmosphere, and climate. These changes result mainly from human activity and are expected to intensify in the near future, with both the magnitude and scope of change likely to continue to increase into the next century (see Chapter 1).

Biological diversity or biodiversity encompasses a suite of scales from species and genetic diversity to landscape diversity. Perhaps of all the changes in biodiversity currently occurring, the most widely recognized are those occurring in species diversity or species richness. However, equally important are those changes in diversity occurring within species (genetic diversity), or at larger scales such as changes in landscape diversity. In addition, changes in the trophic structure and pathways of ecosystems are occurring, and collectively all these changes are referred to as changes in ecological complexity.

Human activity influences both biodiversity and overall ecological complexity, directly and indirectly. Overexploitation of natural resources, such as overgrazing, and the conversion of natural ecosystems into croplands and urban areas results in habitat alteration, destruction, and fragmentation. These changes lead to a global reduction of species diversity and within-species diversity. Indirectly, human activity also affects biodiversity since fossil fuel combustion and land-use change affect the composition of the atmosphere and climate, which in turn result in a global reduction of species diversity.

Changes in species diversity involve not only extinctions but also invasions. Activities such as domestic livestock grazing or crop production may result, at local scales, in net increases or decreases of species diversity while at global scales they almost always result in decreases in species diversity. In summary, at a global scale the terrestrial biosphere is being impoverished and homogenized as a result of human activity.

Whereas human perturbations decrease global species diversity, they tend to increase landscape diversity as a result of the introduction of new landscape units such as croplands in a matrix of native grassland or logged patches in a forest. The intricate patterns of croplands, forests, and grasslands and their spatial distribution affect the functioning of ecosystems, such as the flow of matter, disturbance regimes, and biosphere/atmosphere interactions.

How do anthropogenic changes in land use, chemical composition of the atmosphere, and climate affect biodiversity? How do changes in biodiversity affect the ability of ecosystems to respond to the other changes that are occurring at the global scale such as deforestation, desertification, CO_2 enrichment, and climate change?

The importance of these questions has spawned a series of scientific endeavours that are described in the following pages. This chapter first reviews the causes and consequences of changes in diversity at different scales. It assesses the current understanding of the impact on diversity of changes in land use, disturbance regime, species introductions, climate, and atmospheric composition. Second, it analyses the consequences of those changes for the functioning of ecosystems. This section is organized in a hierarchical fashion, with discussions of the consequences on ecosystem functioning of changes in genetic, species, and landscape diversity. The analyses of causes and consequences of changes in diversity are mostly retrospective and based on the interpretation of the mechanisms of phenomena that have already occurred. The last section attempts to look into the future and to develop biodiversity scenarios. Several other disciplines, from atmospheric science to human demography, provide predictions of trends and patterns of quantities as disparate as CO_2 concentrations in the atmosphere, human population density, cereal production and people at risk from hunger for the next 100–200 years. The biodiversity scenarios described here were based on simulation models of changes in land use, climate, and atmospheric composition and serve as inputs to other models.

11.2   Causes of changes in biodiversity

11.2.1   Impact of changes in land use and disturbance regime

Fragmentation

The total area of available habitat is the primary factor determining species abundance and landscape levels of biodiversity (Noss, 1996). As the habitat within a landscape becomes fragmented by natural or anthropogenic disturbances, or as the quality of ecological resources is modified by climate change, the persistence of species adapted to these landscapes may decline and invasions of exotic species increase (Hobbs, 1989). Time-delays in the response of biota to changes in the heterogeneity and availability of resources may result in several generations elapsing before the consequences of landscape change are realized (Tilman *et al.*, 1994). These interactive effects of the biota with changes in the pattern of habitat fragmentation make changes in diversity at landscape scales difficult to predict.

The degree to which fragmented habitats remain connected is an important factor affecting species abundance and diversity at landscape scales (Fahrig & Merriam, 1985; Gustafson & Gardner, 1996; Noss, 1996). A variety of theoretical models have shown that thresholds exist where incremental reductions in available habitat will result in sudden changes in connectivity (Gardner *et al.*, 1987, 1992, 1993; Lavorel *et al.*, 1995) (Fig. 11.1). Because individual species utilize landscape resources at different spatial and temporal scales, each species experiences different patterns of change depending on the 'dimensions' of their ecological neighbourhoods (Addicott *et al.*, 1987). Knowledge of the resources required by particular species, and information on their dispersal abilities, allows the value of the thresholds to be estimated for each species (O'Neill, 1988; Plotnick & Gardner, 1993). For instance, species that are capable of

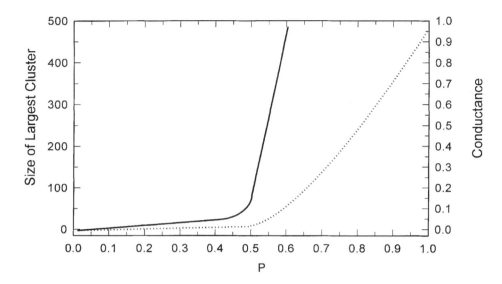

Figure 11.1   Change in conductivity (dashed line) and the average cluster size (solid line) as a function of P, the fraction of the landscape containing suitable habitat. These curves are estimated from gridded landscapes with connectivity between habitat sites established by contact along one or more edges of the four adjacent sites. A connectivity threshold exists at $P = 0.5928$ producing a sudden increase in average cluster size. Assuming that movement is restricted to adjacent habitat sites, conductance below the critical threshold is nearly zero. Above the critical threshold conductance rapidly increases as resistance to movement declines. Adapted from Plotnick & Gardner (1993).

long-distance dispersal and have small neighbourhoods may be able to locate refugia within a disturbed landscape, while species with a larger neighbourhood and poorer dispersal capabilities will be adversely affected by shifting patterns of disturbance (Baker, 1993). Typically, species that survive better in fragmented landscapes are able to either live and reproduce in the matrix of land uses surrounding remnant habitat patches, or have a sufficiently small ecological neighbourhood to persist in small remnant patches, or are mobile enough to integrate many patches into a single interbreeding population (Noss, 1996).

Two empirical studies support these results. The first effect of fragmentation on birds and mammals is the loss of habitat and subsequent reductions in overall levels of abundance (Andren, 1994). As fragmentation continues, the distribution of habitat (i.e. patch size and isolation) becomes progressively more important, with a noticeable threshold in species decline when about 30% of the suitable habitat remains. Metapopulation theory predicts that as habitats become more fragmented and patch sizes become smaller, species will be able to maintain viable populations in fewer and fewer suitable habitats, leading to eventual extinction (Tilman *et al.*, 1994). Hanski *et al.* (1995*b*) have confirmed this result with observed declines of butterfly populations with diminishing fractions of suitable habitat. Although much work remains before there is a general theory that can relate metapopulation dynamics with landscape change, this confirmation of modelling results with field observation is encouraging.

The spatial arrangement of habitat types (e.g. forests, wetlands, riparian vegetation, etc.) within landscapes is a primary factor determining the rate of exchange of material, energy and organisms at scales of kilometres (Turner, 1989; Pickett & Cadenasso, 1995). The amount and spatial arrangement of different habitat types is dynamic (Romme, 1982; Baker, 1989), with disturbances (both natural and anthropogenic) and recovery from disturbance affecting the successional status of individual patches. However, over a sufficient period of time a shifting mosaic of habitat patches occurs within natural landscapes (Bormann & Likens, 1979), maintaining overall levels of habitat diversity. Because habitat patches may be temporarily or permanently lost as the result of disturbance, extinction of species within fragmented landscapes can be expected (Tilman *et al.*, 1994). Therefore, in shifting mosaics frequent exchanges of organisms among isolated habitat patches is essential for maintaining biodiversity at landscape scales (Noss, 1983).

Disturbance

Landscape disturbance is any relatively discrete event external to the system being studied that disrupts ecosystem, community, or population structure and processes (Pickett *et al.*, 1989). It includes events such as fires, storms, outbreaks of pests or pathogens, mass movement of material, and human-induced changes in land use. Rapid shifts in disturbance patterns are now occurring at global scales as humans alter land use (Turner *et al.*, 1990), replacing natural, periodic disturbances with more permanent changes in landscape pattern.

Disturbance regimes can be characterized by four parameters (Sousa, 1984): the *frequency* of disturbance (the number of occurrences within a given time period), the *intensity* of disturbance (the likelihood of spread of disturbance across the landscape), the *severity* of disturbance (the degree of ecological change caused by disturbance) and the *duration* of disturbance (the length of time the disturbance lasts). Changes in these parameters constitute a change in the disturbance regime, which ultimately affects the spatial heterogeneity of ecological resources and species that are adapted to these patterns (Franklin & Forman, 1987; Baker, 1992; Forman, 1995). For instance, if climate change produces warmer and drier conditions, more frequent fires are likely to occur (Romme & Turner, 1991). However, the relationship between changing disturbance regimes and patterns of species abundance is complex. Many species are adapted to existing disturbance regimes, with periodic disturbances providing temporary release from competitors and opportunities for increases in abundance. Disturbances may enhance the invasibility of many species, especially when the availability of a limiting resource (e.g. space or nutrients) changes (Hobbs, 1989). Severe disturbances may also eliminate seeds stored in the soil, causing declines in abundance of native species (Malanson, 1984). Climate-induced changes in the fire regime in Yellowstone National Park, USA, have probably had substantial consequences for the extent and age-class distribution of native forest communities (Romme & Turner, 1991).

The complex patterns generated by changes in landscape heterogeneity and disturbance regimes have been explored with simple probabilistic models (Gardner *et al.*, 1987; Turner *et al.*, 1989*a*) and have suggested that, when susceptible habitats occupy less than 50% of the landscape, disturbance effects are more related to disturbance frequency than intensity. When more than 60% of the landscape is occupied by susceptible habitat, disturbance effects are more related to increased intensity than frequency (Turner *et al.*, 1989*b*). Other spatially explicit models have also indicated that small changes in landscape pattern can cause dramatic shifts in the frequency, duration, and intensity of disturbance events (Franklin & Forman, 1987; O'Neill *et al.*, 1992; Turner *et al.*, 1994).

Shifts in land-use patterns and fire suppression efforts, causing fuels to accumulate, result in an increased risk of larger and more severe fires (Swetnam, 1993). The loss of large areas of forests due to fires may result in significant, long-term changes in the pattern of vegetation regrowth and succession with possible shifts to a new vegetation state (Noble & Slatyer, 1980; Starfield & Chapin, 1996). Results of a landscape-scale fire model have suggested that climate-induced changes in fire regimes produce rapid shifts in the pattern of mature forests (Gardner *et al.*, 1996). Small changes in weather produce significant shifts in the age structure and spatial arrangement of forests. A 'drier climate' results in less mature forest with a greater degree of habitat fragmentation. A 'wetter climate' produces fewer but larger fires. These results are consistent with fire records in giant sequoia stands that show that frequent small fires occurred during warm periods but more widespread fires occurred during cooler climate conditions.

Changing land-use patterns can interact with disturbances to effect changes in the frequency, severity, or extent of the disturbance. For instance, fire suppression in giant sequoia groves has shifted presettlement fire regimes from frequent low-intensity surface fires to infrequent, but increasingly numerous, large catastrophic crown fires (Swetnam, 1993). Increasingly synchronized regional fire regimes may be expected in such human-altered forest landscapes subjected to climatic extremes (Swetnam, 1993). The coexistence of some early colonizing species with more competitively dominant species depends on the temporal phasing (or synchrony) of disturbances, even if the mean rate of disturbance remains constant (Swetnam, 1993). Therefore, it is important that the landscape management regime be formulated such that the historical variation in disturbance regime is maintained if the abundances of species adapted to these landscapes are to persist (Baker, 1992; Forman, 1995).

Many key global change-related questions remain for the study and management of disturbance-prone landscapes. What is the extent, frequency, or severity of disturbance that can be tolerated by the system without irretrievable loss of biotic elements (e.g. species) or processes? What are the ecological effects of a particular disturbance? Could a disturbance qualitatively change the system? How large should a reserve be in a disturbance-prone environment? Should management seek to alter natural disturbance dynamics? Answers to these questions are neither simple nor straightforward and will require methods that verify simulation results with direct observations of landscape change.

11.2.2 Impacts of invasive species

Invasive species pose a considerable threat to biodiversity, particularly on islands. The invasion of the brown snake into Guam has been strongly implicated in the precipitous decline in the populations of ten species of birds (Savidge, 1987). In addition, a decrease in the abundance of a native species of lizard correlated with increased abundance of an invasive lizard species, probably compounded by the effects of the brown snake on lizard predators (Rodda & Fritts, 1992).

Purposeful introductions have often resulted in species extinctions, such as the introduction of the mongoose into islands for the control of rats in sugar cane. These animals subsequently had a large impact on native populations of small rodents and birds. Other biocontrol efforts that have resulted in losses of non-targeted species include the introduction of the *Euglandia rosea*, a carnivorous mollusc from Central America, that was brought on to some Pacific Islands to control the giant African snail, *Achatina fulica*. *Euglandina* also preys on other land snails and has extirpated native snail species. Clarke *et al.* (1984) indicated that 'the number of endemic species that are endangered or already extinct as a result of the introductions must now be well over a hundred'.

Aquatic systems are particularly vulnerable to extinctions. One of the most dramatic cases of species extinctions, due to invasions, has been the introduction of the Nile perch into Lake Victoria. Following its introduction in the late 1950s, over 200 species of haplochromine fish species have been driven to extinction (Witte *et al.*, 1992). The dramatic recent invasions of the Asian clam into San Francisco Bay (Carlton *et al.*, 1990), the zebra mussel into the North American Great Lakes, and a ctenophore into the Black Sea have had enormous impacts on the populations of the native biota. Moyle (1996), in his analysis of

aquatic systems, has noted that species extinctions are most likely to occur
when:

- the successful invader is a top carnivore;
- the invader carries a novel disease organism;
- the invaded ecosystem has a naturally low diversity;
- the invaded ecosystem has been highly disturbed by human or natural
  factors.

Even though such predictions are readily known, new introductions are continuously made into lakes.

As the biota of the world becomes increasingly homogenized, there may be
rather large extinctions, even though in the short term there will be local species
enrichments. The worst-case scenario is given by Wright (1987), using island
biogeographic principles, that indicates the potential loss of species due to the
breakdown of biogeographic barriers will be enormous. Current invasion-
driven extinctions are only the tip of the iceberg of the effects of invaders on the
composition and structure of biotic systems. Thus, invasive species themselves
can be considered a powerful agent of global change.

11.2.3 Impacts of climate and atmospheric composition on diversity

Because of the widespread and dramatic effects of land-use change and species
introductions on species diversity, it is often difficult to recognize the impacts of
gradual changes in climate and atmospheric composition. Nitrogen deposition
has had the most dramatic impacts – eliminating heath species from Dutch
heathlands and increasing dominance by grasses in many heath, meadow, and
forest ecosystems (Berendse & Elberse, 1990).

The paleo-record clearly documents changes in species composition and
diversity in response to past changes in climate (Davis, 1981). Species richness
of Arctic tundra declined 30% with experimental warming and 50% with
warming plus nutrient addition (simulating N deposition) within a decade (Fig.
11.2) (Chapin *et al.*, 1995*b*). Forbs and lichens, which are critical food resources
for caribou and pollinating insects, were most strongly affected, in addition to
mosses, which are critical for insulation of permafrost. Thus, the species most
sensitive to climatic warming have strong feedbacks to both animal diversity and
biogeochemical cycling. Similarly, experimental addition of water and nitrogen
greatly reduced the diversity of plant functional types in a semiarid short-grass
steppe (Lauenroth *et al.*, 1978).

At a larger scale, models that predict changes in distribution of vegetation in
response to climatic change (Prentice *et al.*, 1992, see also Chapter 6) suggest
that there could be major changes in the relative abundance of biomes in

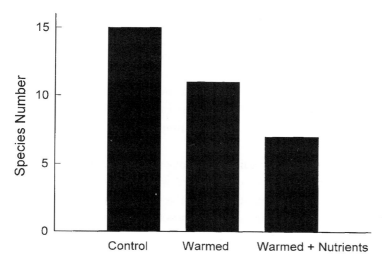

Figure 11.2 The effects of experimental warming and N addition on species richness in the arctic tundra. Adapted from Chapin *et al.* (1995*b*).

response to climatic warming (Chapter 8). For example, in Africa the two biomes that are currently most rare are predicted to decrease by 81% (semideserts) and 69% (broadleaved evergreen forests) in response to a doubled CO_2 climate. If these biome shifts occur, all biomes except hot deserts and tropical rainforests would decrease in abundance in national parks, particularly semideserts, warm grass/shrublands, and broadleaved evergreen forests, substantially reducing the capacity to conserve the diversity of these biomes. These predictions of biome shifts are based on the equilibrium distributions of vegetation with climate and ignore the rates and factors controlling migration of organisms in response to climatic change (Bond & Richardson, 1990). Furthermore, the developing transient dynamic global vegetation models (DGVMs) suggest that the degree of biome shifts will be less than those predicted by equilibrium models (see Chapter 8). There are increasing barriers to migration, as the land between national parks becomes more fragmented and transformed. The wise management of the land matrix between existing protected areas, to allow the movement of organisms, is thus essential to preserving species diversity of protected areas into the future. A priority for planning protected areas in the future is to conserve the connectivity among these areas and to establish new protected areas in places that are presently climatically diverse.

11.3   Consequences of changing biodiversity

11.3.1   Consequences of changes in genetic diversity

Low genetic diversity and the resulting inbreeding depression are among
several processes that increase the probability of extinction in small populations.
Other processes include reduced probability of interaction among individuals
and vulnerability to stochastic variation in environmental and demographic
processes (Hanski *et al.*, 1995*a*).

The reduced genetic diversity in crops is of particular concern because the
world's population relies on only three species (rice, wheat, and maize) for about
60% of its carbohydrate. Approximately 95% of the protein and carbohydrate
consumed comes from about 30 crops. A single variety of wheat accounted for
67% of the area planted to wheat in Bangladesh in 1983. The main crops in the
United States depend on less than nine varieties (NRC, 1994). Catastrophic
outbreaks of disease have occurred in the past (e.g. potato blight in Ireland) and
cannot be dismissed for the future. For example, new varieties of potato blight
recently discovered in Mexico and spreading to the United States are more
virulent than previously known strains and are insensitive to any known
fungicide. In the past, genetic diversity has been an important source of disease
resistance (Holden *et al.*, 1993). Given the large increases in food production
required to meet projected population growth over the next 30 years at least (see
Chapter 9), the emergence of a disease capable of infesting a major wheat or rice
variety could pose a major threat to production. Reductions in genetic diversity
may also reduce the capacity of crops to adjust to changes in climate and
atmospheric composition. Low genetic diversity is of particular concern in
managed forests, where the greater longevity of individuals increases the range
of conditions that each individual is likely to experience during periods of rapid
environmental change.

There are several international consortia that maintain stores of genotypes
of major crops, for example, the International Rice Research Institute. Maintenance of these 'genebanks' has been an important mechanism for conserving
genetic diversity in these crops. Genetic engineering has the potential to
increase diversity at specific loci, for example, resistance to a particular disease,
although development of these varieties will always lag behind the recognition
that a problem exists. Genetic engineering and breeding programmes have
focused primarily on crops and varieties used in intensive agriculture, where
economic and production returns are greatest.

11.3.2 Consequences of changes in species diversity

Effects of species number and relative abundance

ECOSYSTEM IMPACTS OF CHANGES IN DIVERSITY UNDER
FAVOURABLE CONDITIONS.   There is currently no clear understanding
of the relationship between species richness (the number of species present)
and ecosystem processes. For example, along natural productivity gradients,
species richness can be quite high in both unproductive environments (e.g.
Australian heaths) or productive environments (wet tropical forests) (Bond,
1993*a*,*b*; Ricklefs, 1995). Monocultures of some crop varieties are as productive as mixed-cropping systems under favourable conditions (Vandermeer &
Schultz, 1990). Similarly, natural forests dominated by a single tree species do
not differ consistently in productivity and nutrient cycling from more diverse
forests (Rodin & Bazilevich, 1967). Depending on the taxonomic group and
region, species richness can show a variety of relationships with environment
and productivity (Ricklefs, 1995).

Species diversity is a function of both species richness and the relative abundance (evenness) of species (Ricklefs, 1995). This section discusses only the
impacts of species number on ecosystem processes because the impacts of
evenness have not yet been addressed experimentally.

Despite the lack of a clear relationship between species richness and ecosystem processes in natural ecosystems, there is concern that reductions in species diversity below naturally occurring levels might alter ecosystem processes
such as productivity. A decline in species richness could reduce productivity
for at least two reasons: (i) loss of species reduces the probability of there
being at least one species present that is productive under a particular set of
conditions; and (ii) additional species may be able to tap resources that are not
captured by other species, due to differences in rooting depth, phenology,
form of nitrogen utilized, etc. (Tilman, 1988; Chapin *et al.*, 1997*a*). Note that
both explanations depend on traits of species and do not assume a causal link
between species number *per se* and resource capture. For both hypotheses, a
saturating relationship is expected between species richness and ecosystem
parameters (Fig. 11.3) (Vitousek & Hooper, 1993) because the more species
there are in an ecosystem, the more likely it is that a given species that is
gained or lost will be ecologically similar to other species present.

Several recent experiments have shown a positive, saturating relationship
between species richness and various ecosystem processes. Artificial tropical
communities seeded with 0, 1, 2, or 100 species showed greater nitrogen uptake with 100 than with fewer species (Ewel *et al.*, 1991). Manipulation of diversity at four trophic levels (plants, herbivores, parasitoids, and decomposers)
in experimental mesocosms also resulted in a positive correlation between

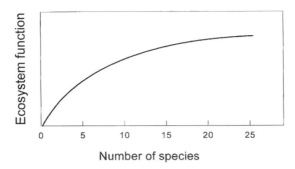

Figure 11.3   Conceptual relationship between species richness and ecosystem functioning. Adapted from Vitousek & Hooper (1993).

species richness and productivity, decomposition, and nutrient retention (Naeem *et al.*, 1994, 1995). However, since only one combination of species was used at each level of species richness in these experiments, it is difficult to separate the effects of the particular species combinations from the effect of species richness.

To minimize the possibility that diversity effects might be a simple consequence of particular species being present in high-diversity treatments, Tilman *et al.* (1996) established an experiment in which species richness was directly manipulated by sowing plots with seven levels of plant species richness (1–24 species). Each replicate was a separate random draw from a pool of 24 experimental species. This experiment also showed a positive, saturating relationship between species richness and aboveground biomass (Fig. 11.4a). Lower concentrations of inorganic nitrogen in the soil of the more diverse plots (Fig. 11.4b) suggested greater nutrient uptake by plants from these plots (Tilman *et al.*, 1996). There was a similar inverse correlation between plot-level diversity and soil nitrate in a nearby native prairie. The conclusion that there is a relationship between species richness and ecosystem processes has been questioned, because as species diversity increased, the probability of including species with large biomass increased simultaneously (Huston, 1997). Since higher diversity treatments have more large species, the reported relationship could be solely the result of higher plant biomass due to the presence of these larger species.

In other experiments plant species richness affected biogeochemical cycling without influencing plant biomass and production. In a greenhouse experiment with Mediterranean grassland species and diversity ranging from 3 to 12, plant species richness had no detectable effect on measured plant parameters but had a strong impact on soil biological activity (Chapin *et al.*, 1997*b*). As a result, organic matter decomposition was enhanced, and the leaching of nitrogen was reduced at high species richness.

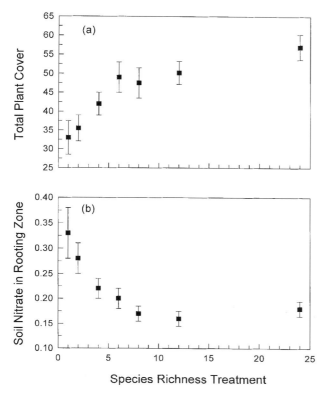

Figure 11.4 The effect of species number on (a) plant cover and (b) total nitrogen in the upper soil layer where most of the roots are concentrated. Results from a manipulative field experiment with grassland species. Adapted from Tilman *et al.* (1996).

Both Ewel's and Tilman's experimental results are consistent with a positive, saturating relationship between species richness and ecosystem processes, but suggest that the instantaneous effect of species richness on the measured parameters may saturate at relatively few (4–10) species. Differences in productivity and nitrogen cycling in these experiments could reflect (i) more complete spatial and/or temporal utilization of space in diverse plots ('more niches occupied'), or (ii) that higher diversity increases the probability of having productive species in the experimental plots. Clearly, additional research is necessary to determine the mechanisms by which species richness influenced ecosystem processes in these experiments. At present, there have been too few experiments conducted to know when, how, and to what extent species richness affects ecosystem functioning. No experiments have explored the role of species richness in natural ecosystems or the role of relative abundance (evenness). As the time scale increases, plants encounter a greater variety of

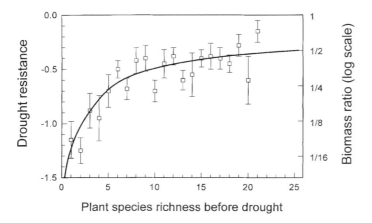

Plant species richness before drought

Figure 11.5   The effect of number of species on resistance to an extraordinary drought. The species richness gradient was experimentally created by adding fertilizer. Drought resistance was estimated as the biomass during the drought/long-term average biomass ratio. Results from a field manipulative experiment in a mesic grassland in tall grass prairie, North America. Adapted from Tilman & Downing (1994).

conditions, suggesting that diversity may be more important than short-term experiments would suggest. There is an urgent need for new experiments, over longer time periods, and in systems of greater diversity to see under what circumstances and by what mechanisms species diversity influences ecosystem processes.

ECOSYSTEM IMPACTS OF DIVERSITY UNDER EXTREME

CONDITIONS.   Species richness may buffer ecosystem processes against extreme events or unanticipated effects of global change (McNaughton, 1993). Indirect evidence for a positive relationship between species richness and ecosystem buffering comes from prairie grassland plots that differed in diversity due to long-term nutrient additions. The more diverse plots showed the least decline in aboveground biomass during a severe drought (Tilman & Downing, 1994) (Fig. 11.5) and recovered more rapidly following drought. However, the less diverse plots were those to which nitrogen fertilizer had been added, so the lower stability of these plots probably reflected a decline in abundance of drought-resistant species (Sala *et al.*, 1995). In addition, the high variability in primary production among years and the reduced primary production during drought in the low diversity–high fertility treatment could be accounted for by the fertilization treatment (Huston, 1997). This explanation is much more parsimonious than the species richness effect. The high fertility (low diversity) treatments were mostly constrained by water availability and therefore the variability in productivity reflected the precipitation pattern. In contrast, low fertility (high diversity) treatments were highly constrained by nutrient availability, particularly when water availability was high, and showed a small response to precipitation variability.

One mechanism by which diversity might confer stability to ecosystem processes is through the contrasting responses of functionally similar species to variations in environment (McNaughton, 1977; Chapin & Shaver, 1985). For example, in response to annual variation in weather or experimental manipulation of environment, ecosystem productivity is much more stable than is that of any individual species (Lauenroth et al., 1978, Chapin & Shaver, 1985). The more species there are in a functional group (group of species with similar effects on ecosystem processes), the lower is the probability that any change in climate or climatic extremes that is severe enough to cause extinction of a species will have serious ecosystem consequences. Thus, genetic diversity and diversity among ecologically similar species may provide insurance against large changes in ecosystem processes in the event of species loss. High diversity might reduce the likelihood of invasions by exotic species. For example, the smaller number of species present on islands than on mainlands may explain the vulnerability of islands to changes in ecosystem processes in response to invasion (Vitousek et al., 1987; Cushman, 1995; Vitousek et al., 1995). If there is greater resistance to invasions in diverse communities, it may reflect the higher probability of having a species in such a community that is ecologically similar to the invader.

Model simulations suggest that species richness could influence ecosystem response to global change. A model community consisting of nine deciduous tree species of differing CO_2 response exhibited a response to elevated CO_2 30% greater than did a model community composed of a single species with the average CO_2 response of the more complex community (Fig. 11.6) (Bolker et al., 1995). Thus, ecosystem models that neglect diversity in simulating responses to CO_2 and other environmental factors may underestimate the response of terrestrial ecosystems to global change. This modelling result contrasts with some of the conclusions of Chapter 7 where the response to elevated CO_2 decreases from leaves to individuals to ecosystems. As the hierarchical level increases and the number of factors controlling the response (from photosynthesis, to leaf growth and primary productivity) increases, the magnitude of the CO_2 effect decreases. In contrast, according to this modelling study, the increase in complexity from a single species to a community and the corresponding inclusion of mechanisms such as competition result in an amplification of the CO_2 effect.

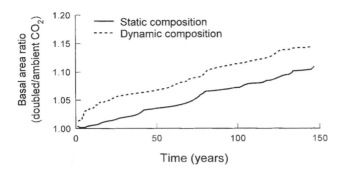

Figure 11.6   The effect of species richness on the ability of ecosystems to respond to global change. Response of a model community made up of nine species each with a different response to CO_2 (dynamic composition) or one species with the average community response to elevated CO_2 (static composition). The dynamic composition allows for competition and successional change, which results in those species with a larger CO_2 response to have a higher importance in the community than in the static composition case with a single CO_2 response. Consequently, the response to elevated CO_2 is larger in the dynamic composition model run. Adapted from Bolker *et al.* (1995).

Effects of species identity

TRAITS OF INDIVIDUAL SPECIES WITH PROFOUND ECOSYSTEM IM-
PACTS.   Underlying the effects of species richness on ecosystem processes (Section 11.3.2) is the assumption that increased richness brings with it a greater number of 'types' of species. A predictive understanding of the conse-quences of changing species diversity or the introduction of exotic species would therefore benefit from knowing which traits most strongly affect ecosystem processes. Species with particularly strong impacts on ecosystem function-ing include those that modify (i) resource availability, (ii) trophic structure, or (iii) disturbance regime (Vitousek, 1990; Chapin *et al.*, 1995*a*).

Resource Supply.   The supply of soil resources is an important 'bottom–up' control that strongly influences the structure and dynamics of all terrestrial ecosystems (Jenny, 1980; Chapin *et al.*, 1996*a*) and consequently their role in regional and global processes. Introduction of exotic species with symbiotic N fixation in Hawaii greatly increased productivity and N cycling and altered the structure and species composition of forests (Vitousek *et al.*, 1987). As another example, Nepalese alder increases N inputs, and bamboo retains newly weathered P and K during the cropping and fallow phases in shifting agricul-ture in India (Ramakrishnan, 1992). Each of these species differs from other species in the community in its effects on nutrient supply, so that introduction or loss of a single species can have profound ecosystem effects. Similarly, introduction of deep-rooted species, such as *Eucalyptus* or *Tamarix*, can tap previously inaccessible water and nutrients (van Hylckama, 1974; Robles & Chapin, 1995). These differences in rooting depth can be important at the regional scale. Model simulations suggest that conversion of the Amazon basin from forest to pasture would cause a permanent warming and drying of South America because the shallower roots of grasses would lead to less evapotranspiration and greater energy dissipation as sensible heat (Shukla *et al.*, 1990).

Plant species also affect resource supply rate through litter-quality effects on nutrient turnover in soils (Melillo *et al.*, 1982; Flanagan & Van Cleve, 1983; Berg & McClaugherty, 1989) and modification of the microenvironment (Wilson & Agnew, 1992; Hobbie, 1995). For example, Arctic mosses, with their low litter quality, low rates of evapotranspiration (leading to water-logging), and effective insulation (preventing soil warming) indirectly inhibit decomposition (Gorham, 1991).

Animals influence the resource base of the ecosystem by redistributing nutrients within an ecosystem or by importing nutrients to oligotrophic ecosystems (e.g. nutrient movement from oceans to streams by migratory salmon). Some soil processes (e.g. nitrification and denitrification) that are critical to nitrogen retention by ecosystems are controlled by relatively few species of microorganisms. Consequently, changes in their abundance could have large effects on N loss from ecosystems (Frost *et al.*, 1995; J. Schimel, 1995).

Trophic structure. Many animal species and microbial diseases (and some plant species) differ strikingly from all other species in the community in the resources that they consume and, therefore, their effects on community structure. These top-down controls are particularly well developed in aquatic systems, where addition or removal of a fish species can have large 'keystone' effects that cascade down the food chain (Carpenter *et al.*, 1992; Power, 1992). Many non-aquatic ecosystems also exhibit strong responses to changes in predator abundance (Hairston *et al.*, 1960; Strong, 1992). For example, removal of elephants or other keystone mammalian herbivores leads to encroachment of woody plants into savannas (Owen-Smith, 1988; Wilson & Agnew, 1992; Zimov *et al.*, 1995). Similarly, epidemic diseases, such as rinderpest in Africa, can act as keystone species by greatly modifying competitive interactions and community structure (Bond, 1993a,b). The top-down controls by predators have greater effect on biomass and species composition of lower trophic levels than on the flow of energy or nutrients through the ecosystem (Carpenter *et al.*, 1985) because declines in producer biomass are compensated by increased productivity and nutrient cycling rates by the remaining organisms. For example, intensely grazed grassland systems such as the southern and southeastern Serengeti Plains (McNaughton, 1985) have a low plant biomass but rapid cycling of carbon and nutrients due to treading and excretion by large mammals, which prevent accumulation of standing dead litter and return nutrients to soil in plant-available forms (McNaughton, 1988). Keystone predators or grazers can thus alter the pathway of energy and nutrient flow, modifying the balance between herbivore-based or detritus-based food chains, but we know less about the effects on overall energy flow and nutrient cycling by the entire ecosystem.

Disturbance regime.   Disturbance is one of the most important ways in which animals affect ecosystem processes (Lawton & Jones, 1995), creating sites for seedling establishment and favouring early successional species (Hobbs & Mooney, 1991; Kotanen, 1995). At the regional scale, disturbances created by overgrazing can alter albedo of the land surface and change patterns of regional temperature and precipitation. For example, in the Middle East, overgrazing by sheep and goats reduced vegetation cover, thereby increasing the albedo and reducing energy absorption (Charney *et al.*, 1977; Schlesinger *et al.*, 1990). Consequently, there was less heating and convective uplift of the overlying air mass and less advection of moisture from the Mediterranean. This reduced precipitation and further contributed to the regional reduction in biomass and production.

Plants alter disturbance regimes through effects on soil stability and flammability. For example, introduction of grasses into forest or shrubland ecosystems can increase fire frequency and cause a replacement of forest by savanna (D'Antonio & Vitousek, 1992). Similarly, boreal conifers are more flammable than deciduous trees because of their large leaf and twig surface area, low moisture content, and high resin content (van Cleve *et al.*, 1991). In early succession, plants reduce disturbance by stabilizing soils and reducing wind and soil erosion. This allows successional development and retains the soil resources that determine the structure and productivity of late-successional stages.

GENERALIZATION OF SPECIES EFFECTS TO REGIONAL AND GLOBAL SCALES.   In contrast to species effects on resource supply, trophic interactions, and disturbance regime, which can only be predicted at present by recognizing individual species, other important effects of species on ecosystem and regional processes are shared in common by many species and their effects can often be generalized, making it possible to model the impacts of changes in species composition on ecosystem processes. Functional types are groups of species that show similar responses to, or effects on, environment (see Chapters 2 and 8). Highly generalized functional types (e.g. grasses, deciduous trees, evergreen trees), defined at the level of growth forms, are often predictable in both their environmental responses and their effects on ecosystem and regional processes (Raunkier, 1934). However, more refined groupings of species (e.g. C3 versus C4 grasses) often fail to show consistent responses to climate and atmospheric composition (see Chapters 7 and 8). In this section we discuss functional types defined with respect to responses to and effects on soil resources (water and nutrients) rather than climate. These provide useful generalizations of some species effects on ecosystem processes.

Plants can be characterized by general 'adaptive strategies' that govern effects on environment (Grime, 1977; Chapin, 1993). A high relative growth rate (RGR), typical of plants from high resource environments, requires high rates (per unit tissue mass) of nutrient absorption, photosynthesis, and water loss and a large leaf area (Chapin, 1980; Lambers & Poorter, 1992). These traits result in high litter quality, which promotes decomposition and nutrient mineralization (Hobbie, 1992), and high rates of water transfer to the atmosphere. A large size, generally conferred by woodiness, allows plants to dominate light resources (Tilman, 1988) and is associated with large amounts of nutrient-absorbing, photosynthetic, and transpiring tissues. The consequences of large size include large annual fluxes of carbon and nutrients, large fluxes of water vapour to the atmosphere (and consequently reduced fluxes to lakes and streams), and a greater radiation absorption (low albedo) in snow-covered landscapes, acting as a feedback to regional climate warming (Bonan et al., 1992; Foley et al., 1994). Together these traits determine many ecological processes with clear ecosystem and global consequences (Chapin, 1993; Hobbie, 1995). Another general axis of plant traits is associated with responses to disturbance and relates to life-history characteristics of plants (Grime 1977; Tilman, 1988). These traits are only weakly related to RGR and size through allocation tradeoffs (Chapin, 1993).

11.3.3 Consequences of altered landscape structure for regional and global processes

Just as knowledge about the species and functional-type diversity can be important for understanding the functioning of individual ecosystems, knowledge about the diversity of ecosystems in a landscape or region can be important for understanding its biogeochemical functioning. The structure and functioning of a landscape can be described at three levels of complexity. In the simplest case, a landscape or region consists of a single homogeneous unit (one ecosystem type), which can be sampled randomly. If the area is a mosaic of different ecosystem types, each of which is biogeochemically independent from it neighbours, sampling among the different ecosystem types must be stratified and each ecosystem type weighted by its relative area. Finally, if the area consists of a mosaic of different ecosystem types that have spatially explicit interactions with their neighbours, these neighbourhood interactions must be clearly recognized. The level of landscape complexity varies among processes. For example, the carbon capital of a region can be estimated without considering spatial interactions among ecosystems whereas spatial pattern is critical to water quality and nitrogen transport to river systems.

Each of the three cases represents a different degree of connection among ecosystems. This biogeochemical view of landscapes and regions conceptualizes diversity as being represented by the number (richness) and relative importance (evenness) of the constituent ecosystems. Examples of each case are described below.

The North American mid-latitude IGBP transect contains excellent examples of cases 1 and 2. The dominant environmental gradient along the transect is the west–east gradient in annual precipitation from 300 to 1200 mm. Regional ecosystem diversity is relatively low at the western and eastern extremes of the transect and highest in the centre. In the west, low precipitation limits land-use alternatives, and most of the area is still in native grassland. In the east, high precipitation and fertile soils have resulted in most of the area being converted to cropland that supports one or two crops. In the centre, ecosystem diversity is high because precipitation is sufficient to support crop production on the best soils and a number of crops are grown. However, on the least productive soils native grasslands still exist and, depending on site conditions, they may be similar to either the dry western types or the wet eastern types. An analysis of the potential sensitivity of the regional carbon balance to climate change and land use first assumed that the entire area was still occupied by native grassland (case 1 – no ecosystem diversity) (Burke *et al.*, 1991). The carbon balance was simulated using the CENTURY model (Parton *et al.*, 1987), and the potential effect of warming as a result of climate change was found to be small compared to the past effect of conversion of a large fraction of the area to cropland. In an ongoing effort, I.C. Burke is considering the effects of land-use diversity using a land-use map as input to CENTURY so that the effects of the various crops and management can be evaluated in terms of positive or negative effects compared to native grasslands (case 2 – regional response as the weighted average of a mosaic of biogeochemically distinct units).

An example of case 3 is given by biogeochemical questions about areas that include riparian ecosystems. These ecosystems have a specific role in landscapes, because they filter sediments and retain nutrients (Gregory *et al.*, 1991). Analysis of an experimental watershed demonstrated that while croplands released most of the nitrogen and phosphorus received during the year, a riparian forest retained most of the nutrients including those from an adjacent cropland (Peterjohn and Correll, 1984). Disturbance to riparian ecosystems may result in severe sedimentation and nutrient loading with the corresponding landscape and regional effects (Burke & Lauenroth, 1995). In addition to being an example of the importance of the spatial arrangement of landscapes in evaluating biogeochemical response of landscapes or regions, riparian ecosystems may be one of the best examples of 'keystone' elements in landscape diversity, because their influence can be disproportionate to their areal extent.

11.4   Societal consequences of altered biodiversity

Biodiversity changes have impacts on all societies through their effect on delivery of ecosystem services, i.e. those products of ecosystems required by humans (Ehrlich & Mooney, 1983). It is not known what fraction of extant species is required to keep ecosystems functioning in a fashion that supports natural rates of carbon and trace-gas exchange with the atmosphere and water and nutrient runoff to aquatic systems. If many of the ecosystem functions of species were redundant, a less diverse world might function 'normally' under average conditions. However, species-poor ecosystems may be more vulnerable to catastrophic disease and/or be less resilient in the face of environmental change than would a natural ecosystem with its original species diversity (Section 11.3.2) (Walker, 1992; Lawton & Brown, 1993). Given the current uncertainty as to the number of species required for normal ecosystem functioning under current and future conditions, policies that conserve species and genetic diversity are a prudent course of action.

Changes in species diversity directly affect the livelihood and culture of traditional societies in the biodiversity-rich regions of the developing tropics. Inhabitants of these regions obtain a variety of resources from the forest such as food, fodder, fuelwood, medicinal plants, and timber for their own use and for cash income. Traditional societies employ a wide variety of food production systems, ranging from shifting to sedentary agriculture. These agro-ecosystem types, including traditional freshwater fishery systems, have close interconnections with natural forests and with complex village societal systems. Together they generate a landscape mosaic that includes (i) natural forests and grasslands, (ii) low-input traditional multi-species shifting agricultural systems, (iii) mid-intensity agro-ecosystems such as agroforestry, alley cropping, rotational cropping systems, etc., and (iv) high-input modern mono-cropping agriculture (Swift *et al.*, 1995). This patchwork mosaic of managed and natural ecosystems may provide a more sustainable livelihood for traditional societies in forested areas than a commitment to a single agricultural system or to a system in which natural reserves are isolated from populated areas (Ramakrishnan, 1992).

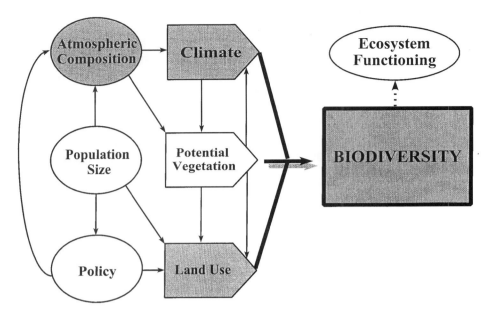

Figure 11.7   The major drivers of change in biodiversity as used in an exercise to develop biodiversity scenarios.

11.5   Implications for the future

Previous sections of this chapter described the patterns of biodiversity change that occur as a result of human activity, the mechanisms underlying the observed changes, and the consequences for ecosystem functioning. Based on this current understanding and the tools already developed, can scenarios of biodiversity change into the next century be developed? Several disciplines from atmospheric sciences to human demography have developed scenarios of such quantities as CO_2 and CH_4 emissions, climate, and population density, based on the current state of knowledge in these fields.

The development of biodiversity scenarios requires input from scenarios of changes in land use, atmospheric composition, and climate, as described in Section 11.2 (Fig. 11.7). In turn, biodiversity scenarios should serve as input to other scenarios. A GCTE project to develop biodiversity scenarios for the major biomes of the world involved scientists familiar with individual biomes and with models that simulate future patterns of the three major drivers of change in biodiversity: land use, climate, and potential natural vegetation. IMAGE 2 (Alcamo, 1994) provided scenarios of land use for the different regions of the world, which in turn were based on scenarios of change in climate and human demography. BIOME 2 (Prentice *et al.*, 1992) provided a scenario of the global distribution of potential vegetation at equilibrium with a double-CO_2 climate.

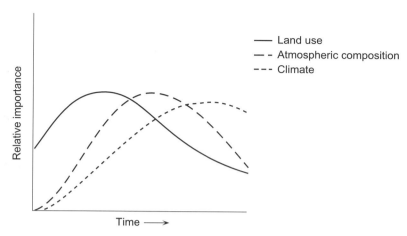

Figure 11.8 Conceptual model of the change through time in the relative importance of the different drivers of change in biodiversity. Relative importance in this conceptual model ranges from 0 to 1 and is formed by the sum of the effects of the three drivers. Consequently, decrease in the relative importance of a driver may result from either a decline of its direct effect or by an increase in the effects of other drivers.

Analysis across biomes from the present to the year 2100 suggested that the relative importance of global change drivers will change with time (Fig. 11.8). Currently, changes in land use have the largest impact on biodiversity at all levels. Conversion of natural ecosystems into croplands, fragmentation of natural ecosystems, and their over-exploitation have far larger impacts on biodiversity than do changes in climate and atmospheric CO_2, which may result in shifts of potential natural vegetation. Changes in land use will be reduced in the short to medium term when most of the arable land is converted to cropland. The direct effect of changes in the atmospheric composition will become an increasingly important driver of biodiversity changes (see Chapter 7). Increases in the atmospheric concentration of CO_2 have already been detected, are substantial, and are expected to continue to increase (see Chapter 1). Because the climate system has considerable inertia, decades will be necessary to observe the effects of current and past changes in atmospheric composition on climate. Furthermore, because ecosystems are adapted to the existing large interannual climate variability, climate change will be the last of the drivers to affect biodiversity significantly.

Cross-biome analysis suggested important spatial patterns for the drivers and responses. The expected temperature changes increase with latitude while precipitation changes, although equally important, have more idiosyncratic patterns. Potential vegetation changes mainly reflect changes in climate since current models do not simulate migration or successional trends. Land-use change is expected to be large in savannas and forests, which are areas where large food demand increases are expected. Therefore, IMAGE 2 predicts some of largest changes to occur in Africa (see also Section 12.4).

Assessment of expected changes in the drivers combined with the differential biome sensitivities yield the patterns of expected biodiversity change. Biomes that are characteristic of extreme environments like tundra, boreal forest, or deserts may be more sensitive to climate change than to land-use change. The difficulties of feasible commercial operations and low population densities may minimize logging of boreal forest or transformation of deserts into croplands before they are impacted by changing climate. The land-use model predicts relatively small changes in land use for these biomes. In addition, warming predictions are larger at higher latitudes.

In contrast to biomes of extreme environments, biodiversity in biomes characteristic of temperate and mesic sites is more likely to be affected more rapidly by changes in land use than by changes in climate. Not only are changes in climate predicted to be smaller but, due to their production potential, these biomes are more likely to be heavily utilized and transformed.

Acknowledgements

We thank Elisabeth Huber-Sannwald and Amy T. Austin for their assistance and insightful comments.

⑫ Interactive and integrated effects of global change on terrestrial ecosystems

B.H. Walker, W.L. Steffen and J. Langridge

12.1   Introduction

The preceding chapters have synthesized what has been learned over the past six years about global change and terrestrial ecosystems. In this chapter these findings, across the GCTE programme, are synthesized in terms of five major global change issues: (i) the terrestrial carbon cycle; (ii) interaction of ecosystem structure and functioning; (iii) vegetation dynamics; (iv) impacts on production systems; and (v) effects on ecological complexity (Section 12.2). An assessment of the net effects of all the global change influences, acting together, is then presented for a few selected regions (Section 12.3). This is done by developing for each region an assessment of what is likely to happen, in an integrated way, under a future climate scenario, including knowledge of disturbance and human activities in the region concerned. The synthesis then returns to the global scale and, using the scenarios of the future potential vegetation of the earth as simulated by DGVMs, discusses how the vegetation cover of the earth may actually change into the future, by incorporating the information on disturbance effects from the regional assessments (Section 12.4). Finally, the chapter touches upon the issue of society's responses to this emerging understanding of global change effects – learning to live with global change (Section 12.5).

12.2   Overall synthesis of research findings

12.2.1   Global change and the terrestrial carbon cycle

This section presents an integration of the various results that have relevance to the terrestrial carbon cycle, and projects the likely role of the terrestrial biosphere in the global carbon cycle over the next 50–100 years.

The current carbon perturbation budget

The starting point is the assessment of the global CO_2 perturbation budget carried out by the Intergovernmental Panel on Climate Change (IPCC, 1994), the terrestrial component of which was amplified by Schimel (1995) and summarized in Table 1.2. This analysis suggests that the terrestrial biosphere is about in balance with regard to the emission and absorption of CO_2, or perhaps is a small sink (see also Keeling *et al.*, 1996*b*). An estimated 1.6 billion tonnes of carbon per year are released through land-use change in the tropics, while about 2.1 billion tonnes of carbon per year are absorbed by terrestrial ecosystems, through the combined effects of forest regrowth, CO_2 fertilization, and nitrogen deposition (Table 1.2).

It is important to note that these estimates are all *average* annual values. There is a very large interannual variability in them and in any one year the net flux can be either positive or negative – i.e. on an annual basis the terrestrial biosphere fluctuates between being a net source and a net sink of carbon (D.S. Schimel, personal communication). Because the biosphere is so closely balanced around a net zero flux, any changes in land use, ecosystem physiology or vegetation structure will have significant effects. It is therefore also important to note that there are significant uncertainties associated with the estimates of all of the budget terms associated with the terrestrial biosphere. Chapter 10 presents a more detailed overview of the terrestrial carbon cycle and its interaction with other biogeochemical cycles.

Can this capability of the terrestrial biosphere to absorb CO_2 be maintained or increased in the future? More specifically:

1. Current absorption of CO_2 by the terrestrial biosphere is about 2.1 Gt C per year, and total releases by human activities are about 7.1 Gt C per year. Thus, the terrestrial biosphere is absorbing about 30% of emissions. Can this fraction be increased or maintained over the next 100 years, or will it drop?
2. Will the source-sink relationship (1.6 Gt C per year released; 2.1 Gt C per year absorbed) be maintained over the next 100 years? If not, how is it likely to change?

This synthesis deals with three important aspects of the terrestrial biosphere's role in the global carbon cycle:

- Land-use/cover change
- Ecosystem physiological responses to environmental change
- Long-term ecosystem structural/compositional responses to environmental change

Land-use/cover change

This component is currently the largest in the budget of terrestrial carbon perturbations and will very likely remain so for the next several decades at least, and perhaps for much of the next century. As noted in Chapters 1 and 9, the human population will increase by nearly one billion per decade for the next three decades at least, and this will require a 2% per year increase in food production for the foreseeable future. This will result both in further conversion of natural ecosystems to agriculture, especially in Africa and Asia (see IMAGE model projections, Alcamo, 1994; Alcamo et al., 1996a,b), and in intensification of production on currently cropped lands. Both of these processes almost always release carbon to the atmosphere (Burke et al., 1993). Losses of carbon include both the initial loss associated with the removal of the natural vegetation and subsequent losses from soil as cropping is established and then intensified. The IMAGE model suggests that net carbon losses associated with all forms of land-use/cover change (currently about 1.1 Gt C per annum when Northern Hemisphere forest regrowth is included) will continue at that level or increase for the first half of the next century at least. Other studies on the long-term carbon dynamics associated with land-use/cover change project large carbon emissions to the atmosphere (Esser, 1990; Cramer & Solomon, 1993). Conversion of natural ecosystems to agriculture and of extensive agricultural systems to intensive ones also reduce the potential of the land cover to sequester carbon via CO_2 fertilization and N deposition (see next section).

The physiological responses of ecosystems to environmental change

Four aspects of ecosystem physiological responses are considered: (i) CO_2 fertilization; (ii) responses to climate change; (iii) interactions among responses; and (iv) the long-term physiological buffering capacity of the terrestrial biosphere.

(I) CO_2 FERTILIZATION. Increasing atmospheric CO_2 concentration leads to an increase in photosynthesis, the CO_2 fertilization effect. However, this increase in initial carbon uptake at the leaf level does not translate directly into increased long-term carbon storage by whole ecosystems; there are many intervening processes that affect the fate of the carbon. As more becomes known about the long-term system-level responses to elevated CO_2 (see Chapter 7), estimates of the magnitude of the CO_2 fertilization effect (and hence its representation in biogeochemical models) have followed a consistent pattern: there has been a continuing reduction in the estimate of the size of the effect. As noted in Chapter 7, when the most recent findings (reduction of stomatal conductance under elevated CO_2 may not be found in mature forests; and increase in C:N ratio observed in live plant tissue is not seen generally in litter), are incorporated into models, the model results show a lower magnitude of the CO_2 response in terms of plant production and soil carbon storage (see Section 7.2.5). This pattern is generally consistent with the results of 'natural experiments' – tree ring studies in forests subjected to the historical increase in atmospheric CO_2

concentration (Schweingruber *et al.*, 1993; Phillips & Gentry, 1994) and growth studies on vegetation around natural CO_2 vents (Hättenschwiler *et al.*, 1997) – which show little or no sustained, long-term increase in above ground growth with higher CO_2 concentrations. (This does not preclude the possibility of increased belowground growth.) Thus, when projecting the future terrestrial carbon budget, the CO_2 fertilization effect, although still potentially significant, may well not be as large as earlier thought.

(II) RESPONSE TO CLIMATE CHANGE.   Many effects of changing temperature and moisture on the terrestrial carbon cycle have been hypothesised, often acting in opposition. For example, increasing temperature and season length in the high latitudes may increase the rate of decomposition of soil organic matter and hence lead to enhanced carbon emissions. At the same time, such increased decomposition will lead to a faster rate of nutrient release, which in turn will stimulate growth and NPP of vegetation. Clearly, the *overall* effect will be the net result of a number of competing, interacting processes (see Chapter 10). So far, an insufficient number of whole ecosystem experiments manipulating temperature and moisture have been done to yield likely patterns of overall response across several biomes. However, a natural experiment is occurring in the high latitudes, where continental regions have been subjected to a temperature increase, as high as 3 °C in some areas, over the past three decades (Chapman & Walsh, 1993). Measurements of the net CO_2 release/ uptake by tundra ecosystems (Oechel *et al.*, 1993; Zimov *et al.*, 1996) in Alaska and Siberia suggest that these systems have gone from being a carbon sink to a source, or are now in approximate balance. These early results are consistent with the projection that enhanced decomposition of soil organic carbon will lead to increased carbon emissions as the climate warms (see Chapter 9). However, much more work is required before these results can be assumed to be more general, both over time and across this and other biomes.

(III) INTERACTIONS.   In reality, all terrestrial responses to the drivers of global change occur simultaneously and interactively. As Schimel (1995) notes, much of the data on the rate of Northern Hemisphere forest regrowth (from abandonment of agriculture) has been taken during the recent period when increasing atmospheric CO_2 and N deposition are also potentially affecting growth rates. Apportioning an overall growth stimulation to these various factors is extremely difficult, but important for future projections of the terrestrial carbon cycle. For example, if more of the overall terrestrial sink is due to the normal successional dynamics of forest regrowth and to N deposition than to CO_2 fertilization (which seems likely given the analysis presented in Section 7.5.1), then the reliability of the terrestrial sink into the next century, from a physiological perspective, is diminished. The N-deposition effect has a maximum (Asner *et al.*, 1997), and it appears that this is now being exceeded for some European forests (Ingestad & Ågren, 1988; Schulze, 1995*b*). The critical issues then become the rate at which agricultural land in the temperate regions of the Northern Hemisphere continues to be abandoned, or at which regrowth forest is converted back to agriculture, and the pattern and amount of N deposition.

(IV) LONG-TERM PHYSIOLOGICAL BUFFERING CAPACITY OF THE
TERRESTRIAL BIOSPHERE (PHYSIOLOGICAL 'SATURATION'). The net uptake of carbon from the atmosphere by terrestrial ecosystems at a site over short (annual) time frames (NEP) is a balance between the assimilation of CO_2 (photosynthesis minus autotrophic respiration) and the loss of CO_2 through decomposition (heterotrophic respiration). As described in detail in Chapter 10, these processes occur at different rates. Carbon assimilation responds positively and almost instantly to increased atmospheric CO_2, whereas decomposition in natural ecosystems responds only indirectly, through changes in temperature, moisture and litter quality – all of which include relatively long delay components. In addition, there are nonlinearities in these processes that limit the rates of long-term (century-scale) carbon sequestration. While the carbon assimilation rate increases with increasing atmospheric CO_2, it does so at a diminishing rate. Decomposition, in contrast, is an exponentially increasing function of temperature. Thus, as global change proceeds, the rate of increase of CO_2 assimilation by terrestrial ecosystems will slow, while rates of decomposition, which releases CO_2 to the atmosphere, will increase (unless litter quality decreases under elevated CO_2; the initial evidence for this is conflicting – see Chapter 7). The net effect is that the ability of the terrestrial biosphere to absorb CO_2 may be diminished as temperature increases. Over the next century, the saturating concentration for CO_2 assimilation will likely not yet be reached in many biomes because CO_2 concentrations will not have risen enough. However, if the early results from three decades of significant warming in the high latitudes are a reliable indication, the effect of warming on carbon release could well become significant in the next few decades and will ultimately limit the overall, long-term capacity of the terrestrial biosphere to absorb carbon.

Long-term ecosystem structural/compositional responses to environmental change
The shifts in biome distributions predicted by equilibrium vegetation models will not occur as smooth transitions. As discussed in Chapter 8, the view on how vegetation structure and composition will change is shifting from one centred on equilibria and changes from one intact biome type to another, to a view that

emphasizes changing ratios of plant functional types, disturbances, lag effects and migration. The emphasis is on transient effects.

According to this emerging view of global vegetation change, the present vegetation assemblages will likely change through increased mortality of some of their component functional types, followed by the establishment and growth of new assemblages of functional types. Changes in natural disturbance regimes, such as fires and insect attacks, will likely be the agents of increased mortality of the existing vegetation. This disturbance-driven mortality, which releases carbon to the atmosphere, is a fast process, while the growth of a new assemblage of vegetation, which absorbs carbon from the atmosphere, is slower. Thus, although the changing biome distribution which will occur in response to a future changing climate may store the same or even more carbon than at present, the process by which this biome change occurs may lead to an increased emission of carbon to the atmosphere as disturbances increase.

Ecosystem structural change has been the subject of much less research than ecophysiological responses to global change, and thus has rarely been included in analyses of the terrestrial carbon cycle. However, its quantitative impact is potentially similar in magnitude to that of physiological processes, especially in the medium term (decades to a century). Evidence from the high latitude regions over the past three decades of warming suggest that this disturbance-driven phenomenon can be significant. Observations of fire frequencies in the Canadian boreal forests over the past century suggest a sharp recent increase, concomitant with rising temperatures (Fig. 12.1). The resulting change in age class structure of the forests indicates that they have gone from being a sink (about 0.2 Gt C per year) to being neutral or a slight source of carbon to the atmosphere (Kurz *et al.*, 1995). If these results are representative of a general pattern as the climate continues to warm, then significant transient releases of carbon may result as disturbance patterns are altered and vegetation assemblages change in response.

Net effects

Two questions were posed at the beginning of this section:

1. Can the current fraction of annual CO_2 emissions absorbed by terrestrial ecosystems (about 30%) be increased or maintained over the next 100 years, or will it drop?
2. Will the present source-sink relationship of the terrestrial biosphere (about neutral with respect to emissions and sequestration) be maintained over the next 100 years?

In attempting to answer these questions, a quantitatively reliable projection of the terms of the perturbation budget for 50–100 years into the future is not

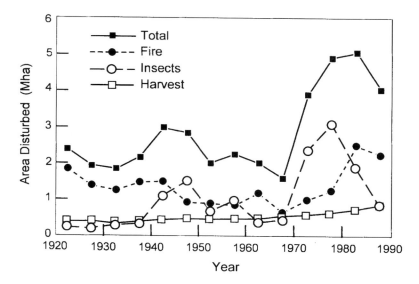

Figure 12.1 The average area of Canadian boreal forest annually disturbed by forest fires, insect-induced stand mortality and clear-cut logging in the period 1920 to 1989. From Kurz *et al.*, 1995.

feasible. There are already large uncertainties associated with the perturbation budget of the 1980s, and even larger uncertainties would apply to specific, quantitative projections for the future. However, it is possible to assess the likely trends in the terms of the budget – whether they will increase or decrease in relative importance – so that an overall trend can be projected.

Based on the analysis presented above, the following trends in the various terms of the perturbation budget are plausible:

■ Emissions from land-use/cover change will almost surely increase (as seems inevitable given the sustained increase in food demand over the next three decades at least).

■ If climate continues to change at the rate it has over the second half of the 20th century, a significant emerging source – the emission of CO_2 associated with the dynamics of large-scale change in ecosystem composition and structure – may add further to the release of carbon from the terrestrial biosphere.

■ The emission of CO_2 from soils as climate warms will become an increasingly important source through the 21st century.

■ Increasing CO_2 fertilization coupled with N deposition will partially offset these expected increases in emissions.

Thus, the answers to the questions are: (1) Given the projected increase in fossil fuel emissions, the fraction of annual anthropogenic CO_2 emissions absorbed by terrestrial ecosystems will almost surely decrease. (2) It is possible that the carbon source-sink relationship for the terrestrial biosphere may remain approximately in balance, especially if the current sinks in the mid-latitudes of the Northern Hemisphere can be increased by policy and management, but a more likely scenario is that the biosphere as a whole will become a net source. This projection has direct implications for the development of strategies to stabilize the concentration of greenhouse gases in the atmosphere.

12.2.2   Interaction of structure and functioning in ecosystem change

Influence of structure on functioning

The interaction between ecosystem structure and functioning is highlighted in the introduction to Chapter 8, where the structural influence on the response of vegetation to increasing atmospheric CO_2 concentration is discussed for various space and time scales (see Table 8.1 and Shugart, 1997). A similar analysis can be applied to a closely related issue of importance to Earth system science, the extrapolation of ecosystem functioning (e.g. photosynthesis, evapotranspiration) from the leaf level to the globe.

Carbon sequestration is an important example of ecosystem functioning where scaling directly from leaf to globe without taking into account the intervening effects of structure is not appropriate, and is no longer done in state-of-the-art DGVMs and biogeochemical models. Higher order structure mediates CO_2 response, and so leaf- or plant-level processes cannot be extrapolated directly to the scale of a whole ecosystem, as shown by reviews such as that of Körner (1993). Projections of long-term responses to elevated CO_2 directly based on sub-plant and plant scales are thus often upwardly biased.

The effects of structure on functioning do not end at the whole ecosystem (stand) level. Although Net Ecosystem Productivity (NEP) is an appropriate measure of the net carbon uptake of an ecosystem, it still does not allow direct extrapolation to the globe over decadal time frames. At these longer time and larger space scales, landscape phenomena, such as disturbance (e.g. fire), become important in controlling the overall productivity of the system. As noted in Table 8.1, landscapes can be viewed as mosaics in different stages of recovery from natural disturbance. Failure to account for the successional stage of the vegetation can lead to unrealistic extrapolations from flux measurements of CO_2 uptake (which estimates NEP) or from elevated CO_2 experiments. For example, eddy correlation measurements of CO_2 uptake of regrowth forests in an early successional stage simply show the strong uptake of carbon associated with rapid growth, with or without elevated CO_2 levels.

The inclusion of these larger scale processes and patterns in analysing terrestrial carbon dynamics has led to the concept of 'Net Biome Productivity (NBP)' (Schulze & Heimann, 1997), described in more detail in Chapter 10 and shown schematically in Fig. 10.2. This is the appropriate concept for an analysis of the global carbon cycle, rather than NEP at the stand level or, even less appropriate, NPP or growth increases. In fact, it is possible, and rather likely under global change, that an ecosystem could have a positive NEP over a multi-year time frame (a response, for example, to increasing rainfall) but that the biome in which it is embedded could have a negative NBP over a multi-decadal time frame (due to increasing fire frequency caused by increased fuel load and higher temperatures).

Changes in the Canadian boreal forests over the past few decades illustrate this effect. There is good evidence from remote sensing data that photosynthesis has increased significantly over the 1981–1991 decade in the high latitudes, including Canada, primarily due to a lengthening of the growing season (Myneni et al., 1997). Detailed process-level studies of gas exchange between Canadian boreal forests and the atmosphere suggest that net carbon uptake (NEP) is positive over the growing season, although there is much variation from site to site (measurements vary from 350 g C m^{-2} yr^{-1} for aspen, an early successional species (Black et al., 1996), to about 50 g C m^{-2} yr^{-1} for jack pine (Baldocchi et al., 1997) to near zero for the northern BOREAS stand of black spruce (P.G. Jarvis, personal communication). However, a continental-scale analysis over the last 100 years of disturbance frequencies in the Canadian boreal forests (Kurz et al., 1995) suggests that over the last few decades they have gone from being a sink of about 0.2 Gt C yr^{-1} to being about neutral in terms of carbon exchange. Thus, for recent years estimates of NEP for the Canadian boreal forest are positive (net carbon uptake at the patch scale over a season), while an estimate of NBP is near zero (no net carbon uptake by the biome at a decadal time scale).

In summary, the uptake or release of carbon by the terrestrial biosphere over decades or a century is the net result of a complex interplay of structure and functioning at a large number of space and time scales. A simple scale-up from leaf or plant to globe is insufficient to understand how the terrestrial carbon cycle is responding to global change; alterations to structure/functioning relationships at a number of intermediate scales must be understood and taken into account.

Influence of functioning on structure

Complementary to the above example, changes in ecosystem functioning induced by global change can lead to changes in structure. For instance, changes in water-use efficiency brought about by elevated CO_2 (Bazzaz et al., 1989; Bazzaz, 1996; Chapter 7) lead to changed competitive relationships among the species in a community, and hence to changes in the relative abundances of the species.

At biome scales changes in temperature and precipitation regimes will lead to changes in water balance, and in boreal forests these projected changes have led to model predictions of marked differences in changes to the vegetation composition of north- and south-facing slopes. In particular, the decline of mixed spruce/birch/aspen forests on south-facing slopes and their potential conversion to cool temperate grassland because of increased drought and fire have been simulated by forest dynamics models (Smith *et al.*, 1995; Shugart, 1997), and recent empirical evidence is lending support to such disturbance-driven change (Jacoby & Darrigo, 1995).

As another example of disturbance-driven dynamics, in Alaska the observed increased temperatures over the recent three decades (Chapman & Walsh, 1993) have led to changed population dynamics of the spruce bark beetle and to increased susceptibility of spruce trees to attack (physiological responses to the temperature increase), and consequently to increased mortality of whole stands of spruce trees to bark beetle attack (a structural effect). There have been observations of spreading spruce bark beetle epidemics in forests on the southern third of the Alaskan transect (south of the Brooks Range) (USDA Forest Service, 1997); and observations, in at least one case, of such insect attacks leading to a conversion of forest to grassland, due to a combination of warming and fire suppression activities (LaBau, 1996), consistent with model predictions (Starfield & Chapin, 1996).

Experimental warming of ecosystems has shown that the effects are mostly indirect (see Chapter 7), via changes in soil moisture, which alter the competitive balance between (for example) shrubs and forbs (Harte *et al.*, 1995), with some empirical evidence from Alaska (Chapin *et al.*, 1995*b*) showing a change in tundra from sedges to shrubs. Chapter 11 has identified a number of lines of evidence that changes in ecosystem functioning caused by global change will result in significant changes in the structure of ecosystems.

Interactive effects

The effects of ecosystem structure on functioning and the effects of functioning on structure do not act independently. The interactions are usually complex, and few generalities can yet be made of the impacts of global change on the nature of this interaction itself. However, as described in Section 7.8.1, one explicit modelling attempt to examine interactive effects indicated that increasing N and air temperatures (as given under one particular scenario of global change) had larger effects on ecosystem properties than increasing levels of CO_2, taken on their own. Increasing N led to increases in plant production, N mineralization and soil C storage while increasing temperature decreased soil C but also increased plant production and N mineralization.

Another example of the interaction of structure and functioning comes from a study on water–use efficiency and productivity along the rainfall gradient on the Northern Australia Tropical Transect (Schulze *et al.*, 1998). For any particular woody species, as rainfall decreased so did the change in the $\delta^{13}C$ ratio, indicating that water-use efficiency declined (a change in functioning in response to an abiotic factor). However, as rainfall declined there was also a change in woody species composition (a change in ecosystem structure), to tree species with higher leaf nitrogen and lower specific leaf areas, and the $\delta^{13}C$ ratio for each succeeding species indicated an increased water use efficiency, such that the community-averaged $\delta^{13}C$ value changed very little down the rainfall gradient. They concluded that changing species composition acted towards maintaining productivity under low nutrient and water availability. A further complication is the impact on the woody plant species composition induced by changes in fire frequencies or herbivory. These effects will limit the extent to which the species replacement along the gradient is able to dampen the change in overall $\delta^{13}C$ ratio, and therefore the water use efficiency and productivity of the vegetation. This latter effect highlights again the important difference between NEP and NBP, outlined above.

Bonan *et al.*'s (1992) modelled sensitivity study of boreal forest deforestation showed a very strong potential interaction. Loss of forest caused a sufficient cooling in the summer (through increased albedo) to prevent new forest becoming established. In an analogous way Xue and Shukla's (1993) model experiments showed that reducing vegetation cover in the Sahel, to desert-like conditions, resulted in reduced rainfall, and thereby shifted the region further towards desert.

In summary, these examples show that ecosystem structure and functioning are intricately linked. Although global change studies usually consider one or the other independently, they operate together as a system. At present many of their interactions can be untangled and explained after the fact, but they are difficult to predict *a priori*.

12.2.3 Vegetation dynamics under global change

One of the first 'global change products' from the terrestrial ecology community was a group of models of vegetation change at the global scale (e.g. Prentice *et al.*, 1992). These 'equilibrium' models assumed that vegetation is in equilibrium with its abiotic environment. Though these models did not simulate the dynamics through which vegetation would actually respond to a changing environment, they were nevertheless useful in indicating how the distribution of biomes might look under a different climate. The models quickly found applications in a wide range of impact studies.

Extensive use of equilibrium models, however, has perhaps unconsciously led to a paradigm of vegetation change based on a rearrangement of present biomes, a type of vegetative plate tectonics. The paradigm suggests that there are beginning and end points (i.e. 'when climate change occurs...'), resulting in a transition from one equilibrium distribution of biomes to another.

As described in Chapter 8, reality will be quite different. Global change-driven impacts on vegetation composition and structure, at scales from the stand to the globe, are occurring now, are continuous, will likely accelerate, and have no identifiable or predictable end-point. These nonequilibrium, transient dynamics of changing vegetation composition and structure include several important features (Steffen *et al.*, 1996; Cramer & Steffen, 1997):

- Biomes will not shift as intact entities. Species respond differently in competitive abilities (e.g. relative growth rates), migration rates, recovery from (response to) disturbance and in other ways. Thus, new combinations of species will arise.
- Palaeo studies and model simulations suggest that many plant species can migrate fast enough to keep up with projected climatic change, but only if they can migrate through continuous, relatively undisturbed natural ecosystems (see Chapter 8). This emphasizes the important consequences of fragmentation of natural ecosystems (see Chapters 9 and 11) as a global change phenomenon.
- Invasion of alien species into natural ecosystems is already a problem and is worsening under global change (see Chapter 11). It will probably be exacerbated by the trends in land-use/cover change and by the globalization of trade.
- Disturbances (e.g. fire, dieback due to insect attacks) are increasing in some regions, leading to increasing mortality and to more systems in early successional states.
- Because direct physiological effects (e.g. response to elevated CO_2) lead to changing competitive abilities at the species level (see Chapter 7), they are important for structure and composition as well as for functioning.
- Three of the four DGVMs used for the transient run of vegetation dynamics in response to the derived transient climate through to 2100 (Chapter 8) showed a total global vegetation biomass increase of between 0.7 and 3.5 Pg C yr^{-1}, with an overall increase, over the hundred years of climate change, in NPP of between 44% and 71%. The big enhancements in NPP were in the high latitudes, with increasing CO_2 driving additional increases in NPP in the semi-arid regions.
- The shifts in vegetation distribution in the models were driven by changes in climate, with limited influence from CO_2, except in the arid regions.

Taken together, these factors, and the trends which are beginning to emerge in some regions (see transect-based analysis in the next section), suggest some generalizations about vegetation dynamics in the 21st century:

- Rates of biological invasions of alien species and of biodiversity loss are currently high and are not likely to diminish over the next several decades (see Chapters 1 and 11). Thus, the terrestrial biosphere of the 21st century will probably be further impoverished in terms of species richness and ranges, and substantially 'reorganized' in terms of species composition, with as yet unknown consequences for ecosystem functioning.
- More natural ecosystems will be in an early successional state or will be converted into human-dominated terrestrial production systems. A good example of the latter is the conversion of tropical forest into oil palm plantations and of mangrove forests into prawn farms, two important land-use changes in Southeast Asia. These trends will result in a generally 'weedier', structurally simpler biosphere with fewer systems in a more, ecologically complex old-growth state.
- Many forests appear to be sensitive to projected climate change on a time scale of centuries. Decadal responses to the shifting climate are likely to be muted and masked due to lag effects in demographic processes. However, associated secondary effects due to changes in disturbance regimes will very probably be detected on decadal time scales.

These broad-scale generalizations mask important regional differences, which are discussed in Section 12.3.

At finer scales, a recurring outcome of studies is that the future state of the vegetation, under any particular global change scenario, is heavily dependent on the previous disturbance regime. For example, Sykes and Prentice (1996) modelled the composition of the mixed conifer/northern hardwood forests of northern Europe predicted under a doubled CO_2 climate change scenario. The major changes represented continuing successional (non-climatic) responses to the cessation of disturbance 150 years ago. Eventually, the model indicates that climate change allows *Fagus* to appear and compete successfully with *Picea*, but it takes some hundreds of years for the effects of the past disturbance to work through.

Interaction of climate with other factors is very important in the dynamics of vegetation. For example, soil type strongly mediates the direct effect of climate. In a simulation of temperate forest dynamics in North America (Post & Pastor, 1996), there was no decrease in soil water availability predicted under a doubled CO_2 climate on a silty clay-loam, and replacement of mixed spruce-fir/northern-hardwood with a more productive northern-hardwood forest was projected. On an adjacent sandy soil there were increases in the proportion of the year with soil moisture below wilting point and the mixed forest was replaced by a stunted pine-oak forest.

12.2.4   Impacts of global change on managed production systems

Although intensive cropping areas and nature reserves can readily be classified as 'managed' and 'natural' ecosystems, respectively, much of the Earth's land cover, such as forests, rangelands and complex agroecosystems, cannot be so easily classified. The vegetation dynamics of these areas are subject to both direct human management and to natural processes. This section pulls together research across all four GCTE Foci, combining physiology, structure and complexity, from the perspective of production in two of these types of managed ecosystems. First, however, the impacts of global change on crop production are briefly reviewed.

Major grain crops

The conclusion of the work presented in Chapter 9 is that the impacts of global change on crop production will have a strongly regional character. It is likely that in the major temperate grain-producing areas – the 'bread baskets' of Europe and North America – yields of important crops will increase under rising temperatures and atmospheric CO_2. The future of crop yields in the tropical regions under global change is far less certain, with some projections suggesting static or declining yields. The GCTE synthesis has highlighted two key issues that affect these generalizations. First, as noted in both Chapters 7 and 9, the actual impact of elevated atmospheric CO_2 on crop growth, and especially on yields, is likely to be significantly less than earlier estimates, and will probably be less than 20% for doubled current CO_2 under most agronomic situations. Second, the yields of many cropping systems are sensitive to the moisture regime. Projections of changes to moisture regimes are not yet reliable enough to determine whether current agricultural infrastructure and management options can cope with them.

The likely impacts of global change on the world's major grain crops are described in more detail in Chapter 9 and summarized in the global analysis in Section 12.4.

Complex agroecosystems

Most global change research to date on agriculture has focused on intensively managed, single-crop systems. Much of the world's population, however, relies on more complex agricultural systems for food and fibre, systems that are increasingly being pushed towards intensification and simplification (Chapter 9). As noted in Section 12.5 below, there is an inescapable trade-off between production and resilience. The critical question for complex agroecosystems is how much and what kinds of complexity can be lost while still maintaining an acceptable level of resilience and productivity in the face of accelerating change.

The early results of work on biodiversity and ecosystem functioning (Chapter 11) show that both productivity and resilience appear to be related to species richness for species-poor systems (e.g. Naeem *et al.*, 1994, 1995; Tilman *et al.*, 1996; McNaughton, 1993), but that this function seems to saturate relatively quickly. On the other hand, the impacts of species loss (simplification) in more complex systems can be highly dependent on the nature of the species rather than simply on the total number (Section 11.3). Agroecosystems are normally modified from their natural precursors to favour certain types of species – those that provide a desired product, retain and recycle nutrients, or control a pest, for example. Thus, the effects of simplification of these systems are more likely to be expressed in a series of sharp changes in functioning as key species are lost or thresholds reached, rather than as a smoothly varying function.

Managed forests

Work across GCTE suggests that global change is bringing both benefits and hazards to forest production. As noted in Chapter 7, research on *in situ* forests located next to CO_2 springs shows that, although there is no long-term increase in biomass, there is stimulation of growth for the first 30 years or so (Hättenschwiler *et al.*, 1997). Also, N deposition has clearly led to enhanced growth of trees in Europe (Section 7.5: Assmann & Franz, 1963). Tree growth may also be stimulated by the effects of elevated CO_2 on water use efficiency, especially in semi-arid forests, although the early results from mature trees do not show a strong effect (Chapter 7). In the longer term, changes in species composition may also work to enhance the productivity of managed forests. A model study of the effects of elevated CO_2 on the composition of temperate forests shows a shift in composition to those species with higher growth rates and hence to increased C uptake by the system as a whole (Section 11.3: Bolker *et al.*, 1995).

Other system-level responses to global change suggest possible reductions in forest productivity. As noted in Section 12.2.1, enhanced growth does not necessarily lead to longer-term carbon uptake; changes in disturbance regimes of Canadian boreal forests over the past 30 years have led to a reduction in net C uptake of about 0.2 Gt C yr^{-1} (Kurz *et al.*, 1995). Both model studies (Smith *et al.*, 1995; Shugart, 1997; Starfield & Chapin, 1996) and observations in Alaska (Jacoby & Darrigo, 1995; USDA Forest Service, 1997) suggest species compositional changes in and dieback of boreal forests due to a warmer, drier climate.

The net effect of global change on timber yield from managed forests will depend on the interaction of these positive and negative effects on productivity, on their relative time scales, and on appropriate management strategies. For example, increased fire frequency in the boreal zone may lead to net loss of forest production, despite increased growth rates, unless fire suppression activities are maintained or enhanced. Fire suppression activities, however, can create problems of their own. In southern Alaska, suppression of fires has led to stands of unusually old spruce forests, which have become more susceptible to attack by bark beetles, which in turn have experienced rapid increases in population because their life cycles can be completed in one year instead of two because of the recent increase in temperature in this region. The result has been the conversion of some forests to grasslands (LaBau, 1996).

The lesson from these two examples is that, to grasp the opportunities of global change and avoid the hazards, management of terrestrial production systems needs to be proactive, based increasingly on an understanding of the dynamics of the system rather than on past experience. The abiotic environment can no longer be assumed to be constant or to be varying at historical rates of change, nor can system responses to a changing environment be readily predicted on past knowledge. Much more needs to be known about vegetation dynamics under global change, and about the integration of this knowledge into management strategies (e.g. Campbell *et al.*, 1997).

12.2.5   The effects of global change on ecological complexity

In terms of significance to ecosystem functioning and the 'goods and services' it provides, ecological complexity is much more than just the number of species. It is the product of genetic diversity within species, the diversity of species, the diversity of functional types of species and the diversity (number, sizes and spatial pattern) of ecosystems in the landscape. Chapter 11 provides evidence from a range of studies to support the importance of the last of these, and given the accelerating rate of land-use change and fragmentation of broadscale ecosystems, landscape disturbance emerges as the primary cause of change in ecosystem functioning.

This statement is backed up by the results of a 'biodiversity futures scenarios' exercise run by GCTE's Focus 4. The analysis across all biomes (see Chapter 11) indicates quite clearly that changes in land use are the most important drivers that will impact on biodiversity at all levels over the next several decades. The introduction of alien species and changing atmospheric composition comes next, before climate change assumes a dominant effect.

Another emerging generalization is that biotic change is almost invariably brought about by changes in disturbance regime, which in turn is a function of the frequency, intensity, severity and duration of disturbances (Sousa, 1984). If climate change produces warmer and drier conditions, changes in ecosystem composition will likely be through increased severity and frequency of fires rather than through the direct effects of the climate itself (Romme & Turner, 1991).

Turning to the functional consequences of changes in diversity, experiments suggest that it is both the functional kinds of species and the diversity of species *per se* that are important in controlling (for example) nitrogen uptake in an ecosystem (Ewel *et al.*, 1991; Tilman *et al.*, 1996). The instantaneous effect of species richness saturates at a relatively few species (4–10). An important aspect of ecosystem functioning is the constancy (or conversely, the variability over time) in ecosystem processes. One mechanism whereby diversity might confer stability to processes is through the diversity of responses of functionally similar species to variation in the environment (McNaughton, 1977; Chapin & Shaver, 1985). Such diversity provides insurance against large changes in ecosystem processes in the face of environmental change and species loss. Chapter 11 provides a number of examples of the effects on ecosystem processes through changes (often introductions) of particular kinds of species.

Trophic structure is the final aspect of ecosystem complexity that deserves mention. Top-down controls on ecosystem structure by top carnivores or keystone herbivores have been well documented (e.g. Rasmussen, 1941; Owen-Smith, 1988), but the ecological consequences of such effects, in terms of changes in ecosystem processes, has not been sufficiently well researched to make any generalizations. It is an area that requires considerably more attention.

In summary, there are numerous examples of changes in ecological complexity at all scales, through a range of global change drivers, primarily change in land use. Although there is evidence that such changes have direct effects on ecosystem processes, one of the potentially most important consequences, about which we know the least, is the effect they may be having on the resilience of ecosystems in the face of environmental change (where resilience is defined as the capacity of a system to continue functioning in its current manner in the face of environmental or anthropogenic stress or disturbance).

12.3   Global change at regional scales: net impacts in four major zones

Research into global change effects on terrestrial ecosystems has thus far been focused at either a very local scale (the plant community, or patch) or the global scale. Interest in the consequences of global change, however, is increasingly directed towards regional scale assessments of the combined effects of all the changes acting together to produce a net effect. Such a perspective, of the relative magnitudes of the various effects and the nature of their overall impact, is needed to guide such processes as the IPCC Working Group II assessments and the planning activities of national governments. There is no formal methodology for conducting such complex analyses, and this section is an attempt to provide initial ideas by examining in some detail what is occurring, and what might occur in the future, in a number of selected regions in the world.

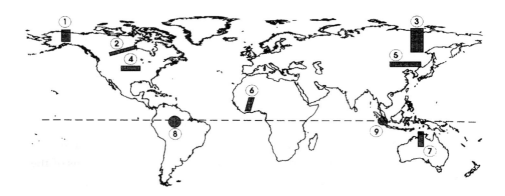

Figure 12.2   Locations of the IGBP Terrestrial Transects selected for the regional analysis of global change issues (Section 12.3). 1 = Alaska Latitudinal Gradient (ALG – 71°N 152°W, 71°N 145°W, 62°N 145°W, 62°N 152°W). 2 = Boreal Forest Transect Case Study (BFTCS – 54°N 112°W, 59°N 93°W, 55°N 93°W, 50°N 112°W). 3 = Siberia Far East Transect (SIB-E (FEST) – 74°N 125°E, 74°N 135°E, 55°N 135°E, 55°N 125°E). 4 = North American Mid-Latitude Transect (NAMER – 42°N 105°W, 42°N 90°W, 38°N 90°W, 38°N 105°W). 5 = North East Chinese Transect (NECT – 46°N 108°E, 46°N 132°E, 42°N 132°E, 42°N 108°E). 6 = Savannas in the Long-Term (SALT – 16°N 4°W, 16°N 0, 5.5°N 3°W, 5.5°N 7°W). 7 = Northern Australia Tropical Transect (NATT – 11°S 129°E, 11°S 134°E, 24°S 134°E, 24°S 129°E). 8 = Amazon (LBA – The Large-scale Biosphere–Atmosphere Experiment in Amazonia). 9 = Southeast Asian Integrated Study (Jambi Province, Sumatra, transect). Note: Transects 8 and 9 are based on a conceptual gradient of land-use change intensity.

This analysis is based on four biogeographic zones and, within each, two contrasting regions each centred on one of the IGBP Terrestrial Transects (Fig. 12.2; see Chapter 4). The intention is that this preliminary attempt will serve as a platform for more intensive studies of the regions, and provide an impetus and some guidance for extending such analyses to other regions.

For each transect, the major drivers are examined in terms of their impacts on ecosystem functioning (see Chapter 7), ecosystem structure and change (see Chapter 8), and the consequences for the production of food and fibre (see Chapter 9), with some brief comments on changes in ecological complexity (see Chapter 11). The analysis first describes the present situation, including what are considered to be the important issues in the region concerned. The differences between the potential present vegetation cover along the transect and what actually exists are analysed in terms of the disturbance regime, including land use. The emphasis then moves to a projection of the future vegetation cover along the transect, based on likely changes in the disturbance regime, considered together with predictions of future potential vegetation based on DGVMs run with a future climate scenario.

For the present and future potential vegetation the output from two transient DGVMs, the Sheffield DGVM (F.I. Woodward, personal communication) and IBIS (J. Foley, personal communication), as described in Chapter 8, were used. The predicted present potential vegetation can be compared in each case to available maps of potential vegetation produced by local experts (see Plates 8–10). The latter are at much finer scales than the coarse GCM scale used for the DGVM runs, which makes the comparison difficult. However, given the present state of GCMs the scale is appropriate, as it does not give a false sense of accuracy. The present and future land use comparison is based on runs of the IMAGE model (Alcamo, 1994; Alcamo *et al.*, 1996*a,b*), produced by R. Leemans (personal communication) for the years 1990 and 2090, using the IMAGE 2.1 version. The IMAGE climate dataset differs from the transient climate data used for the DGVMs, and the predicted present vegetation cover produced by the DGVMs and by IMAGE are thus different on two counts – different climate sets and different vegetation models. It must be remembered that the DGVMs do not depend on a correlation between present vegetation and climate, but predict the proportions of grass and of deciduous and evergreen woody biomass from basic principles, as a function of climate and soils. A set of translation rules has been derived to convert these simulated ratios of functional types into vegetation types. A modified IGBP-DIS classification system (Woodward & Steffen, 1996) has been adopted as the standard for this analysis. The translation rules have relied mainly on the biomass of the functional types and to a lesser extent on the ratios of leaf area index (LAI – for IBIS) or ratios of per cent cover (for SDGVM), rather than on the per cent cover of each of the three functional types and canopy height of the overstorey, as used in the IGBP-DIS classification.

The four selected zones are:

  (i)   the high latitudes (tundra and boreal forest biomes);
 (ii)   the mid-latitudes (grassland and deciduous temperate forest biomes);
(iii)   the sub-humid to semi-arid tropics (tropical savanna biome);
(iv)   the humid tropics (tropical rainforest biome).

12.3.1   The high latitudes (tundra–boreal forest biomes)

The selected regions are northern North America and the Far East of Siberia (Fig. 12.2). The North American analysis is based on a set of two transects, which together encompass the tundra and boreal forest biomes and the transition between them. The transects are located in Alaska and Canada (Boreal Forest Transect Case Study – BFTCS). The Alaska transect runs south from Prudhoe Bay to the Brooks Range through tundra vegetation, while the Canadian transect lies along a NE–SW axis through the provinces of Alberta,

Saskatchewan and Manitoba and encompasses primarily boreal forest systems and the transition to the prairie biome to the south-west. The Far East Siberian Transect (FEST) is located roughly along $135°$ E meridian, from Tiksi on the Arctic Ocean to south of Yakutsk.

The present

CLIMATE.   The climate of the high latitude transects is dominated by a strong north–south gradient in temperature. It is characterized by short growing seasons and long, cold winters where the vegetation is inactive and the ground is frozen and snow-covered. There is also a gradient of moisture regimes; the climate of the Alaskan transect is moister than that of the BFTCS, and the FEST climate is drier and colder than that of the bulk of northern Eurasia to the west.

POTENTIAL VEGETATION.   The existing and the predicted (modelled) potential vegetation cover under present climate on the transects are presented in Plate 8. The overall patterns are controlled by the north–south temperature gradient. The Alaskan transect is typified by tundra systems, with shrub-dominated systems in the south tending to tussock tundra and moss/lichen dominated systems in the north. The BFTCS is a predominantly forested transect; forest cover is greater than 75% for much of the area, with wetlands accounting for much of the remainder and some tundra patches occurring at the extreme northeast end. The southwest end of the transect lies in the North American prairie biome and thus incorporates the aridity-driven transition between forest and grassland.

FEST is dominated by taiga (boreal forest), with tundra prevalent along the Lena River mouth in the far north and in the upland areas. *Larix* is the dominant tree with much of the transect covered by monospecific stands (Schulze *et al.*, 1995). The Siberian transect is topographically more complex than its North American counterparts; it incorporates much of the Lena River basin, which is bordered by ranges carrying areas of mountain tundra.

An important feature of these high latitude transects is the existence of extensive zones of permafrost. Both the Alaskan transect and FEST are underlain by continuous permafrost, which can be hundreds of metres thick with an active (thaw) layer of < 1 m during the short growing season. Permafrost dynamics play a crucial role both in the hydrological regimes of the high latitudes and in vegetation composition and functioning.

At the landscape scale, the hydrological cycle – in particular, the effects of permafrost (e.g. active layer depth), the precipitation regime, and the spatial pattern of areas subject to inundation and anaerobic soil conditions – is a major factor controlling ecosystem structure and functioning. Surface hydrology largely determines the finer scale mosaic of bogs, meadows and forests of different compositions (Kelliher *et al.*, 1997).

ACTUAL VEGETATION: LAND USE AND NATURAL DISTURBANCES. Fire, insect attacks (both natural disturbances) and timber harvesting are the major disturbances that control the successional dynamics and strongly influence the structure of the forests in both regions. Although the present biome boundaries correspond well to those predicted by equilibrium vegetation models, a simple biome classification hides variations in age class structure of different stands as determined by the disturbance regime. The mean return interval of forest fires along the FEST is the subject of considerable conjecture but little long-term data; according to one study in the Yakutsk area (Osawa *et al.*, 1994), the return interval could be as low as 30 years. In the southern ends of both the BFTCS and Siberian transect, fire suppression activities over past decades have probably resulted in fewer and smaller fires.

Logging activities are occurring in pockets along the BFTCS, although most of the area is still unlogged forest. Timber harvesting has not yet reached a level where extensive areas are being logged for the second time. Timber harvesting has been a major activity in the southern parts only of the FEST.

Agriculture is an important activity only at the south-western (prairie) end of the BFTCS, where extensive wheat production occurs. There is virtually no cropping in the FEST, with some very small areas of conversion of forest to grassland for grazing limited to isolated patches along the Lena River. By comparison to most parts of the Earth's land surface, human population density is very low along these high latitude transects and is not predicted to increase significantly.

Thus, in general the actual vegetation of the high latitude transects corresponds well with the potential, although the age class structure has been modified somewhat by logging and changes to natural disturbance regimes.

The big issues

The overriding issue for both transects is the effect on ecosystem structure, composition and functioning of increasing temperature predicted under global change (and observed in some continental areas over the past three decades, see below). The impact of warming on disturbance patterns will likely be the dominant factor. The consequences of temperature-related effects for the global carbon cycle is of major concern.

Changes in logging patterns are also an increasingly important issue, but any attempt at designing sustainable forestry practices will have to be carried out in the context of a rapidly changing abiotic environment.

The future

The potential future vegetation and possible land-use patterns for both tran-sects are shown in Plate 8. In general, the DGVMs predict the northward expansion of woody systems (forest or woodland) into tundra for the Alaskan transect and FEST, in response to increasing temperature. For the BFTCS, IBIS predicts no change from the present distribution of forest, while SDGVM predicts increasing forest at the expense of shrub- and woodland. Direct, human-driven land-cover change is expected to be small for all three transects.

As previously mentioned, the high latitude transects have already been subjected to a strong temperature increase over the past several decades. In fact, the temperature record shows an increase of up to 1 °C per decade for the past three decades over much of Siberia, north-western Canada and much of Alaska (Chapman & Walsh, 1993). Thus, trends that are beginning to build now in the high latitude regions may be used as an indicator of the effects forecast by GCM scenarios of global warming. The major findings beginning to emerge include:

VEGETATION GROWTH. An increase in the length of growing season has been observed (see Section 12.2.2).

FOREST PRODUCTIVITY. Measurements of CO_2 uptake at boreal forest sites in both the FEST and BFTCS and results from the BOREAS project (see Section 12.2.2) show that these forests are, in general, of low productivity. Measurements of CO_2 fluxes and estimates of NPP at leaf, stand and regional levels for a *Larix* forest near Yakutsk (FEST) showed that the Net Ecosystem Productivity (NEP) of Siberian larch forests is small and positive, but that the Net Biome Productivity (NBP) (incorporating the effects of fires) is probably zero (Hollinger *et al.*, 1995, 1997).

TRACE GAS EMISSIONS. There is some evidence that the recent warming has led to an apparent increase in carbon emissions from soils (see Section 12.2.1).

PERMAFROST DYNAMICS. Studies in northern Alaska show that the upper layers of permafrost are very sensitive to changes in soil surface tempera-ture (Osterkamp & Romanovsky, 1996) and preliminary calculations suggest that the depth of the active (thaw) layer could double for a 4 °C increase in temperature (Kane *et al.*, 1992). Such an increase in active layer depth would probably lead to exposure of carbon to release as either CH_4 or CO_2, and would also significantly increase rooting depth and available water and nutrients. The latter would increase potential ecosystem productivity. However, thawing of permafrost often leads to subsidence of the ground surface, producing an irregular topography of channels, pits, ponds, lakes and 'drunken forests' (trees leaning in random directions).

DISTURBANCE REGIMES.   There has been an apparent increase in fire frequencies and extents over continental Canada during the past three decades, as well as an increase in damage due to insect attacks in Alaskan boreal forests (Section 12.2.2).

Management issues in the high latitudes are primarily focused on the sustainable development and maintenance of the forestry industry. This translates into (i) the development of strategies to deal with the apparent change in natural disturbance regimes and (ii) the design of sustainable logging regimes.

Increasing fire frequency is the most important of the likely changes in disturbance. In contrast, fire suppression activities are likely to decrease, both in Siberia and in Canada, due to a re-evaluation of the ecological role of fire and to economic constraints. This may well lead to a change in management strategies, towards protection and management of smaller, high productive areas.

In regard to logging regimes, subtle changes can lead to significant changes in ecosystem structure and composition. For example, at the southern end of the FEST slightly shortened logging rotations coupled with increasing fire changes the subsurface hydrology significantly, leading to waterlogging of soils and subsequently to a thick, permanent cover of grasses – a process known in Russia as 'green desertification' (D. Efremov, personal communication).

The dynamics of the extensive *Larix* forests further north are subject to changes in both fire regimes and the dynamics of the underlying permafrost in ways that are not yet well understood. Global change impacts on these 'fire and ice' dynamics will be a crucial factor to consider in developing sustainable management strategies.

Although changing climate may allow the expansion of agriculture northwards, this will likely not occur in the Canadian wheat belt due to soil limitations (i.e. the granitic shield). In East Siberia, which was not glaciated, there are no such soil constraints, but low population density, distance to markets and poor infrastructure may act as equally powerful limitations. Economic constraints are weaker in West Siberia, which may experience a northwards expansion of agriculture.

ECOLOGICAL COMPLEXITY   In terms of global change and ecological complexity, little is known about the high latitudes. In general, these systems are relatively species poor, and changes in complexity are likely to have significant impacts on ecosystem functioning (see Chapter 11). One 10-year study in Alaska, undertaken during the recent period of rapid warming, shows an intriguing response in tundra biodiversity, with an increase in deciduous shrubs and a decrease in sedges (Chapin *et al.*, 1995*b*). Such changes may have important consequences for regional water and energy fluxes, thus providing a link between changing complexity and ecosystem functioning (Weller *et al.*, 1995).

The high latitudes – summary

Climate change, especially increases in temperature, will likely be the dominant force driving changes to the composition, structure and functioning of high latitude ecosystems. Already some continental areas of the high latitudes, including the regions covered by the transects used in this analysis, have experienced significant temperature increases over the past three decades. As described above, the net effect is that these high latitude regions will probably change from being carbon sinks to net sources of carbon over much of the twenty-first century.

The sustainable management of boreal forests is the critical issue in terms of human use of the high latitudes. Again, climate change will probably be the dominant effect, and the new reality of increasing disturbance regimes and warmer and drier conditions requires a reappraisal of logging methods to avoid practices that could lead to unacceptable regrowth or to conversion to grassland.

12.3.2   The mid-latitudes (grassland–temperate forest biomes)

The two selected regions are northeast China and the midwest USA (Fig. 12.2). Each features an east–west moisture gradient from temperate forest at the eastern, wetter end to short grassland at the western, arid end (in the case of China, through to desert). The North East China Transect (NECT) runs for 1600 km from Chaingbai Mountain in the east to the Inner Mongolian steppe in the west, and the midwest USA transect (NAMER) extends 1200 km from the edge of the Mississippi River in the east to the foothills of the Rocky Mountains in the west.

The present

CLIMATE.   Present rainfall along the NECT varies from 800 mm yr^{-1} to 170 mm yr^{-1} in Inner Mongolia. The climate is continental in the west, with a coastal influence in the mountainous eastern section, and ranges from arid to humid. The NAMER has a gradient of annual rainfall from 1200 mm to 300 mm. The climate is continental throughout the transect and described as temperate, ranging from semi-arid to humid.

POTENTIAL VEGETATION.   The existing and the predicted present potential vegetation cover on the transects are presented in Plate 9. The dominant potential vegetation type of both in the east is temperate forest – deciduous broadleaf forests and mixed broadleaf and coniferous forests in the NECT and oak–hickory forest and a forest/grassland mosaic in the NAMER – through tall grass steppe or prairie in the central parts, into short grass steppe or prairie in the west. The NECT is more variable in the central region, with extensive potential shrubland, and continues into desert grasslands further west.

The primary determinant of vegetation type is the moisture regime (rainfall coupled with atmospheric demand for water). For the NAMER potential evapotranspiration is 1500 mm yr^{-1}, exceeding rainfall for the entire length of the transect. Soil type is a very important secondary determinant. On both transects, extensive areas of sand deposits/dunes, with marked differences in vegetation and ecosystem functioning from the finer textured soils, are prominent features. Both DGVMs predict more evergreen forest than potentially occurs on both transects, but the overall trend is reproduced, though not so well for NAMER. The SDGVM prediction of woodland rather than prairie in the western NAMER is a function of the cut-off levels used in the translation from functional types to vegetation classes.

ACTUAL VEGETATION: LAND–USE AND NATURAL DISTURBANCE. For both transects the difference between actual and potential vegetation is accounted for almost entirely by land-use change. Agriculture is the dominant land use, with cropping the major activity in the central part of the transects and livestock production in the west.

For the NECT, the central part of the transect, mainly the Liao River Basin, is the most intensively used, with rice and maize grown in the moister steppe region and spring wheat and millet in the drier steppe. Generally low levels of fertilization are used, and crop yields are thus fair (e.g. an average of 6750 kg ha^{-1} for maize). Approximately 50% of the land potentially suitable for cultivation has so far been converted to agriculture.

In the east, the major disturbance is logging of the forest. Apart from a few small reserves, all the forest has now been logged to some extent, and most of it has been changed to secondary oak woodlands, with around 10% of them, in the river valleys, converted to rice paddies.

The western, arid steppe region of the NECT is used for pastoralism, and high grazing pressures are a major disturbance causing significant change in vegetation, a loss of plant cover, and significant soil erosion, primarily by wind. The sandy areas in their undisturbed state consist of a temperate savanna dominated by *Ulmus pumila* and *Prunus davidiana* (the wild apricot) on stabilized soils. The effects of grazing have been to trigger shifting sands and dune development, with the loss of the savanna, and its replacement, mostly by sagebrush (*Artemisia* spp).

For NAMER there is very little land left under natural vegetation in the central and eastern parts; virtually all that can be converted to cultivation has already been converted. In terms of current land use, the eastern part is used for maize, the central part for winter wheat, and the western, uncultivated part for livestock grazing. Crop yields along the transect are generally high (e.g. around 7000 kg ha^{-1} in Weld County, Colorado).

The two transects contrast sharply in their population densities. For the NECT, densities are high, especially in the central part of the transect, and rising. The population is distributed widely over a large number of villages (i.e. high densities in the rural areas), with a few large cities. NAMER also includes two large cities (Denver and Kansas City), but the rural areas of the transect have low population densities (0–5 persons km^{-2}), which are static or declining.

The most important natural disturbance is drought. Several have occurred in the NAMER region over the past 50 years, but none of them have included two or more successive years of drought. The last time such a succession of droughts occurred was in the 1930s, and it had a very severe impact on both the ecology and economy of the region.

Because most of the ecosystems on both of these transects are so intensively managed, other natural disturbances like fire and insect outbreaks are not important. The exception to this generalization is the western steppe region of the NECT, which experiences periodic plagues of locusts and mice (*Mus musculus*), both of which are encouraged by heavy grazing.

Desertification is also an important issue in this heavily grazed region. For instance, the desertified area in Xilingele County is about 9 575 000 ha, about half the total grassland area according to a 1985 survey. The desertification resulted in a change in the structure of the grassland ecosystem, including invasion by weed species and a decrease in grasses palatable to livestock, and the modified structure was accompanied by plagues of locusts and mice.

An important actual and potential biotic disturbance along the NAMER is invasion of alien species. As one example, leafy spurge (*Euphorbia esula*) lowers livestock production (it is poisonous and toxic in a tactile manner), increases costs of crop production and leads to a decline in biodiversity.

The big issues

The major issue all along the NECT is sustainable land use. The population's growing food demands must be met without reducing the production potential. A second issue, of a global nature, is that the non-sandy soils in this region have very high carbon content, which is rapidly lost through cultivation (e.g. Burke *et al.*, 1993) and through increase in temperature.

The big issue for the NAMER transect is continued, economically viable, high agricultural production, both for domestic consumption and export.

The future

The potential future vegetation and possible land-use patterns of both transects are depicted in Plate 9. Potential vegetation is a function of the future climate scenario only. The future land cover depicted by the IMAGE model is a function of the future climate scenario, the scenario of human population growth, and the assumptions about human consumption and production.

For northeast China potential vegetation shows a slight increase in evergreen forest at the expense of savanna (i.e. increased woodiness). For NAMER the 2090 potential vegetation shows more deciduous forest. The IMAGE model suggests that much of the NAMER will revert to natural vegetation (for economic reasons). Local opinion (W.K. Lauenroth, personal communication) is that the most likely future pattern of land cover is probably similar to the existing pattern, but with further intensification in some areas. There are still large areas of land in the western region that could be converted to cultivation if suitable crops are found. The combination of technology (crop breeding) coupled with increasing water use efficiency under higher CO_2 may well make this possible. Whether or not it occurs will be determined by economic factors.

For the NECT the likely pattern is an intensification of the existing patterns in the three disturbance zones. Both the IMAGE model and local authorities (Xin-shi Zhang, personal communication) suggest that forest cover will decrease in the east, though conversion to cultivated land will be small (a few percent more, limited by topography and suitable soil). Significantly more land will be converted to agriculture in the central zone, and continued heavy grazing will keep the arid steppe region in a low-cover, degraded state. A combination of improved technology and an increase in rainfall could lead to a western extension of cultivation on non-sandy soils. Future crop yields under the above climate change scenario and improved technology could increase significantly (e.g. nearly double for maize).

In terms of future agricultural production the most significant concern is drought. A single drought year is manageable economically (at least for the USA; possibly not for China) and does not cause irreversible ecological change. However, two successive droughts would have a devastating effect, ecologically and on the farming system as a whole. The capability of elevated CO_2 to ameliorate the effects of drought on these agroecosystems through increased water use efficiency is still largely unknown.

A particular future issue in the western part of the NAMER concerns the extensive deposits of sands in Nebraska. The sand is currently stabilized, with only about 0.1% of it in a shifting form. However, the vegetative cover is very sensitive to any changes in climate, as it is close to the point where any increase in aridity would lead to a significant reduction in cover. Two thousand years ago the western region of the NECT was in the same state. Denudation of the vegetative cover there led to the start of shifting sand, and the formation of the moving dunes. If a combination of climate change and grazing pressure in the western USA prairies leads to loss of cover, the sand will start to blow in an easterly direction. Given the very large amount of sand involved, this would have major consequences, not only for the western region but also for the central crop-producing region to the east. The drier southwestern grasslands on the NAMER transect are close to the threshold for change to desert, and an increase in aridity and temperature would probably lead to this change occurring.

In terms of carbon budgets, the NAMER region is currently a source, owing to loss of soil carbon following cultivation. By the middle of the next century the rate of loss will probably stabilize and then reverse, leading to a carbon sink. Reversion of cultivated land in the eastern region to forest would add to the size of this sink. For the NECT region, the combined effects of global change (especially the extension and intensification of cultivation) mean that it will continue to be a net carbon source to the atmosphere.

The mid-latitudes – summary

Land-use change is the dominant issue and the dominant component of global change in these regions. In NECT, land use will intensify and ecological sustainability will be the big issue. In NAMER, depending on economic forecasts, agricultural land use could well decrease with economic sustainability being the big issue.

These regions are highly vulnerable to rare, climatic events, and a double drought in the regions of these two transects would have serious consequences for the world's food supply. The DGVMs suggest a slight overall increase in potential forest cover under climate change.

In both regions, wind erosion of sandy soils is a major concern, though it is still under control in NAMER. For NECT, erupting populations of locusts and mice are a serious problem. In NAMER, some grasshopper activity occurs, but such plagues are presently not a problem. Under future climate changes, and with the possibility of introduced species gaining a foothold, this issue may become more important.

In terms of their influence on the global environment, these mid-latitude grassland to forest regions are presently a source of carbon. NECT will continue to be so into the middle of the next century; NAMER will probably become a small sink.

12.3.3   The sub-humid to semi-arid tropics (tropical savanna biome)

The two selected regions are west Africa and northern Australia (Fig. 12.2), both of which involve strong rainfall gradients with little difference in temperature. In west Africa the SALT transect (Savanne a Long Terme) constitutes a gradient extending north from the edge of the humid coastal forest for 2000 km into the semi-desert of Mali's Sahelian zone. In northern Australia the NATT (Northern Australian Tropical Transect) extends south for just about 2000 km from near Darwin on the northern Australian coast to west of Alice Springs in the centre of the continent.

The present

CLIMATE. The SALT gradient of annual rainfall is from 1200 mm to < 200 mm, occurring in the summer, and the climate is described as hot and humid at the southern end and hot and arid in the north. The climate of the NATT is monsoonal tropical, ranging from humid to semi-arid, with an annual rainfall of 1600 mm in the north to less than 300 mm in the south.

POTENTIAL VEGETATION. Actual and predicted present vegetation cover are presented in Plate 10. The vegetation along the SALT grades from closed forest through humid grass savannas, savanna woodlands, open savanna woodlands, shrub and grass savannas to arid desert vegetation. The major primary determinant of natural savanna structure and function is the moisture regime, with the two main features being the total annual amount and the seasonal distribution. A subsidiary primary determinant is soil nutrient status, varying at a much finer scale than does rainfall. Net primary productivity is strongly controlled by the moisture regime, and varies by a factor of 10 from the moist south to the arid north. However, there is much variability at the landscape scale, as much of the SALT region is covered by 'tiger bush', or 'brousse-tigree', with dense strips of woody vegetation alternating with more sparsely vegetated strips. The two main secondary determinants of vegetation structure and functioning are fire and grazing. West African savannas are 'but a fallow of secondary stages of recovery in a system of cultivation and cattle breeding' (J.-C. Menaut, personal communication).

Vegetation type along the NATT varies from open *Eucalyptus*-dominated forests (*c.* 20 m tall) with an annual grass dominated understorey at the wetter end through *Eucalyptus* savannas with a vigorous perennial tussock grass layer and varying amounts of woody plants, to mostly open grassland on either black cracking clay soils (at various places along the transect) or sandy soils at the southern, dry end, dominated by hummock grasses. Below 500 mm rainfall, *Acacia*-dominated savannas and perennial hummock grasses (*Triodia* spp. and *Plectrachne* spp.) prevail. (This pattern of annual grasses at the wet end and perennial at the dry end (Alice Springs) is unique in the world's savannas).

The DGVMs reproduce the potential vegetation quite well (given the scale

difference) though the SDGVM predicts rather too much woody vegetation in
the centre of the NATT.

ACTUAL VEGETATION: LAND USE AND NATURAL DISTURBANCE.
In both transects land use is the dominant factor causing the actual vegetation
to vary from the potential. The magnitude of the effects, however, contrast
sharply due to the great differences in human population densities along
SALT and NATT.

For the NATT human densities are very low, less than 0.13 per km², and
concentrated at the northern end. For SALT human densities are high, and
mostly higher than considered sustainable (Beets, 1990*a,b*). An analysis of sustainable population densities based on estimates of sustainable cropping intensities, livestock rates and fuelwood supplies suggests that the Sahelian zone
(centre of the transect) is at its limit, the arid Sahelo–Saharan region (northern
end of the transect) is already overpopulated by 700% (mainly due to depleted
fuelwood), while the southern end of the transect (the Sahelo–Sudanian region) is currently overpopulated by 33% (Beets, 1990*a,b*). The population in
the Sahelian zone has doubled between 1960 and 1988 (3% growth rate) and is
destined to double again by 2011. Livestock numbers have roughly followed
the increase in human numbers (Le Houerou, 1993; Valentin *et al.*, 1994)

For SALT the major disturbances are a suite of direct human-driven
processes – cultivation, harvesting of woody vegetation for firewood, and
grazing by domestic livestock. Cultivation occurs along the transect down to
where rainfall is limiting, at around 400 mm. The major crops are maize and
cassava in the wetter regions, sorghum, millet and ground nuts in the middle
regions, and millet in the drier northern region. As an example of land-use effects on vegetation cover, in the Niamey region (70 km east, around the town
of Banizoumbou) the present actual vegetation consists of just over 20% cultivated lands, and just under 40% each for fallow lands and still uncleared
land, mainly on iron pan zones occurring on hillslopes, and the tiger-bush
plateaux (d'Herbes & Valentin, 1997). Livestock densities increase three-fold
from north to south. In the drier north cattle make up just over 50% of the
biomass with goats, sheep, camels and donkeys making up the rest. In the
south cattle make up over 73% of the biomass (Le Houerou, 1989).

Although direct human effects are the dominant factors in controlling vegetation structure and functioning along SALT, they interact with several
natural disturbances. Fire is a regular feature of the wetter end of the gradient,
and the present trend is for an increase in the number of fires, but with the
areal extent of the fires becoming smaller, therefore creating a more spatially
heterogeneous vegetation. Present fire frequencies are around once every one
or two years at the wet end (*c*. 80% of the area is burned every year), one in

six years in the central region, down to nil in the dry Sahel. In the drier regions fire is strongly suppressed, and is considered to be a 'pest'.

Droughts are a feature of the region and cause significant changes in vegetation, particularly the loss of perennial grasses, and trees in the Sahel. Wind with associated dust (the Harmattan) is also a regular stress event (rather than a disturbance as such), resulting in a dense haze and reduced radiation, soil erosion, and soil deposition. There is a net loss of carbon and nitrogen, via wind erosion, from the region. In the northern Sahelian region locust outbreaks are a serious problem.

In contrast to the SALT, the differences between potential and actual vegetation along the NATT are small, and are due primarily to the effects of cattle grazing and changes in natural fire regimes. They are not reflected in major changes in vegetation type, but rather in the proportions of the component species. Fire is a dominant feature of this region, ranging from about once every two years at the northern end to once every 20 years at the southern end. Changes in fire frequencies constitute a significant disturbance. Since the 1970s control of fires and reduced use of fire as a management tool, together with reduced grass cover through grazing pressure, have been accompanied by a significant increase in woody plants (Bastin & Andison, 1990). The IMAGE prediction of 'pasture' is a land use description rather than a vegetation structure description, and all of the designated pasture on the NATT is, in fact, savanna (a wooded rangeland) that is used for grazing.

Cattle grazing is the major disturbance affecting composition of the herbaceous vegetation. Livestock densities are moderate by world standards, of the order of 3 to 4 per km² in the central part of the transect (the Victoria River District). Native wildlife do not constitute a significant biomass, but introduced feral herbivores (water buffalo, donkeys, camels, horses, goats) do. Light grazing pressure does not lead to significant or irreversible change, but on hard, lateritic soils sustained high grazing pressure leads to loss of cover and the development of a sealed soil surface (known locally as a 'scald') (Mott, 1985, 1987). This in turn leads to increased run-off and less soil water. Once in this state, it is very difficult to reverse the process and the system can remain degraded for many years.

The impacts of alien species constitute a major disturbance along some parts of this transect. The invasive woody shrub *Mimosa pigra* has invaded about 800 km² of clay plains in the wetter northern end. *Acacia nilotica*, a major woody 'weed' in other Australian tropical grasslands on heavy soils, remains a threat but is not yet in the region. Various introduced grasses (*Andropogon gayanus*, *Pennisetum polystachion*, *Brachiaria mutica*) are problems in high rainfall areas and *Cenchrus ciliaris* has invaded in the drier end. Feral animals are having significant negative effects on biodiversity in some areas.

The big issues

Along the SALT the overriding issue is the ability of the region to sustain the increasing numbers of people who live in it and who depend on maintaining crop yields and sufficient livestock and fuelwood. Possible desertification in the Sahelian region is the dominant issue in the north, while crop production and increasing pressure on the land is an increasing issue in the south and middle zones. Human diseases are also an issue in West African savannas.

In northern Australia the big issue is one of economics, sustainable pastoralism, rather than the ability to sustain the population. Land occupancy by private graziers on the NATT is declining, but ownership of pastoral properties by large, often multinational, corporations is increasing. There is also an increasing ownership of land by Aboriginal people, who may or may not maintain commercial cattle enterprises. A potential increase in irrigated crops is also an issue.

Feedbacks to climate/atmosphere are less of an issue in the savannas than in the boreal and tropical rainforests, but wind erosion and long-range transport of dust is an important feature of the northern SALT region, and has been implicated in fertilization of the Atlantic Ocean (Prospero & Nees, 1977).

The future

The potential future vegetation, under the particular climate change scenario, is presented in Plate 10. For both regions it suggests relatively small changes from the present potential vegetation compared to the changes expected at higher latitudes. An increase in biomass is expected due to increased water use efficiency under higher CO_2, resulting in an increase in woody vegetation – savanna replacing grassland, and woody savanna replacing open savanna. Higher biomass will likely be matched by higher fire intensities, and perhaps by increased fire frequencies, though frequency will depend on the length of the dry season, grazing intensity and human decisions. The vast majority of fires are lit by people and this is unlikely to change.

A particular climatic effect in the SALT region (and to some extent on the NATT) is the potential change in the grass composition and cover through high temperature effects on seedling emergence on sandy soils. Present DGVMs do not take this into account. It is already a limiting factor in some seasons, as this is one of the hottest regions in the world, with soil temperatures rising to 59 °C at 2 cm depth in the Niamey region (Stomphe, 1990).

As illustrated by the IMAGE scenarios for future land cover (Plate 9), the major departure from the potential vegetation along the SALT will continue to be through land use, which will intensify (see Table 9.9). IMAGE suggests that virtually all the savanna and woodland will be replaced by pasture and cropland. At the present growth rate (3%), the human population will double twice by the second half of the next century. Total crop yields need to match, at least, the growth in human population, although this seems unlikely to be achieved. There is some small potential for increase through increased water use efficiency, but it is management practices and technology that will have most effect. Changes in types of crops under projected climate changes are difficult to forecast. It is likely that socio-economic factors will be more important than climate in influencing the types of crops grown. Livestock numbers are currently close to carrying capacity, and therefore unlikely to increase except in the moist savanna zone. Wood production for domestic fuel supplies is dwindling rapidly and at present rates of use will be depleted by the middle of next century.

The intensification process appears to be occurring now and gaining momentum. The annual transmigration of people and cattle between the northern and southern Sahelian zones, which has gone on for hundreds of years, is now stopping, and the present trend is for settled populations with a net southward migration due to political influences. This trend is driving increasing crop production and pressure on the land in areas of rapid population increase. In the north the question of whether or not desertification in the Sahelian region is occurring is unresolved. Periodic degradation of the pastoral areas occurs. Ten years after the last major drought, which caused massive mortality of livestock in the north with a resulting southward migration of people, the vegetation has recovered.

Human diseases on SALT are getting worse, and any increases in temperature will intensify this through effects on faster life cycles of the disease organisms and their vectors, and an increased potential for being spread. Meningitis, malaria and AIDS (so far only in the bigger towns) are major problems in human health. Bilharzia is an increasing problem, though onchocerciasis, previously very widespread and a major health problem, is now under control. Tsetse flies, causing livestock deaths through the disease ngana, are a problem in the south.

Future trends in land use along the NATT depend primarily on the pattern and extent of the demand from Southeast Asia. If the present demand for live cattle continues, it is likely that there will be an increasing pressure on the land, with some increase in degradation, and also an increase in land under irrigated pastures. The extent of irrigated pasture will depend on competition for water from cotton, sugar cane and horticultural products. The future of both cotton and sugar cane depends on world market prices; the future of horticulture depends on Australian and Southeast Asian demand. The IMAGE scenario indicates virtually no change from present land cover, except for an increase in shrubs in the southern parts due to climate change.

In terms of atmospheric feedbacks, the SALT region as a whole is currently a net source of CO_2, and this will increase. Land use (cultivation, firewood harvesting, grazing, fire) will override any increases in productivity due to higher water use efficiency. The Harmattan is a function of wind and soil cover. The trend is for winds to increase (if anything) and for soil cover to decrease, and the net effect is therefore likely to be increased soil erosion and loss of carbon and nitrogen. The NATT is currently probably about in balance in terms of carbon fluxes, and in future could become either a sink or source, depending on the economics of pastoralism.

The sub-humid to semi-arid tropics – summary

The two savanna regions differ markedly in terms of global change issues and responses, owing to differences in human population densities and economics. Nevertheless, in both Africa and Australia, land use is likely to override effects of CO_2 and climate change for several decades. The regions are currently sensitive to extremes of climate because of high levels of grazing and/or cultivation, so any increase in frequency or intensity of droughts will lead to resource depletion.

In terms of atmospheric feedbacks, the SALT region is currently a net source of CO_2, and the magnitude of this source will increase. The NATT is currently probably about in balance in terms of carbon fluxes, and in future could become either a sink or source, depending on the economics of pastoralism.

12.3.4   The humid tropics (tropical rainforest biome)

The selected regions are the Amazon Basin of South America and the humid tropics of Southeast Asia, with a focus on the Jambi Province region of Sumatra (Indonesia). These transects differ from the others in this analysis in that they are conceptual gradients of land-use intensity rather than contiguous physical gradients based on temperature or moisture (see Chapter 4; Koch *et al.*, 1995*a*,*b*). Thus, the transects do not lie in a straight line in geographical space but rather consist of a set of sites located within the regions depicted in Figure 12.2. The broader Amazon Basin region is the subject of an intensive regional global change study (LBA – The Large Scale Biosphere–Atmosphere Experiment in Amazonia – LBA Science Planning Group, 1996), of which the land-use intensity transect is a component.

The compact land-use intensity transect in Jambi Province, stretching through central Sumatra in a east–west direction for approximately 200 km between the cities of Jambi and Padang, encompasses a wide range of both natural vegetation (from coastal wetlands to montane forests) and agricultural systems (from intensive rice cultivation to complex 'jungle rubber' systems). Much of the work on this transect is carried out under the auspices of the Alternatives to Slash-and-Burn Programme (van Noordwijk *et al.*, 1995).

The present

CLIMATE.   The climate is typified by high rainfall and temperature in both areas. Precipitation is generally higher than potential evapotranspiration so moisture deficits are usually not a problem. Seasonality of precipitation is more pronounced at the boundaries of these humid tropical regions – to the south and east of the Amazon Basin, where the humid forests change into tropical seasonal forests and then into savannas, and to the northwest (northern Thailand) and southeast (eastern Indonesia) boundaries of Southeast Asia, where tropical seasonal forests are found. Seasonality in Southeast Asia is very pronounced at these boundaries due to the strength of the Asian monsoon system.

There are two important differences between the regions. First, Southeast Asia is an archipelago of islands and peninsulae, and its climate is dominated by the surrounding oceans. Amazonia lies in the centre of a large continental land mass, and it is estimated that half or more of its rainfall is recycled water from forest evapotranspiration (e.g. Lean & Warrilow, 1989; Nobre *et al.*, 1991). Thus, it strongly influences its own climate. Second, Southeast Asia is more topographically complex, with high mountain ranges forming the backbones of most of the larger islands. Thus, altitudinal zonation is a prominent feature of the climate in many areas of the Southeast Asian region. Significant mountains (the Andes) are found only on the western boundary of Amazonia; the basin itself is flat.

POTENTIAL VEGETATION.   In both regions humid tropical evergreen forests are dominant. In Southeast Asia the finer scale patterns are more complex due to the sharp altitudinal gradients; vegetation types range from coastal wetlands through lowland tropical evergreen forest to montane forests and alpine grasslands at higher elevations.

ACTUAL VEGETATION: LAND USE AND NATURAL DISTURBANCES. Direct, human-driven land-cover conversion, primarily for agriculture, has the most significant effect on vegetation composition and structure in both regions. In large parts of Southeast Asia, commercial timber harvesting and conversion of natural forests to plantations are also important, as are hurricanes, volcanoes and landslides. In both regions under natural conditions (i.e. with little or no human influence), fires are rare, owing to the high moisture content of potential fuel.

In the Amazon Basin, land-cover conversion ('deforestation') has been a major trend for several decades, originally driven by the internal migration of rural people from the heavily populated regions to the south and northeast.

More recently, international economic forces are playing an increasingly important role in driving land-use/cover change (Skole *et al.*, 1994). The deforestation process, over decadal time scales, produces complex landscape patterns caused by clearing followed by cultivation, then often followed by conversion into pasture or abandonment into secondary regrowth (see Fig. 1.1 – Skole *et al.*, 1994). Thus, the vegetation in the areas affected by the deforestation process is a mosaic of primary forest, secondary forest, pastures, and cultivated land.

There appears to have been a sharp increase in deforestation rates in the 1980s, which led to the outbreak of international concern, but the rates have probably slowed and stabilized in the 1990s (Fearnside, 1993). Estimates of the overall fraction of Amazonia's humid tropical forests that have been subjected to clearing and the subsequent land-cover trajectories vary, but it is generally thought to be less than 10% (Fearnside, 1993; INPE, 1992; Skole & Tucker, 1993).

There is a much longer history of cultivation in Southeast Asia than in the Amazon Basin, but recent population growth, industrialization and urbanization are leading to new and complex patterns. In the Jambi Province transect in Sumatra, population density in neighbouring Java is driving some of the land-cover conversion process through an official transmigration programme. There are large-scale logging operations in the region as well as the conversion of primary forests into oil palm plantations. In addition, spontaneous migration (as opposed to planned, transmigration) is encroaching on the forest margins opened up by the infrastructure associated with the officially approved developments. The resulting mosaic of land-cover types is even more complex than that in the Amazon Basin. It typically contains most of the following types in various patterns and areal ratios: permanent/continuous food crops, alang-alang (*Imperata cylindrica*) grasslands, shifting cultivation (slash-and-burn), fallow rotation, rubber/oil palm plantations, logged-over forests, jungle rubber/secondary forests, agroforestry systems, and primary forests (van Noordwijk *et al.*, 1995).

The rate of conversion of primary forests to agricultural systems varies greatly among the Southeast Asian countries. For example, in the Philippines, less than a quarter of the original primary forest remains, while substantially more than half of Malaysian and Indonesian primary forests remain (Sayer & Collins, 1991). In the Jambi Province transect, over half of the original forest cover remains, although the rate at which the remainder is being converted is large (van Noordwijk *et al.*, 1995)

The big issues

In both regions the big issue is land-use/cover change, especially its rate, driving forces, and consequences for nutrient dynamics, soil structure, biodiversity and the long-term sustainability of terrestrial production systems. In Southeast Asia, the impact of land-cover conversion in upland regions on the coastal zone, through transport of sediments and nutrients downstream, is also an important issue.

From a global perspective, the big issues are (i) the potential direct effect of large-scale, rapid conversion of forests to grasslands and agriculture on regional and global climate through changes in land surface characteristics (e.g. Henderson-Sellers *et al.*, 1993); and (ii) the large carbon losses associated with land-cover conversion and subsequent land uses, currently estimated to be about 1.6 Gt per year for the tropics as a whole (Schimel, 1995).

The future

Land use is predicted to remain the dominant component of global change in the humid tropics for at least the first half of the twenty-first century. Climate will likely continue to play a secondary role in influencing the future structure of vegetation; there is little predicted change in potential vegetation with changing climate over the next 100 years or so (see DGVM results in Chapter 8).

This generalization masks strong differences between the two regions in terms of overall rates of land-cover conversion in the future. Although both will experience a continuing overall loss of primary forest according to the IMAGE model, the loss appears to be surprisingly modest in the Amazon Basin and perhaps even stabilizes with half or more of the primary forest still intact at the end of the twenty-first century. However, in the IMAGE simulation Southeast Asian primary forests are both reduced significantly in overall extent and further fragmented as the region attempts to keep pace with its growing food demand. In addition, the two most populous countries in the world, China and India, border on the Southeast Asian region and will probably have a strong impact on land-use/cover change as their economies more fully integrate into the regional economy and transmit their food demands into it.

There is also some concern about climate change, especially in Southeast Asia, where the Asian monsoon is the dominant feature of the regional climate. Significant changes in the intensity and reliability of the monsoon, coupled with land-use practices, could change the region's major natural disturbance regimes. The combination of more frequent and extended droughts, increased slash-and-burn agriculture, conversion of primary forests to plantations, and more intensive logging of primary forests could induce a strong rise in fire frequencies. The extensive fires in Kalimantan in the early 1990s and the 1997 fires in Sumatra and Kalimantan, which both affected regional air quality and disrupted traffic at major Southeast Asian airports, were attributed to such combinations.

Although temperature rise is not predicted to be large in the humid tropics, any increase above 32 °C would pose problems for rice production in Southeast Asia. As noted in Chapter 9, yields drop by 5% for every degree rise above this threshold.

The most important consequences of global change, particularly land-use/ cover change, over the next century will occur in the following:

SUSTAINABLE DEVELOPMENT.   The predicted continuing conversion of primary forest to agriculture could confound efforts to move towards sustainable management of terrestrial ecosystems in the humid tropics. Much more needs to be known about the changes in pools and fluxes of carbon and nutrients associated with the land-cover conversion process and the subsequent alternative land uses. For example, there is already some evidence that the land-use/ cover change trajectories in the Amazon Basin are far from sustainable on a decadal time frame. Agricultural plots are often abandoned on a 3–5 year timeframe, usually due to rapidly declining fertility and problems with pests and weeds, and more forest is cleared (e.g. Nepsted *et al.*, 1991). Alternatively, farmers abandon agriculture entirely and pursue small-scale mining, leading to a large number of areas of spontaneous development and deforestation scattered through the Basin (LBA Science Planning Group, 1996). In Southeast Asia, development will be driven by a mix of local factors (e.g. market gardening versus subsistence farming), regional pressures (e.g. transmigration) and international economic drivers (e.g. the global price of palm oil), with the last playing an increasingly important role as the 'tiger economies' of the region integrate more fully into the global economy.

GLOBAL CARBON CYCLE.   The IMAGE model projections for the Amazon Basin reflect a forecast improved economic situation, leading to a changed pressure on the land. The IMAGE scenario notwithstanding, the underlying pressures for land-use change will probably intensify in the Amazon Basin for the next several decades at least; populations are projected to grow substantially and there are plans for a more extensive road network throughout the region.

In Southeast Asia, net conversion from forest to agricultural cover types will probably continue. The implication for the global carbon cycle is that this region will continue to be a significant net source of carbon to the atmosphere. Given the magnitude of required increased food production (see Chapter 9), it will take a rapid and major improvement in management techniques for carbon losses to be held at their current levels.

The humid tropics – summary

Land-use/cover change is the dominant component of global change in the humid tropical regions now and for the foreseeable future. The pressures for conversion of forests to agricultural cover types and for commercial harvesting of forests will intensify through much of the twenty-first century. Climate change is secondary, although significant changes to the Asian monsoon, interacting with land-use/cover change, could be important.

The predicted consequences of continuing loss of primary forests in the humid tropics are:

- Achieving sustainable development of the terrestrial biosphere in a broad regional sense will become more difficult unless understanding of the biogeochemical changes and impacts on resources associated with land-use/cover change improves rapidly and appropriate management strategies are devised and adopted. Macro-economic decisions about large scale transformation of humid tropical forests will be particularly important.
- The humid tropics will probably continue as a net source of carbon to the atmosphere, with a possibility that the magnitude of carbon loss will increase with increasing food production.

12.3.5 Summary of transect analysis

From the eight regions, an overall synthesis confirms that global change will be very different, on a regional scale, around the world.

- Of the eight, six (all but NAMER and NATT) will be net sources of carbon in the future.
- Land-use change rather than climate change is the dominant component of global change in all but the high latitude regions.
- Secondary effects (wind erosion, rodent and locust plagues, insect attacks on trees, fire frequencies, 'weed' invasions. . .) were raised by regional experts in almost all cases as potentially very serious but as yet unpredictable consequences of the interactive effects of climate and land-use change.
- Food production per unit area will potentially decline in the tropics, will be maintained or increased in the mid latitudes, and will only very slightly increase in the high latitudes (for up to $2 \times CO_2$ climate scenarios).

12.4 Global overview

12.4.1 Scenarios of global potential vegetation

By drawing together the results from (i) earlier equilibrium vegetation models; (ii) the transient DGVM outputs in Chapter 8; (iii) the synthesis of vegetation dynamics under global change in Section 12.2.3; and (iv) the regional analyses in the previous section, a number of inferences (if not conclusions) about the future potential vegetation cover of the world (i.e. vegetation excluding the effects of direct human disturbance and land use change) can be made.

- In the high latitude regions of the northern hemisphere, a northward extension of deciduous trees into the present boreal forests is likely, as is the incursion of woody vegetation into the tundra.
- Most models predict that the open savanna regions in the tropics will become woodier, though as biomass increases a potential increase in frequencies of fires may oppose this trend. The major limitation in projecting this scenario is the inadequate prediction of regional scale rainfall in current GCMs.
- In the mid-latitudes, where both increased temperatures and increased precipitation occur, the differences in vegetation on contrasting soil types is likely to increase; sites with high water-holding capacities will 'advance' in succession, while those where water-holding capacity is currently the limiting determinant of vegetation development will revert to drier or 'lower' successional states.

Changes in disturbance regimes, and the transient dynamics associated with them, will play a strong role in modifying the general projections outlined above. Capturing these disturbance-related effects is a major challenge facing the developers of DGVMs.

- Changes in plant community species composition in many biomes will most probably occur within the timeframe of a human life-span, owing to changes in disturbance regimes.
- As an example, vegetation changes induced through increases or changes in climate variability have occurred in the boreal forest, where a significant temperature rise has occurred over many areas over the past three decades. Increases in fire frequency and insect attack are causing changes in the age structure of the forests, and these disturbances will likely trigger changes in the proportional composition of species. The purely 'climatic' potential vegetation is thus quite different from the potential vegetation modified by natural biotic disturbances.
- Past disturbances will play an important role in determining the future composition of long-lived plant communities. Most of the world's 'natural' vegetation is not in equilibrium with the present climate, and, so long as it is not subject to altered disturbance, will continue to lag as climate changes in the future.
- The transient pattern of biomass and NPP suggests that there will be an increase in NPP (of perhaps 50% over the next century), but the actual biomass increase will depend on disturbance regimes.

12.4.2 Scenarios of global land cover

Plate 11 is a difference map in land cover between 1990 and 2090, produced by the IMAGE 2.1 model (Alcamo *et al.*, 1996*b*). Vegetation change in IMAGE is based on an equilibrium approach (see the BIOME model of Prentice *et al.*, 1992) and so the projected extent of change in natural vegetation is likely to be somewhat greater than that produced by a transient DGVM.

The driving forces behind the land cover projections in Plate 11 are population growth and economic growth in various sectors, which together drive energy and food consumption. Land-use change is based on per capita food consumption (which in turn is based on income, land productivity and the preference of meat over grain). IMAGE does not take into account effects of future land degradation or potential new irrigated agricultural land, and it also assumes that the projected human population increases in all regions will in fact occur (i.e. no resource limitations to the forecast population growth rates).

The forecast major changes in land cover will be due largely to conversion to agriculture in subtropical and tropical Africa and in Asia (except the high latitudes). Significant climate-driven change is indicated for the high latitude northern hemisphere.

The IMAGE projections include the time course of changes in the variables (Alcamo *et al.*, 1996*a,b*), and details of changes in crop types due to climate change (Leemans & van den Born, 1994). The biggest changes are likely to occur within the next few decades. For example, cattle numbers tend to increase up to about 2030, and then decline. In OECD countries this is due to demographic changes and increasing production efficiency, leading to stabilized numbers after 2030, while in Asia and Africa increased demand for land for agricultural crops to feed the population results in a decline in livestock numbers after 2030. In Latin America there is a very rapid increase in agricultural land up to 2030, and a decline thereafter due to improvements in crop and pasture yields.

Overall, IMAGE suggests that, using the UN intermediate population estimates adopted by the IPCC, about one-third of the earth's land cover will change in the next hundred years, and that the biggest changes will occur within the next three decades.

About 15% of current land cover is now intensive agriculture. By how much will that (can that) change? IMAGE suggests that this may double, approximately, but, as indicated by the more detailed assessment of the transects in this chapter (see the account of arable land on the SALT transect), soil type will be a limiting factor at local scales.

The global scale projections of economically driven land-use change in the IMAGE scenario will clearly have regional scale variations driven by regional and local economics. The differences between the midwest USA and China transects, and between the West African (SALT) and northern Australian (NATT) tropical transects, strongly emphasize this point. What happens to the vegetation along the NATT, for example, will depend very much on the future economic state of Southeast Asia, and the political and social attitudes in those countries to the continued importation of live cattle.

In terms of biodiversity effects, the IMAGE scenario shows that more forest land is converted to another land cover type under the combined effects of climate change and land use than any other type of biome. Climate change adds significantly to the changes due to land use effects alone, and the threat to biodiversity is therefore greater.

12.4.3   Futures for food and fibre

What does GCTE's research effort so far suggest for the future of managed production systems under a rapidly changing global environment?

- Technology, management expertise, and economic flexibility in the developed world are sufficient to cope with global change, and perhaps could even benefit from it in the higher latitudes. This is not true, however, in the tropics, where providing adequate food is currently a problem. This synthesis concurs with other studies that global change will exacerbate, rather than ameliorate, the food-supply problem in the tropics (see Chapter 9).
- Pests, diseases and weeds will become a greater problem under global change, as suggested by several recent simulation studies (e.g. Box 9.3), as well as by emerging evidence in the high latitudes (see Section 12.2.2).
- Irreversible damage to soil physical structure is widespread now and likely to increase, especially as cultivation moves into more marginal lands in the tropics. The probable increase in soil erosion caused by increasing cultivation coupled with changing climate is a major potential constraint on achieving the required increases in food production.

Increasing the production of food and fibre will be the dominant theme for the coming decades. Nevertheless, increasingly complex, additional demands are being made of terrestrial production systems. In addition to being more efficient, productive and robust in the face of accelerating environmental change, terrestrial production systems of the twenty-first century will simultaneously have to (i) reduce local and regional pollution from agricultural sources; (ii) emit less carbon to the atmosphere, and, wherever possible, actively sequester carbon; and (iii) be more effective in conserving biodiversity.

The overall conclusion of the synthesis is that terrestrial production systems will show mixed results in attempting to achieve all of these goals. They will probably be able to meet the food and fibre demands of the significantly larger human population by the middle of next century, but (i) global change will make that task more difficult, rather than easier, especially in the tropics where food demand is the greatest (see Chapter 9); and (ii) the Earth's natural ecosystems will pay a heavy price in terms of biodiversity loss and ability to provide ecosystem goods and services, most notably in the tropics.

12.4.4 Uncertainties and surprises

At all levels of this synthesis, from the detailed studies of CO_2 effects on plants through the interactive effects of global change drivers on ecological communities and ecosystem functioning, to the scenarios for production systems and land cover change, uncertainty prevails. We can't be sure what will happen, and therefore some big surprises are almost certainly in store. In addition to the uncertainties we have at least recognized, there are also likely to be direct effects of global change on human societies – diseases, climatic disasters (sea level rises, storms, droughts, exceptional high temperature periods), water resources and refugees (with the possibility of resource wars) – that may be almost impossible to predict.

Virtually by definition, there is no deterministic solution to this uncertainty. The most effective approach is one of successive approximation, as we improve our models of global change and of ecosystem and agricultural response and feedbacks. The important part in this process is recognizing the various levels of uncertainty and couching any policy implications accordingly.

12.5   Living with global change

Global change is occurring now, will continue for the foreseeable future and is likely to intensify in many aspects. It is an emerging reality that will increasingly impact on the political process, on regional strategic planning and on the daily lives of resource managers. Learning to live with global change, to avoid the worst of the hazards and capitalize on opportunities as they arise, requires that societies develop creative and innovative strategies. These must be built upon a sound scientific understanding of terrestrial ecosystem interactions with global change. Although GCTE and similar efforts have made good progress in the last six years, large challenges remain, both in the basic understanding of the science and in the development of research tools to improve that understanding.

12.5.1   Socio-political context

The single biggest issue that will determine how terrestrial ecosystems, globally, will change is outside the scope of a biophysical synthesis, and lies squarely in the domain of the 'human dimensions' of global change. It is the socio-political

Box 12.1   Emerging questions and challenges

The synthesis has highlighted a number of significant challenges that need to be resolved before we are able to achieve the predictive capability we require. In particular:

- *The integration of natural and social sciences.*   How can closer collaboration between the natural and social sciences improve understanding of the complex impact-feedback loops involving socio-economic and political systems and institutions and the natural environment? In the immediate future, closer collaboration is needed to improve our capability to undertake integrated assessments of global change impacts.

- *Sustainable development and global change.*   Sustainable development and global change are closely related. The continuing build-up of anthropogenically generated trace gases in the atmosphere and the rapid loss of biodiversity suggest that past and present rates of development are not sustainable in a biophysical sense. These global environmental changes are now producing biophysical constraints on sustainable development strategies for the future. How can we more effectively merge these closely related research-applications efforts to deliver more effective outcomes for policy and resource management?

- *Rates of change.*   The Earth's environment is constantly changing, but anthropogenically driven global change appears to be much more rapid than natural, 'background' change, causing serious problems for terrestrial ecosystems to adapt without human intervention. What are 'safe' rates of global change that avoid dangerous disruption of natural ecological processes and cycles, and what needs to be done to slow global change to these rates?

- *Interactive effects of global change drivers.*   The components of global change are not independent but interact strongly. How can we devise more appropriate methodologies for studying the interactions of terrestrial ecosystems with combinations of global change drivers, as opposed to the linear 'driver–response–impact' chain of reasoning?

- *Climate scenarios.*   Although much progress has been made in our ability to model climate processes, the present capability still falls well short of what is required to study impacts on ecological systems. Can climate prediction capability be improved to produce realistic regional-scale scenarios of water availability and of the nature and frequencies of extreme events?

- *The interaction between physiology and structure.*   Failure to include the interactive effects of ecosystem physiology and structure continues to confound much global change research. How can we more effectively integrate these two strands of research to gain a whole ecosystem perspective of global change interactions, over multiple space and time scales?

- *Landscape processes.*   Much of the 'action' in terms of disturbance dynamics and direct human impacts on terrestrial ecosystems occurs at the landscape scale. How can we improve our understanding of these phenomena, both to assist resource managers to 'live with global change' and to facilitate the scaling between patch and globe?

- *Ecological complexity and resilience.*   How can we gain a better quantitative understanding of the relationships between ecological complexity (including biological diversity) and ecosystem resilience, and the ways in which global change will affect this relationship?

issue of how to resolve the developing world/developed world disparity – the legitimate aspirations of the citizens of developing countries for a better quality of life – and the implications of this disparity for land-use and -cover change and the use of terrestrial resources. The way in which this issue develops will have an overriding effect on what happens in the future to the terrestrial biosphere. Isolationist policies on the part of any region or nation that fail to account for the biological, physical and socioeconomic teleconnections that link widely separated regions of the globe, or for the accelerating changes to the Earth system itself, are doomed to failure. Global change is truly a global problem.

12.5.2   Prepare for the future

Based on what we already know, there are a number of preparatory actions that societies can begin to take now to lessen the likely negative effects of global change. For example:

Crop breeding.   Development of new crops should take into account more variable water and nutrient availability, and include adaptation for climatic extremes (e.g. high temperature for rice).

Agricultural systems planning and management.   Forecast changes in land suitability for different crops (including tree plantations) and for livestock enterprises will identify required future infrastructure changes, and imbalances between such things as irrigation water supply and demand.

Conservation strategies.   Reserve network designs should be based on environmental gradients, which allow for spatial shifts in environmental domains.

In addition to taking proactive measures such as these, societies must also face the question of taking action now versus taking action later (or taking no action), for example, on proposals to limit greenhouse gas emissions. There are global processes that have lag times of decades or even centuries. The consequences of not taking action now may not be felt until the middle of next century, but when these consequences do occur, they will be serious and very difficult to avoid. An example is the diminishing ability of the terrestrial biosphere to absorb carbon as both atmospheric CO_2 concentration and temperature increase; a lack of action now could lead to a large, unavoidable, additional CO_2 release a century or so from now, through increasing decomposition of soil organic matter (about twice as much carbon is stored belowground than above in terrestrial ecosystems).

It is important, however, to put such biophysical considerations into the global socio-political context. As described above, the foremost global change issue is land-use change, especially in the developing world, and its effects on the terrestrial biosphere. Thus, efforts to slow carbon build-up in the atmosphere by planting trees or improving agricultural practices in the developed world amount to fiddling at the margins. Unless the developing world/developed world disparity is tackled directly, then there is little that can be done in terms of managing terrestrial ecosystems in the developed world to slow global change significantly; it may just provide a few years of temporary lag effect.

12.5.3   Design for resilience – expect the unexpected

As noted in Section 12.4.4, we will probably never be able to predict, with a high degree of certainty, precisely how terrestrial ecosystems will interact with accelerating environmental change. Thus, the analogy that ecosystems can be 'managed' in the same way that much simpler human-designed industrial systems can is misleading and dangerous. In terms of terrestrial ecosystem interactions with global change, we must expect the unexpected (and unpredictable), and keep open as many response options as possible. There is an inescapable trade-off between resilience and production in managed terrestrial ecosystems: the most productive agricultural systems are often the simplest, but they are the least resilient to disturbance and perturbation. Highly productive systems are required to feed an expanding population; complex, resilient systems are required to be able to respond to future shocks and disturbances, and to continue providing the ecosystem 'goods and services' we need.

The simple, but very important, policy message is that, under global change, we need to design resource-use systems (e.g. agriculture, forestry) for resilience as well as for production. A policy aimed at maximum yield carries with it an increasing probability of production failure and irreversible damage (such as from stabilized to mobile sands – see the contrast between the midwest USA and China transects, Section 12.3.2). What we still need to know is how much, and what kinds of, complexity are needed to design into resource-use systems. Striking the right balance in the production/resilience dichotomy is the biggest environmental challenge facing humanity in the twenty-first century.

Acknowledgements

We thank Ian Woodward and his research group at the University of Sheffield, UK, for the use of output from SDVM in the transect-based analysis, Jon Foley and his group at the University of Wisconsin, USA, for the use of output from IBIS, and Rik Leemans and his group of RIVM, The Netherlands, for the use of output from IMAGE 2.1. We are also grateful to the following people who provided data, background information and expert opinion on the present and future state of vegetation on the transects used in analysis: Mike Apps, Terry Chapin, Qiong Gao, David Halliwell, Joe Landsberg, Bill Lauenroth, Jean-

Claude Menaut, Daniel Murdiyarso, Bill Parton, Vladislav Rojkov, Detlef Schulze, Anatoly Shvidenko, Christian Valentin, Dick Williams and Xinshi Zhang.

References

Aber, J. D. & Melillo, J. M. (1982). Nitrogen immobilisation in decaying hardwood leaf litter as a function of initial nitrogen and lignin content. *Canadian Journal of Botany*, **60**, 2263–9.

Addicott, J. K., Aho, J. M. Antolin, M. F. Richardson, J. S. & Soluk, D. A. (1987). Ecological neighborhoods: scaling environmental patterns. *Oikos*, **49**, 340–6.

Agbu, P. A. & James, M. E. (1994). *The NOAA/NASA Pathfinder AVHRR Land Data Set User's Manual*. Goddard Distributed Active Archive Center, NASA, Goddard Space Flight Center.

Aggarwal, P. K. & Kalra, N. (eds.), (1994). *Simulating the Effect of Climatic Factors, Genotype and Management on Productivity of Wheat in India*, New Delhi, India: Indian Agricultural Research Institute.

Ågren, G. I., McMurtrie, R. E., Parton, W. J., Pastor, J. & Shugart, H. H. (1991). State-of-the-art of models of production–decomposition linkages in conifer and grassland ecosystems. *Ecological Applications*, **1**(2), 118–38.

Aguiar, M. R., Paruelo, J. M., Sala, O. E. & Lauenroth, W. K. (1996). Ecosystem responses to changes in plant functional type composition – an example from the Patagonian steppe. *Journal of Vegetation Science*, **7**(3), 381–390.

Ahmad, W., Hill, G., Williams, D. & Cook, G. (1996). Role of SPOT and AIRSAR digital data for identifying and mapping different land cover types in the tropical savannas of Northern Australia. Report to CEOS – IGBP-DIS High Resolution Data Pilot Project. Presented at CEOS Plenary.

Ajtay, G. L., Ketner, P. & Duvigneaud, P. (1979). Terrestrial primary production and phytomass. In *The Global Carbon Cycle*, ed. B. B. Bolin, E. T. Degens, S. Kempe & P. Ketner, pp.129–71. New York: John Wiley and Sons. SCOPE.

Alcamo, J. (ed.) (1994). *IMAGE 2.0: Integrated Modeling of Global Climate Change*. Dordrecht, The Netherlands: Kluwer Academic Publishers. 318 pp.

Alcamo, J., Kreileman, E. & Leemans, R. (eds.) (1996a). *Integrated Scenarios of Global Change*. London: Pergamon Press.

Alcamo, J., Kreileman, G. J. J., Bollen, J. C., van den Born, G. J ., Gerlagh, R., Krol, M. S., Toet, A. M. C. & de Vries, H. J. M. (1996b). Baseline scenarios of global environmental change. *Global Environmental Change*, **6**, 261–303.

Alfonssi, D., Sandroni, S. & Viarengo, S. (1991). Tropospheric ozone in the nineteenth century: the Moncalieri series. *Journal of Geophysical Research*, **96**, 17349–52.

Amir, J. & Sinclair, T. R. (1991). A model of the temperature and solar radiation effects on spring wheat growth and yield. *Field Crops Research*, **28**, 59–69.

Amthor, J. S. (1995). Terrestrial higher-plant response to increasing atmospheric CO_2 in relation to the global carbon cycle. *Global Change Biology*, **1**, 243–74.

Anderson, L. A. & Sarmiento, J. L. (1994). Redfield ratios of remineralisation determined by nutrient data analysis. *Global Biogeochemical Cycles*, **8**, 65–80.

Andreae, M. O. (1991). Biomass burning: its history and distribution, and its impact on environmental quality and global climate. In *Global Biomass Burning*, ed. J. S. Levine. pp. 3–21. Cambridge (USA): MIT Press.

Andren, H. (1994). Effects of habitat fragmentation on birds and mammals in landscapes with different proportions of suitable habitat: a review. *Oikos*, **71**, 355–66.

Archer, S., Schimel, D. S. & Holland, E. A. (1995). Mechanisms of shrubland expansion: land use, climate or CO_2? *Climatic Change*, **29**, 91–9.

Arnold, J. G., Engel, B. A. & Srinivasan, R. (1993). A continuous time grid cell model. In *ASAE Spokane, WA, USA, June 17–19*.

Arnold, J. G., Williams, J. R., Nicks, A. D. & Sammons, N. B. (1990). *SWRRB: A Basin Scale Simulation Model for Soil and Water Management*. Texas A&M University.

Arnold, R. W., Szabolcs, I. & Targulian, V. O. (1990). *Global Soil Change*. Laxenburg, Austria: International Institute for Applied Systems Analysis.

Arnone, J. A., III & Körner, Ch. (1995). Soil and biomass carbon pools in model communities of tropical plants under elevated CO_2. *Oecologia*, **104**, 61–71.

Arnone, J. A., III, Zaller, J. G., Ziegler, C., Zandt, H. & Körner, Ch. (1995). Leaf quality and insect herbivory in model tropical plant communities after long-term exposure to elevated atmospheric CO_2. *Oecologia*, **104**, 72–8.

Arp, W. J., Kuikman, P. J. & Gorrisen, A. (1997). Climate change: the potential to affect ecosystem functions through changes in the amount and quantity of litter. In *Driven by Nature: Litter Quality and Decomposition*, ed. G. Cadisch & K. E. Giller, pp. 187–200. Wallingford: CAB International.

Ash, A. & Stafford Smith, D. M. (1996). Evaluating stocking rate impacts in rangelands: animals don't practice what we preach. *Rangelands Journal*, **18**, 216–43.

Asner, G. P., Seastedt, T. R. & Townsend, A. R. (1997). The decoupling of terrestrial carbon and nitrogen cycles. *Bioscience*, **47**, 226–34.

Assmann, E. & Franz, F. (1963). Vorlaeufige Fichtenertragstafeln für Bayern. Forstwissenschaftliches. *Centralblatt*, **84**, 13–14.

Atkinson, D. (1993). *Global climate change: its implications for crop protection*. 102 pp.

London: British Crop Protection Council.

Austin, R. B. & Jones, H. G. (1975). The physiology of wheat. In *Report of the Plant Breeding Institute*, pp. 20–73. Cambridge: Cambridge University Press.

Bacastrow, R. & Keeling, C. D. (1973). Atmospheric carbon dioxide and radiocarbon on the natural carbon cycle. In *Carbon and the Biosphere*, ed. G. M. Woodwell & E. V. Pecan, pp. 86–135. US Atomic Energy Commission.

Bak, P., Chen, K. & Tang, C. (1988). Self-organized criticality. *Physical Review*, **A38**, 364–74.

Baker, B. B., Bourdon, R. M. & Hanson, J. D. (1992). FORAGE: a model of forage intake in beef cattle. *Ecological Modelling*, **60**, 257–79.

Baker, W. L. (1989). A review of models of landscape change. *Landscape Ecology*, **2**, 111–31.

Baker, W. L. (1992). The landscape ecology of large disturbances in the design and management of nature reserves. *Landscape Ecology*, **7**, 181–94.

Baker, W. L. (1993). Spatially heterogeneous multi-scale response of landscape to fire suppression. *Oikos*, **66**, 66–71.

Baldocchi, D. D., Vogel, C .A. & Hall, B. (1997). Seasonal variation of carbon dioxide exchange rates above and below a boreal jack pine forest. *Agricultural and Forest Meteorology*, **83**, 147–70.

Ball, A. S. & Drake, B. G. (1997). Short-term decomposition of litter produced by plants grown in ambient and elevated CO_2 concentration. *Global Change Biology*, **3**, 29–35.

Ball, J. T., Woodrow, I. E. & Berry, J. A. (1987). A model predicting stomatal conductance: its contribution to the control of photosynthesis under different environmental conditions. In *Progress in Photosynthesis Research*, Vol. IV, ed. I. Biggins, pp. 221–4. Dordrecht: Martinees Nijhof.

Barkstrom, B. R. (1984). The Earth Radiation Budget Experiment (ERBE). *Bulletin of the American Meteorological Society*, **65**, 1170–85.

Barnes, P. W, Flint, S. D. & Caldwell, M. M. (1995). Early-season effects of supplemented solar UV-B radiation on seedling emergence, canopy structure, simulated stand photosynthesis and competition for light. *Global Change Biology*, **1**, 43–53.

Bartholomew, J. C., Christie, J. H., Ewington, A., Geelan, P. J. M., Lewisobe, H. A. G., Middleton, P. & Winkleman, H. (eds.) (1988). *The Times Atlas of the World*, (7th Ed.), 228 pp. London: Times Books Limited.

Bartlett, K. B. & Harriss, R. C. (1993). Review and assessment of methane emissions from wetlands. *Chemosphere*, **26**, 261–320.

Barton, C. V. M., Lee, H. S. J. & Jarvis, P. G. (1993). A branch bag and CO_2 control system for long-term CO_2 enrichment of mature Sitka spruce (*Picea sitchensis* (Bong.) *Carr*). *Plant, Cell and Environment*, **16**, 1139–48.

Bascompte, J. & Sole, R. V. (1996). Habitat fragmentation and extinction thresholds in spatially explicit models. *Journal of Animal Ecology*, **65**(4), 465–73.

Bastiaans, L. (1993) Understanding yield reduction in rice due to leaf blast. 127 pp. PhD Thesis, Wageningen Agricultural University, Wageningen, The Netherlands.

Bastin, G. & Andison, R. T. (1990). Kidman Springs Country – 10 years on. *Range Management Newsletter*, **90/1**, Australian Range Management Society, May 1990.

Batjes, N. H. & Bridges, E. M. (1994). Potential emissions of radiatively active gases from soil to atmosphere with special reference to methane – development of a global database (WISE). *Journal of Geophysical Research – Atmospheres*, **99**(D8), 16479–89.

Bauer, G., Schulze, E.-D. & Mund, M. (1997). Nutrient status of the evergreen conifer *Picea abies* and the deciduous heartwood *Fagus sylvatica* along an European transect. *Tree Physiology*, **17**, 777–86.

Bayley, S. E. & Schindler, D. W. (1991). The role of fire in determining stream water chemistry in northern coniferous forests. In *Ecosystem Experiments*, eds. H. A. Mooney, E. Medina, D. W. Schindler, E.-D. Schulze & B. H. Walker, Vol 45, pp. 141–68. Chichester: SCOPE, John Wiley and Sons.

Bazilevich, N. I., Rodin, L. Y. & Rozov, N. N. (1971). Geographical aspects of biological productivity. *Social Geography Review Translations*, **12**, 293–317.

Bazzaz, F. A. (1996). *Plants in Changing Environments*. Cambridge: Cambridge University Press.

Bazzaz, F. A., Garbutt, K., Reekie, E. G. & Williams, W. E. (1989). Using growth analysis to interpret competition between a C_3 and a C_4 annual under ambient and elevated CO_2. *Oecologia*, **79**, 223–35.

Becker, A. & Bugmann, H. (1997). Predicting global change impacts on mountain hydrology and ecology: integrated catchment hydrology/altitudinal gradient studies. 59 pp. Workshop report. IGBP Report No. 43, International Geosphere-Biosphere Programme, Stockholm.

Beer, T. & Williams, A. (1995). Estimating Australian forest fire danger under conditions of doubled carbon dioxide concentrations. *Climatic Change*, **29**(2), 169–88.

Beerling, D. J. (1993). The impact of temperature on the Northern distribution limits of the introduced species *Fallopia japonica* and *Impatiens glandulifera* in North-West Europe. *Journal of Biogeography*, **20**(1), 45–53.

Beets, W. C. (1990a). *Raising and Sustaining Productivity of Ecosystems*, ed. G. W. Koch & H. A. Mooney, pp. 197–214. San Diego: Academic Press.

Beets, W. C., (1990b). *Raising and Sustaining Productivity of Smallholder Farming System in the Tropics*. Alkmaar, Holland: AgB Publishing. 738 pp.

Belward, A. & Loveland, T. (1995). The IGBP-DIS 1-km land cover project. In *Remote Sensing in Action*, ed. P. J. Curran & Y. C. Robertson, pp. 1099–106. Southampton: University of Southampton.

Belward, A. & Loveland, T. (1996). The DIS 1-km land cover data set. *IGBP Newsletter 1996*, (**27**), 7–9.

Benson, M. L., Landsberg, J. J. & Borough, C. J. (1992). The biology of forest growth experiment: an introduction. *Forest Ecology and Management*, **52**, 1–16.

Berendse, F. & Elberse, W. T. (1990). Competition and nutrient availability in heathland and grassland ecosystems. In *Perspec-*

tives on Plant Competition, ed. J. B. Grace and D. Tilman, pp. 93–116. San Diego: Academic Press.

Berg, B. & McClaugherty, C. (1989). Nitrogen and phosphorus release from decomposing litter in relation to lignin. *Canadian Journal of Botany*, **67**, 1148–56.

Berthelot, B., Dedieu, G., Cabot, F. & Adam, S. (1994). Estimation of surface reflectances and vegetation index using NOAA/AVHRR: methods and results at global scale. In: 6th International Symposium 'Physical Measurements and Signatures in Remote Sensing', pp. 33–40 Val d'Isère, France. 17th–21st January 1994: ISPRS/CNES.

Bessie, W. C. & Johnson, E. A. (1995). The relative importance of fuels and weather on fire behaviour in subalpine forests. *Ecology*, **76**(3), 747–62.

Betts, R. A., Cox, P. M., Lee, S. E. & Woodward, F. I. (1997). Contrasting physiological and structural vegetation feedbacks in climate change simulations. *Nature*, **387**, 796–9.

Bierne, B. P. (1975). Biological control attempts by introductions against pest insects in the field in Canada. *Canadian Entomologist*, **107**, 225–36.

Billings, W. D., Godrey, P. J., Chabot, B. F. & Bourque, D. P. (1971). Metabolic acclimation to temperature in arctic and alpine ecotypes of *Oxyria digyna*. *Arctic and Alpine Research*, **3**, 277–89.

Black, T. A., den Hartog, G., Neumann, H. H., Blanken, P. D., Yang, P. C., Russell, C., Nesic, Z., Lee, X., Chen, S. G., Staebler, R. and Novak, M. D. (1996). Annual cycles of water vapour and carbon dioxide fluxes in and above a boreal aspen forest. *Global Change Biology*, **2**, 219–29.

Blaxter, K. L. & Clapperton, J. L. (1965). Prediction of the amount of methane produced by ruminants. *British Journal of Nutrition*, **19**, 511–22.

Boardman, J., Evans, R., Favis-Mortlock, D. T. & Harris, T. M. (1990). Climate change and soil erosion on agricultural land in England and Wales. *Land Degradation & Rehabilitation*, **2**(2), 95–106.

Boardman, J. & Favis-Mortlock, D. T. (1998). *Modelling Soil Erosion by Water*.

NATO-ASI Series I-55. Heidelberg: Springer-Verlag.

Boerner, R. E. J. & Rebbeck, J. (1995). Decomposition and nitrogen release from leaves of three hardwood species grown under elevated O_3 and/or CO_2. *Plant and Soil*, **170**, 149–57.

Bolin, B. B., Degens, E. T., Duvigneaud, P. & Kempe, S. (1979). The global biogeochemical carbon cycle. In *The Global Carbon Cycle*, ed B. B. Bolin, pp. 1–56. New York: John Wiley and Sons. SCOPE.

Bolker, B. M., Pacala, S. W., Bazzaz, F. A., Canham, C. D. & Levin, S. A. (1995). Species diversity and ecosystem response to carbon dioxide fertilization: conclusions from a temperate forest model. *Global Change Biology*, **1**, 373–81.

Bonan, G. & van Cleve, K. (1992). Soil temperature, nitrogen mineralisation and carbon source-sink relationships in boreal forests. *Canadian Journal of Forest Research*, **22**, 629–39.

Bonan, G. B. (1993). Do biophysics and physiology matter in ecosystem models? *Climatic Change*, **24**(4), 281–5.

Bonan, G. B. & Sirois, L. (1992). Air temperature, tree growth, and the northern and southern range limits to *Picea mariana*. *Journal of Vegetation Science*, **3**(4), 495–506.

Bonan, G. B., Pollard, D. & Thompson, S. L. (1992). Effects of boreal forest vegetation on global climate. *Nature*, **359**, 716–18.

Bond, W. J. (1993a). Keystone species. *Ecological Studies*, **99**, 237–54.

Bond, W. J. (1993b). Keystone species. In *Biodiversity and Ecosystem Function*, ed. E. D. Schulze & H. A. Mooney, pp. 237–53. Berlin, Heidelberg, New York: Springer-Verlag.

Bond, W. J. (1997). Functional types for predicting changes in biodiversity: a case study in Cape fynbos. In *Plant Functional Types: Their Relevance to Ecosystem Properties and Global Change*, ed. T. M. Smith, H. H. Shugart & F. I. Woodward, pp.174–94. Cambridge: Cambridge University Press. IGBP Book Series Vol. 1.

Bond, W. J. & Richardson, D. M. (1990).

What can we learn from extinctions and invasions about the effects of climate change? *South African Journal of Science*, **86**, 429–33.

Boote, K. J., Jones, J. W., Mishoe, J. W. & Berger, R.D. (1983). Coupling pests to crop growth simulators to predict yield reductions. *Phytopathology*, **73**, 1573–600.

Bormann, F. H. & Likens, G. L. (1979). *Pattern and Process in a Forested Ecosystem*. New York: Springer-Verlag.

Botkin, D. B., Janak, J. F. & Wallis, J.R. (1972). Some ecological consequences of a computer model of forest growth. *Journal of Ecology*, **60**, 849–73.

Bottner, P. & Couteaux, M. M. (1991). Effect of plant activity on decomposition: soil–plant interactions in response to increasing atmospheric CO_2 concentration. In *Decomposition and Accumulation of Organic Matter in Terrestrial Ecosystems: Research Priorities and Approaches*, ed. N. van Breemen, pp. 39–45. Netherlands: Doorwerth.

Bowen, H. J. M. (1966). *Trace Elements in Biochemistry*. London: Academic Press.

Box, E. O. (1981). *Macroclimate and Plant Forms: An Introduction to Predictive Modeling in Phytogeography*. 258 pp. The Hague: Dr. W. Junk Publishers.

Box, E. O. (1996). Plant functional types and climate at the global scale. *Journal of Vegetation Science*, **7**(3), 309–20.

Bradley, R. S., Diaz, H. F., Eischeid, J. K., Jones, P. D., Kelly, P. M. & Goodess, C. M. (1987). Precipitation fluctuations over northern hemisphere land areas since the mid-19th century. *Science*, **237**, 171–75.

Bradshaw, R. H. W. & Zackrisson, O. (1990). A two thousand year record of a northern Swedish boreal forest stand. *Journal of Vegetation Science*, **1**(4), 519–28.

Brasier, C. M. & Scott, J. K. (1994). European oak declines and global warming: a theoretical assessment of the activity of *Phytophthora cinnamomi*. *OEPP/EPPO Bulletin*, **24**, 221–32.

Brinkman, R. (1987). Agro-ecological characterization, classification and mapping. Different approaches by the international agricultural research centres. In *Agricultural Environments: Characterization, Classification and Mapping*, ed A. H. Bunting, pp. 31–42. Wallingford, UK: CAB International.

Brooking, I. R., Jamieson, P. D. & Porter, I. (1995). The influence of day length on final leaf number in spring wheat. *Field Crops Research*, **41**, 155–65.

Brown, S., Gillespie, J. R. & Lugo, A. E. (1991). Biomass of tropical forests of South and Southeast Asia. *Canadian Journal for Science*, **21**, 111–17.

Brunnel, J. P., Walker, G. R., Walker, C. D., Dighton, J. C. & Kennett-Smith, A. (1991). Using stable isotopes of water to trace plant water uptake. In *Stable Isotopes in Plant Nutrition, Soil Fertility and Environmental Studies*. Proceedings of an International Symposium, October 1990, pp. 543–51. Vienna: IAEA and FAO.

Buchmann, N., Schulze, E.-D. & Gebauer, G. (1995). ^{15}N-nitrogen and ^{15}N-nitrate uptake of a 15-year-old *Picea abies* plantation. *Oecologia*, **102**, 361–70.

Bugmann, H.K.M. (1994). On the Ecology of Mountainous Forests in a Changing Climate: A Simulation Study. PhD Thesis, No. 10638, Swiss Federal Institute of Technology, Zurich, Switzerland.

Bugmann, H. K. M. (1996a). Functional types of trees in temperate and boreal forests: Classification and testing. *Journal of Vegetation Science*, **7**(3), 359–70.

Bugmann, H. K. M. (1996b). A simplified forest model to study species composition along climate gradients. *Ecology*, **77**(7), 2055–74.

Bugmann, H. (1997a). Sensitivity of forests in the European Alps to future climatic change. *Climate Research*, **8**(1), 35–44.

Bugmann, H. K. M. (1997b). Gap models, forest dynamics and the response of vegetation to climate change. In *Past and Future Rapid Environmental Changes: The Spatial and Evolutionary Responses of Terrestrial Biota*, ed. B. Huntley, W. Cramer, A. V. Morgan, H. C. Prentice & J. R. M. Allen, Berlin: Springer-Verlag. NATO ASI Series, Vol. 47.

Bugmann, H. K. M. & Fischlin, A. (1996). Simulating forest dynamics in a complex topography using gridded climatic data. *Climatic Change*, **34**, 201–11.

Bugmann, H. K. M. & Martin, P. (1995). How physics and biology matter in forest gap models. *Climatic Change*, **29**(3), 251–7.

Bugmann, H. K. M., Yan, X., Sykes, M. T., Martin, P., Lindner, M., Desanker, P. V. & Cumming, S. G. (1996). A comparison of forest gap models: model structure and behaviour. *Climatic Change*, **34**, 289–313.

Buol, S. W., Sanchez, P. A., Kimble, J. M. & Weed, S. B. (1990). Predicted impact of climate warming on soil properties and use. In *Impacts of Carbon Dioxide, Trace Gases and Climate Change on Global Agriculture*, ed. B.A. Kimball. *Agronomy Society of America Special Publication*, **53**, 71–82.

Burdon, J. J. (1993). The role of parasites in plant populations and communities. *Ecological Studies*, **99**, 165–80.

Buringh, P. & Dudal, R. (1987). Agricultural land use in space and time. In *Land Transformation in Progress*, ed. M. G. Wolman & F. G. A. Fournier, pp. 9–43. New York: John Wiley & Sons. SCOPE.

Burke, I. C., Kittel, T. G. F., Lauenroth, W. K., Snook, P., Yonker, C. M. & Parton, W. J. (1991). Regional analysis of the Central Great Plains: sensitivity to climate variability. *Bioscience*, **41**, 685–92.

Burke, I. C. and Lauenroth, W. K. (1995). Biodiversity at landscape to regional scales. In: *Global Biodiversity Assessment*, ed. UNEP, pp. 304–11. Cambridge: Cambridge University Press.

Burke, I. C., Lauenroth, W. K., Parton, W. J. & Cole, C. V. (1993). Interactions of land use and ecosystem structure and function: a case study in the Central Great Plains. In *Integrated Regional Models*, ed. G. E. Likens & P. Groffman, pp. 193–216. New York: Chapman and Hall.

Caldwell, M. M. & Flint, S. D. (1994). Stratospheric ozone reduction, solar UV-B radiation and terrestrial ecosystems. *Climate Change*, **28**, 375–94.

Caldwell, M. M., Teramura, A. H. & Terini, M. (1989). The changing solar ultraviolet climate and the ecological consequences for higher plants. *Trends in Ecology and Evolution*, **4**, 363–7.

Caldwell, M. M., Teramura, A. H., Tevini, M., Bornman, J. F., Bjorn, L. O. & Kulandaivelu, G. (1995). Effects of increasing solar ultraviolet radiation on terrestrial plants. *Ambio*, **24**, 166–73.

Campbell, B. D. & Stafford Smith, M. (1993). Defining GCTE modelling needs for pastures and rangelands. *Proceedings of the XVII International Grassland Congress*, 1249–53.

Campbell, B. D., McKeon, G. M., Gifford, R. M., Clark, H., Stafford Smith, M., Newton, P. C. D. & Lutze, J. L. (1996). Impacts of atmospheric composition and climate change on temperate and tropical pastoral agriculture. In *Greenhouse: Coping With Climate Change*, ed. W. J. Bouma, G. I. Pearman & M. R. Manning, pp. 171–89. Melbourne: CSIRO.

Campbell, B. D., Stafford Smith, D. M. & McKeon, G. M. (1997). Elevated CO_2 and water supply interactions in grasslands: a pastures and rangelands management perspective. *Global Change Biology*, **3**, 117–87.

Canadell, J., Jackson, R. B., Ehleringer, J. R., Mooney, H. A., Sala, O. E. & Schulze, E.-D. (1996a). Maximum rooting depth of vegetation types at the global scale. *Oecologia*, **108**, 583–95.

Canadell, J. G., Pitelka, L. F. & Ingram, J. S. I. (1996b). The effects of elevated CO_2 on plant–soil carbon below-ground: A summary and synthesis. *Plant and Soil*, **187**, 391–400.

Cannell, M. G. R. (1995). *Forests and the Global Carbon Cycle in the Past, Present and Future*. 66 pp. *European Forest Institute. Report 2.*

Cao, M., Ma, S. & Han, C. (1995). Potential productivity and human carrying capacity of an agro-ecosystem: an analysis of food production potential of China. *Agricultural Systems*, **47**, 387–414.

Cardon, Z. G. (1996). Effects of root exudation and rhizo deposition on ecosystem carbon storage under elevated CO_2. *Plant and Soil*, **187**, 277–88.

Carlton, J. T., Thompson, J. K., Schemel, L. E. & Nichols, F. H. (1990). Remarkable invasion of San Francisco Bay (California, USA) by the Asian clam

Potamocorbula amurensis I. Introduction and dispersal. *Marine Ecology Progress Series*, 66, 81–94.

Carpenter, S. R., Fisher, S. G., Grimm, N. B. & Kitchell, J. F. (1992). Global change and freshwater ecosystems. *Annual Review of Ecology and Systematics*, 23, 119–39.

Carpenter, S. R., Kitchell, J. F. & Hodgson, J. R.(1985). Cascading trophic interactions and lake productivity. *BioScience*, 35, 634–9.

Ceulemans, R. & Mousseau, M. (1994). Tansley Review No. 71 Effects of elevated atmospheric CO_2 on woody plants: a review. *New Phytologist*, 127, 425–46.

Chakraborty, S. (1997). *How will Plant Diseases Impact on Pasture Production under Climate Change?* XVIII International Grasslands Congress, Calgary, Canada.

Chapin, F. S. III (1993). Functional role of growth forms in ecosystem and global processes. *Scaling Physiological Processes: Leaf to Globe*, ed. J. Ehleringer & C. Field, pp. 287–312. San Diego: Academic Press.

Chapin, F. S., III (1980). The mineral nutrition of wild plants. *Annual Review of Ecology and Systematics*, 11, 233–60.

Chapin, F. S., III (1983). Direct and indirect effects of temperature on arctic plants. *Polar Biology*, 2, 47–52.

Chapin, F. S., III, Bret-Harte, M. S., Hobbie, S. E. & Zhong, H. L. (1996). Plant functional types as predictors of transient responses of arctic vegetation to global change. *Journal of Vegetation Science*, 7(3), 347–58.

Chapin, F. S., III, Lubchenco, J. & Reynolds, H. L. (1995a). Biodiversity effects on patterns and processes of communities and ecosystems. In *Global Biodiversity Assessment*, ed. UNEP, pp. 23–45. Cambridge: Cambridge University Press.

Chapin, F. S., III, Moilanen, L. & Kielland, K. (1993). Preferential use of organic nitrogen for growth by non-mycorrhizal arctic sedge. *Nature*, 361, 150–53.

Chapin, F. S., III, Sala, O. E., Burke, I. C., Grime, J. P., Hooper, D. U., Lauenroth, W. K., Lombard, A., Mooney, H. A.,

Mosier, A. R., Naeem, S., Pacala, S. W., Roy, J., Steffen, W. & Tilman, D. (1997a). Ecosystem consequences of changing biodiversity. *BioScience*, 48, 45–52.

Chapin, F. S., III & Shaver, G. R. (1996). Physiological and growth responses of arctic plants to a field experiment simulating climatic change. *Ecology*, 77, 822–40.

Chapin, F. S., III & Shaver, G. R. (1985). Individualistic growth response of tundra plant species to environmental manipulations in the field. *Ecology*, 66, 564–76.

Chapin, F. S., III, Shaver, G. R., Giblin, A. E., Nadelhoffer, K. G. & Laundre, J. A. (1995b). Response of arctic tundra to experimental and observed changes in climate. *Ecology*, 76, 694–711.

Chapin, F. S., III, Walker, B. H., Hobbs, R. J., Hooper, D. U., Lawton, J. H., Sala, O. E. & Tilman D. (1997b). Biotic controls over the functioning of ecosystems. *Science*, 277, 500–4.

Chapman, L. J., Chapman, C. A. & Wrangham., R. W. (1992). *Balanites wilsoniana*: Elephant dependent dispersal. *Journal of Tropical Ecology*, 8, 275–83.

Chapman, W. L. & Walsh, J. E. (1993). Recent variations of sea ice and air temperatures in high latitudes. *Bulletin of the American Meteorological Society*, 74, 33–47.

Charney, J. G. (1975). Dynamics of deserts and droughts in the Sahel. *Quarterly Journal of the Royal Meteorological Society*, 101, 193–202.

Charney, J. G., Quirk, W. J., Chow, S.-H. & Kornfield. J. (1977). A comparative study of effects of albedo change on drought in semiarid regions. *Journal of Atmospheric Science*, 34, 1366–85.

Chen, J. M. & Cihlar, J. (1996). Retrieving leaf area index of boreal conifer forests using landsat TM images. *Remote Sensing of Environment*, 55, 152–62.

Chen, Z. Z., Huang, D. H. & Zhang, H. F. (1988). A model of the relation between below ground biomass of Aneurolepidium chinese and *Stipa grandis* grasslands and precipitation, in the Xilin River watershed of Inner Mongolia. In *Research on Grassland Ecosystems* (2), ed. S. Jiang, pp. 20–5 (in Chinese). Beijing: Science Press of China.

Ciais, P., Tans, P. P., Denning, R., Francey, R., Trolier, M., Meijer, H. J., White, J. W., Berry, J. A., Randall, D., Collatz, J. G., Sellers, P. J., Monfrey, P. & Heimann, M. (1997). A three dimensional synthesis of d[18]O in atmospheric CO_2. Part 2: Simulations with the TM2 transport model. *Journal of Geophysical Research*, **102**, 5873–84.

Ciais, P., Tans, P. P., Trolier, M., White, J. W. C. & Francey, R. J. (1995). A large northern hemisphere terrestrial CO_2 sink indicated by the $^{13}C/^{12}C$ ratio of atmospheric CO_2. *Nature*, **269**, 1098–102.

Clarke, B., Murray, J. & Johnson, M. S. (1984). The extinction of endemic species by a program of biological control. *Pacific Science*, **38**, 97–104.

Claussen, M. (1994). On coupling global biome models with climate models. *Climate Research*, **4**, 203–21.

Claussen, M. (1996). Variability of global biome patterns as a function of initial and boundary conditions in a climate model. *Climate Dynamics*, **12**, 371–9.

Coakley, S. M. (1988). Variation in climate and prediction of disease in plants. *Annual Review of Phytopathology*, **26**, 163–81.

Coakley, S. M. (1995). Biospheric change: will it matter in plant pathology? *Canadian Journal of Plant Pathology*, **17**, 147–53.

Coffin, D. & Lauenroth, W. K. (1989). Disturbances and gap dynamics in a semiarid grassland: a landscape-level approach. *Landscape Ecology*, **3**(1), 19–27.

Coffin, D. P. & Lauenroth, W. K. (1990). A gap dynamics simulation model of succession in a semi arid grassland. *Ecological Modelling*, **49**, 229–66.

Coffin, D. P. & Lauenroth, W. K. (1996). Transient responses of North-American grasslands to changes in climate. *Climatic Change*, 1996, 269–78.

Coffin, D. P. & Urban, D. L. (1993). Implications of natural history traits to system-level dynamics – comparisons of a grassland and a forest. *Ecological Modelling*, **67**(2–4), 147–78.

Comins, H. N. & McMurtrie, R. E. (1993). Long-term response of nutrient limited forests to CO_2 enrichment: equilibrium behaviour of plant–soil models. *Ecological Applications*, **3**, 666–81.

Conroy, J. P. (1992). Influence of elevated atmospheric CO_2 concentrations on plant nutrition. *Australian Journal of Botany*, **40**, 445–56.

Conway, T. C., Tans, P. P., Waterman, L. S., Thoning, K. W., Kitzis, D. R., Masarie K. A. & Zhang. N. (1994). Evidence for interannual variability of the carbon cycle from the NOAA/CMDL global air sampling network. *Journal of Geophysical Research*, **99**, 22831–55.

Coops, N. C., Waring, R.H. & Landsberg, J.J. (1997) Assessing forest productivity in Australia and New Zealand using a physiologically-based model driven with averaged monthly weather data and satellite derived estimates of canopy photosynthetic capacity. *Forest Ecology and Management* (in press).

Cornell, S., Rendell, A. & Jickells, T. (1995). Atmospheric inputs of dissolved organic nitrogen to the oceans. *Nature*, **376**, 243–6.

Cotrufo, M. F. & Ineson, P. (1995). Effects of enhanced atmospheric CO_2 and nutrient supply on the quality and subsequent decomposition of fine roots of *Betula pendula* Roth. and *Picea sitchensis* (Bong.) Carr. *Plant and Soil*, **170**, 267–77.

Cotrufo, M. F., Ineson, P. & Rowland, A. P. (1994). Decomposition of tree leaf litters grown under elevated CO_2: effect of litter quality. *Plant and Soil*, **163**, 121–30.

Coughenour, M. B. & Chen, D. X. (1997). Assessment of grassland ecosystem responses to atmospheric change using linked plant soil process models. *Ecological Applications*, **7**, 802–27.

Coughenour, M. B. & Parton, W. J. (1996). Integrated models of ecosystem function: a grassland case study. In *Global Change and Terrestrial Ecosystems*, ed. B. H. Walker & W. L. Steffen, pp. 93–114. Cambridge: Cambridge University Press.

Couteaux, M. M., Bottner, P. & Berg, B. (1995). Litter decomposition, climate and litter quality. *Trends in Ecology and Evolution*, **10**, 63–6.

Cramer, W. (1997). Using plant functional types in a global vegetation model. In

Plant Functional Types: Their Relevance to Ecosystem Properties and Global Change. ed. T. M. Smith, H. H. Shugart & F. I. Woodward, pp. 271–88. Cambridge: Cambridge University Press. IGBP Book Series Vol. 1.

Cramer, W. & Fischer, A. (1996). Data requirements for global terrestrial ecosystem modelling. In *Global Change and Terrestrial Ecosystems*, ed. B. Walker & W. Steffen. Cambridge: Cambridge University Press. IGBP Book Series Vol. 2. pp. 530–65.

Cramer, W. & Leemans, R. (1993). Assessing impacts of climate change on vegetation using climate classification systems. *In Vegetation Dynamics and Global Change*, ed. A. M. Solomon & H. H. Shugart, pp. 190–217. New York: Chapman & Hall.

Cramer, W., Moore, B., III & Sahagian, D. (1996). Data needs for modelling global biospheric carbon fluxes – lessons from a comparison of models. *IGBP Newsletter 1996*, (27), 13–15.

Cramer, W. & Steffen, W. (1997). Forecast changes in the global environment: What they mean in terms of ecosystem responses on different time-scales. In *Past and Future Rapid Environmental Changes: The Spatial and Evolutionary Responses of Terrestrial Biota*, ed. B. Huntley, W. Cramer, A. V. Morgan, H. C. Prentice & J. R. M. Allen, pp. 415–26. Berlin: Springer-Verlag. NATO ASI Series, Vol. I No. 47.

Cramer, W. P. & Solomon, A. M. (1993). Climate classification and future global redistribution of agricultural land. *Climate Research*, 3, 97–110.

Crawford, T. L., Dobosy, R. J., McMillen, R. T., Vogel, C. A. & Hicks, B. B. (1996). Air–surface exchange measurements in heterogenous regions: extending tower observations with spatial structure observed from small air craft. *Global Change Biology*, 2, 275–86.

Crosson, P. & Anderson, J. R. (1994). Demand and supply: trends in global agriculture. *Food Policy*, 19, 105–19.

Crutzen, P. J., Aselmann, I. & Seiler, W. (1986). Methane production by domestic animals, wild ruminants, other her-

bivorous fauna and humans. *Tellus*, 38B, 271–84.

CSIRO. (1995). The PLOT – what is it? *The Futures Gazette*, November 1995, pp. 5,11.

Cumming, S. G. & Burton, P. J. (1996). Phenology-mediated effects of climatic change on some simulated British Columbia forests. *Climatic Change*, 34, 213–22.

Curtis, P. S. (1996). A meta-analysis of leaf gas exchange and nitrogen in trees grown under elevated carbon dioxide. *Plant, Cell and Environment*, 19, 127–37.

Curtis, P. S., Drake, B. G., Leadley, P. W., Arp, W. J. & Whigham, D. F. (1989). Growth and senescence in plant communities exposed to elevated CO_2 concentrations on an estuarine marsh. *Oecologia*, 78, 20–6.

Curtis, P. S., Klus, D. J., Kalisz, S. & Tonsor, S. J. (1996). Intraspecific variation in CO_2 responses in *Raphanus raphanistrum* and *Plantagolanceolata*: Assessing the potential for evolutionary change with rising atmospheric CO_2. In *Carbon Dioxide, Populations, and Communities*, ed. Ch. Körner & F. A. Bazzaz, pp. 13–22. San Diego: Academic Press.

Cushman, J. H. (1995). Ecosystem-level consequences of species additions and deletions on islands. In *Islands*, ed. P. M. Vitousek, pp. 135–47. Berlin: Springer-Verlag.

Dai, A. & Fung, I. (1993). Can climate variability contribute to the 'missing' CO_2 sink? *Global Biogeochemical Cycles*, 7, 599–609.

Dale, V. H. & Rauscher, H. M. (1994). Assessing impacts of climate change on forests: the state of biological modeling. *Climatic Change*, 28(1–2), 65–90.

D'Antonio, C. M. & Vitousek, P. M. (1992). Biological invasions by exotic grasses, the grass/fire cycle, and global change. *Annual Review of Ecology and Systematics*, 23, 63–87.

Davidson, E. A., Hart, S. C. & Firestone, M. K. (1992). Internal cycle of nitrate in soils of a mature coniferous forest. *Ecology*, 73, 1148–56.

Davis, M. B. (1981). Quaternary history and the stability of forest communities. In

Forest Succession: Concepts and Application, ed. D. C. West, H. H. Shugart & D. B. Botkin, pp. 132–153. New York: Springer-Verlag.

Davis, M. B., Schwartz, M. W. & Woods, K. (1991). Detecting a species limit from pollen in sediments. *Journal of Biogeography*, **18**, 653–68.

Dawson, T. E. (1993a). Water sources of plants as determined by xylem-water isotopic composition: perspectives on plant competition, distribution, and water relations. In *Stable Isotopes and Plant Carbon–Water Relations*, ed. J. R. Ehleringer, A. E. Hall & G. D. Farquhar, pp. 465–96. San Diego: Academic Press.

Dawson, T. E. (1993b). Hydraulic lift and water use by plants: implications for water balance, performance, and plant–plant interactions. *Oecologia*, **95**, 565–74.

Dawson, T. E. & Ehleringer, J. R. (1991). Streamside trees that do not use stream water. *Nature*, **350**, 335–37.

De Fries, R. S. & Townshend, J. R. G. (1994). NDVI-derived land cover classifications at a global scale. *International Journal of Remote Sensing*, **15**(17), 3567–86.

De Pury D. G. G. & Farquhar, G. D. (1997). Simple scaling of photosynthesis from leaves to canopies without the errors of big-leaf models. *Plant, Cell and Environment*, **20**, 537–57.

De Roo, A. P. J., Wesseling, C. G., Jetten, V. G. & Ritsema, C. J. (1997). In *Modelling Soil Erosion by Water*, ed. J. Boardman & D. Favis-Mortlock. Berlin: Springer-Verlag.

Debevec, E. M. & MacLean, S. F. (1993). Design of greenhouses for the manipulation of temperature in tundra plant communities. *Arctic and Alpine Research*, **25**, 56–62.

Deevey, E. S., Jr. (1960). The human population. *Scientific American*, **203**, 195–204.

Denslow, J. S. (1997). The effects of functional group diversity on disturbance ecology of tropical moist forests. In *Diversity and Processes in Tropical Forest Ecosystems*, ed. G. H. Orians, J. Dirzo & J. H. Cushman. Berlin: Springer-Verlag.

Dent, J. B. (1993). Potential for systems simulation in farming systems research. In *Systems Approaches for Agricultural Development*, ed. F. Penning de Vries, P. Teng & K. Metselaar, pp. 325–40.

Desanker, P. V., Frost, P. G. H., Justice, C. O. & Scholes, R.J . (eds.) (1997) Causes and consequences of land use and land cover change in central African miombo ecosystems: Prospectus for an IGBP LUCC Miombo Network. Workshop report. IGBP Report No. 41,109 pp, International Geosphere–Biosphere Programme, Stockholm.

Deschamps, P.-Y., Bréon, F.-M., Leroy, M., Podaire, A., Bricaud, A., Buriez, J.-C. & Sèze, G. (1994). The POLDER Mission: instrument characteristics and scientific objectives. *IEEE Transactions on Geoscience and Remote Sensing*, **32**(3), 598–615.

Dewar, R. C. (1996). The correlation between plant growth and intercepted radiation: an interpretation in terms of optimal nitrogen content. *Annals of Botany*, **78**, 125–36.

Dewar, R. C. (1997). A synthesis of several empirical plant growth phenomena in terms of the concept of quasi equilibrium. *Plant, Cell and Environment* (submitted).

d'Herbes, J. M. & Valentin, C. (1997). Land surface conditions of the Niamey region – ecological and hydrological implications. *Journal of Hydrology*, **189**, 18–42.

Dhillion, S. S., Roy, J. & Abrams, M. (1996). Assessing the impact of elevated CO_2 on soil microbial activity in a Mediterranean model ecosystem. *Plant and Soil*, **187**, 333–42.

Dickinson, R. E., Henderson-Sellers, A. & Kennedy, P. J. (1993). Biosphere–Atmosphere Transfer Scheme (BATS) Version 1e as coupled to the NCAR Community Climate Model. National Center for Atmospheric Research. 80 pp. *NCAR Technical Note*: **NCAR/TN-387**.

Dickinson, R. E., Henderson-Sellers, A., Kennedy, P. J. & Wilson, M. F. (1986). Biosphere-Atmosphere Transfer Scheme (BATS) for the NCAR Community Climate Model. National Center for Atmospheric Research. 69 pp. *NCAR Technical Note*: **NCAR/TN-275**.

Donigian, A. S. Jr, Patwardhan, A. S., Jackson, R. B. IV, Barnwell, T. O. Jr, Weinrich, K. B. & Rowell, A. L. (1995). Modeling the impacts of agricultural management practices on soil carbon in the central U. S. In *Soil Management and Greenhouse Effect*, ed. R. Lal, J. Kimble, E. Levine & B. A. Stewart, pp. 121–35. Boca Raton, FL, USA: Lewis Publishers.

Donnelly, J. R. (1984). The productivity of breeding ewes grazing on lucerne or grass and clover pastures on the tablelands of southern Australia. III. Lamb mortality and weaning percentage. *Australian Journal of Agricultural Research*, **35**, 709–21.

Drake, B. G. & Leadley, P. W. (1991). Canopy photosynthesis of crops and native plant communities exposed to long-term elevated carbon dioxide. *Plant, Cell and Environment*, **14**, 853–860.

Drake, B. G., Peresta, G., Beugeling, E. & Matamala, R. (1996). Long-term elevated CO_2 exposure in a Chesapeake Bay wetland: ecosystem gas exchange, primary production, and tissue nitrogen. In *Carbon Dioxide and Terrestrial Ecosystems*, ed. E. W. Koch & H. A. Mooney. San Diego: Academic Press.

Drenth, H., ten Berge, H. F. M. & Riethoven, J. J. M. (eds). (1994). *ORYZA Simulation Modules for Potential and Nitrogen Limited Rice Production*. SARP Research Proceedings. Wageningen, The Netherlands: AB-DLO.

Dufrène, E., Pontailler, J. -Y. & Saugier, B. (1993). A branch bag technique for simultaneous CO_2 enrichment and assimilation measurements on beech (*Fagus sylvatica* L.). *Plant, Cell and Environment*, **16**, 1131–8.

Durka, W., Schulze, E.-D., Gebauer, G. & Voerkelius, S. (1994). Effects of forest decline on uptake and leaching of deposited nitrate determined from ^{15}N and ^{18}O measurements. *Nature*, **372**, 765–7.

Dye, D. G. (1992). Satellite estimation of the global distribution and interannual variability of photosynthetically active radiation. 180 pp. Ph.D. thesis, University of Maryland.

Dyer, J. M. (1995). Assessment of climatic warming using a model of forest species migration. *Ecological Modelling*, **79**(1–3), 199–219.

Dyson, T. (1996). *Population and Food*. London: Routeledge.

Eamus, D. & Jarvis, P. G. (1989). The direct effects of increase in the global atmospheric CO_2 concentration on natural and commercial temperate trees and forests. *Advances in Ecological Research*, **19**, 1–55.

Ehleringer, J. R. & Dawson, T. E. (1992). Water uptake by plants: Perspectives from stable isotopes. *Plant, Cell and Environment*, **15**, 1073–82.

Ehleringer, J. R., Hall, A. E. & Farquhar, G. D. (1993). *Stable Isotopes and Plant Carbon–Water Relations*. 555 pp. San Diego: Academic Press.

Ehleringer, J. R., Phillips, S. L., Schuster, W. F. S. & Sandquist, D. R. (1991). Differential utilization of summer rains by desert plants: implications for competition and climate change. *Oecologia*, **88**, 430–34.

Ehrlich, P. R. & Mooney, H. A. (1983). Extinction, substitution, and ecosystem services. *BioScience*, **33**, 248–54.

Ellenberg, H. Jr. (1986). Vearnderungen der Flora Mitteleuropas unter dem Einfluá von Düngung und Immissionen. *Schweiz. Z. Forstwesen*, **136**, 19–36.

Ellery, W., Scholes, M. C. & Scholes, R. J. (1996). The distribution of sweetveld and sourveld in South Africa's grassland biome in relation to environmental factors. *African Journal of Range and Forage Science*, **12**, 38–45.

Ellsworth, D. S., Oren, R., Huang, C., Phillips, N. & Hendrey, G. R. (1995). Leaf and canopy responses to elevated CO_2 in a pine forest under free-air CO_2 enrichment. *Oecologia*, **104**, 139–46.

Eltahir, E. A. B. & Bras, R. L. (1994). Sensitivity of regional climate to deforestation in the Amazon basin. *Advances in Water Resources*, **17**, 101–15.

Emanuel, W. R., Shugart, H. H. & Stevenson, M. P. (1985a). Climatic change and the broad-scale distribution of terrestrial ecosystems complexes. *Climatic Change*, **7**(1), 29–43.

Emanuel, W. R., Shugart, H. H. & Stevenson, M. P. (1985b). Response to comment: climatic change and the broad-scale distribution of terrestrial ecosystems complexes. *Climatic Change*, **7**, 457–60.

Enting, I. G. & Mansbridge, J. V. (1991). Latitudinal distribution of sources and sinks of CO_2: results of an inverse study. *Tellus*, **43**, 156–79.

Esser, G. (1989). Global land-use changes from 1860 to 1980 and future projections to 2500. *Ecological Modelling*, **44**, 307–16.

Esser, G. (1990) Modelling global terrestrial sources and sinks of CO_2 with special reference to soil organic matter. In *Soils and the Greenhouse Effect*, ed. A. F. Bouwman, pp. 247–61. New York: John Wiley & Sons Ltd.

Evans, L. T., Wardlaw, I. F. & Fischer, R. A. (1975). Wheat. In *Crop Physiology – Some Case Histories*, ed. L. T. Evans. pp. 101–50.

Ewel, J. J., Mazzarino, M. J. & Berish. C. W. (1991). Tropical soil fertility changes under monocultures and successional communities of different structure. *Ecological Applications*, **1**, 289–302.

Fahrig, L. & Merriam, G. (1985). Habitat patch connectivity and population survival. *Ecology*, **66**, 1762–68.

Falkenmark, M. (1997). Meeting water requirements of an expanding world population. *Philosophical Transactions of the Royal Society of London Series B- Biological Sciences*, **352**, 929–36.

FAO (1979). *FAO Production Yearbook*. UN Food and Agriculture Organization. 32.

FAO (1991*a*). *Agrostat PC – Land Use*. UN Food and Agriculture Organization. 1/7.

FAO (1991*b*). *The Digitized Soil Map of the World* (Release 1.0). UN Food and Agriculture Organization. 67/1.

FAO (1993*a*). *Agriculture: Towards 2010*. Conference of FAO, November 1993. FAO, Rome 320 pp.

FAO (1993*b*). Forest Resources Assessment 1990 – Tropical countries. *FAO Paper*, 112.

FAO/UNESCO (1974). *Soil Map of the World*, 1 : 5,000,000. UN Food and Agriculture Organization.

Farquhar, G. D. (1989). Models of integrated photosynthesis of cells and leaves. *Philosophical Transactions of the Royal Society, London*, **72**, 245–50.

Farquhar, G. D., Ehleringer. J. R. & Hubick, K. T. (1989). Carbon isotope discrimination and photosynthesis. *Annual Review of Plant Physiology and Plant Molecular Biology*, **40**, 503–37.

Farquhar, G. D. & Lloyd, J. (1993). Carbon and oxygen isotope effects in the exchange of carbon dioxide between terrestrial plants and the atmosphere. In *Stable Isotopes and Plant Carbon–Water Relations*, ed. J. R. Ehleringer, A. E. Hall & G. D. Farquhar, pp. 47–70. San Diego: Academic Press.

Farquhar, G. D., Lloyd, J., Taylor, J., Flanagan, L. B., Syversten, J. P., Hubick, K. T, Wong, S. C. & Ehleringer, J. R. (1993). Vegetation effects on the isotope composition of oxygen in atmospheric CO_2. *Nature*, **363**, 439–43.

Farquhar, G. D. & Sharkey, T. D. (1982). Stomatal conductance and photosynthesis. *Annual Review of Plant Physiology*, **33**, 317–45.

Farquhar, G. D., Von Caemmerer, S. & Berry, J. A. (1980). A biochemical model of photosynthetic CO_2 assimilation in leaves of C_3 species. *Planta*, **149**, 78–90.

Fearnside, P. M. (1993). Deforestation in the Brazilian Amazon: The effect of population and land tenure. *Ambio*, **22**, 537–45.

Feddes, R.A., Kabat, P., van Bakel, P. J. T., Bronswijk, J. J. B. & Halbertsma, J. (1988). Modelling soil water dynamics in the unsaturated zone – State of the art. *Journal of Hydrology*, **100**, 69–111.

Field, C. B. & Mooney, H. A. (1986). The photosynthesis–nitrogen relationship in wild plants. In *On the Economy of Plant Form and Function*, ed T. J. Givnish, pp. 25–56, Cambridge: Cambridge University Press.

Field, C. B., Chapin, F. S., III, Matson, P. A. & Mooney, H. A. (1992). Responses of terrestrial ecosystems to the changing atmosphere: a resource-based approach. *Annual Review of Ecology and Systematics*, **23**, 201–35.

Field, C. B., Chapin, F. S., III, Chiariello, N. R., Holland, E. A. & Mooney, H. A. (1996). The Jasper Ridge CO_2 experiment: design and motivation. In *Carbon Dioxide and Terrestrial Ecosystems*, ed. G. W. Koch & H. A. Mooney, pp. 212–45. San Diego: Academic Press.

Field, C. B., Gamon, J. A. & Penuelas, J. (1993). Remote sensing of terrestrial photosynthesis. *Ecological Studies*, **100**, 511–28.

Field, C. B., Jackson, R. B. & Mooney, H. A. (1995). Stomatal responses to CO_2: implications from the plant to the global scale. *Plant, Cell and Environment*, **18**, 1214–25.

Firbas, F (1949). *Spät- und nacheiszeitliche Waldgeschichte Mitteleuropas nördlich der Alpen*. Vol 1, 480 pp. Jena: Gustav Fischer Verlag.

Fischer, A. (1997). Seasonal features of global NPP models for the terrestrial biosphere. In *Past and Future Rapid Environmental Changes: The Spatial and Evolutionary Responses of Terrestrial Biota*, ed. B. Huntley, W. Cramer, A. V. Morgan, H. C. Prentice & J. R M. Allen, pp.469–83. Berlin, Heidelberg, New York, London, Paris, Tokyo: Springer-Verlag. NATO ASI Series, Vol. 47.

Fischer M. J., Rao I. M. & Ayarza M. A. (1994). Carbon storage by introduced deep-rooted grasses in the South American savannas. *Nature,* **371**, 236–8.

Fischer, A., Louahala, S., Maisongrande, P., Kergoat, L. & Dedieu, G. (1996). Satellite data for monitoring, understanding and modelling of ecosystem functioning. In *Global Change and Terrestrial Ecosystems*. ed. B. Walker & W. Steffen, Vol. 2, pp. 566–91, Cambridge: Cambridge University Press.

Fischlin, A., Bugmann, H. K. M. & Gyalistras, D. (1995). Sensitivity of a forest ecosystem model to climate parameterization schemes. *Environmental Pollution*, **87**(3), 267–82.

Flanagan, D. C. & Nearing, M. A. (1995). USDA Water Erosion Prediction Project: Hillslope Profile and Watershed Model Documentation. NSERL. 10.

Flanagan, P. W. & Van Cleve, K. (1983). Nutrient cycling in relation to decomposition and organic matter quality in taiga ecosystems. *Canadian Journal of Forest Research*, **13**, 795–817.

Flanagan, P. W. & Veum, K. (1974). Relationships between respiration, weight loss, temperature, and moisture in organic residues on tundra. In *Soil Organisms and Decomposition in Tundra*, ed. A. J. Holding, O. W. Heal, S. F. Maclean Jr & P. W. Flanagan, pp. 249–77. Stockholm: Tundra Biome Steering Committee.

Flannigan, M. D. F. & Van Wagner, C. E. (1991). Climate change and wildfire in Canada. *Canadian Journal of Forest Research*, **21**, 66–72.

Flint, S. D. & Caldwell, M. M. (1996). Scaling plant ultraviolet spectral responses from laboratory action to field spectral weighing factors. *Journal of Plant Physiology*, **148**, 107–14.

Foley, J. A., Kutzbach, J. E., Coe, M. T. & Levis, S. (1994). Feedbacks between climate and boreal forests during the Holocene epoch. *Nature*, **371**, 52–4.

Foley, J. A., Levis, S., Prentice, I. C., Pollard, D. & Thompson, S. L. (1998). Coupling dynamic models of climate and vegetation. *Global Change Biology*, (in press).

Foley, J. A., Prentice, I. C., Ramankutty, N., Levis, S., Pollard, D., Sitch, S. & Haxeltine, A. (1996). An integrated biosphere model of land surface processes, terrestrial carbon balance, and vegetation dynamics. *Global Biogeochemical Cycles* **10**(4), 603–28.

Forman, R. T. T. (1995). Land mosaics: the ecology of landscapes and regions. Cambridge: Cambridge University Press.

Francey, R. J. & Tans, P. P. (1987). Latitudinal variation in oxygen-18 in atmospheric CO_2. *Nature*, **327**, 495–7.

Franck, V. M., Hungate, B. A., Chapin, F. S., III & Field, C. B. (1997). Decomposition of litter produced under elevated CO_2: dependence on plant species and nutrient supply. *Biogeochemistry*, **36**, 223–37.

Frankland, J. C., Magan, N. & Gadd, G. M. (1994). *Fungi and Environmental Change*. Cambridge: Cambridge University Press, 351 pp.

Franklin, J. F. & Forman, T. T. (1987). Creating landscape patterns by forest cutting: ecological consequences and principles. *Landscape Ecology*, **1**, 5–18.

Franz, F., Rohle, H. & Meyer, F. (1993). Wachstums- und Ertragsleistung der Buche. Aggemeine. *ForstZeitschrift*, **6**, 262–7.

Freer, M., Moore, A. D. & Donnelly, J. R. (1997). Grazplan – Decision support systems for Australian Grazing Enterprises. 2. The animal biology model for feed intake, production and reproduction and the GrazFeed DSS. *Agricultural Systems*, **54**, 77–126.

Friedlingstein, P., Fung, I., Holland, E., John, J., Brasseur, G., Erickson, D. & Schimel, D. (1995). On the contribution of CO_2 fertilisation to the missing biospheric sink. *Global Biogeochemical Cycles*, **9**, 541–56.

Friend, A. D. (1991). Use of a model of photosynthesis and leaf micro environment to predict optimal stomatal conductance and leaf nitrogen partitioning. *Plant, Cell and Environment*, **14**, 895–905.

Friend, A. D. (1997). Parameterisation of a global daily weather generator for terrestrial ecosystem and biogeochemical modelling. *Ecological Modelling*, (in press)

Friend, A. D. & Cox, P. M. (1995). Modelling the effects of atmospheric CO_2 on vegetation–atmosphere interactions. *Agricultural and Forest Meteorology*, **73**(3–4), 285–95.

Friend, A. D., Shugart, H. H. & Running, S. W. (1993). A physiology-based gap model of forest dynamics. *Ecology*, 74(3), 792–7.

Friend, A. D., Stevens, A. K., Knox, R. G. & Cannell, M. G. R. (1997). A process-based, terrestrial biosphere model of ecosystem dynamics (Hybrid v3.0). *Ecological Modelling*, **95**, 249–87.

Frost, T. M., Carpenter, S. R., Ives, A. R. & Kratz, T. K. (1995). Species compensation and complementarity in ecosystem function. In *Linking Species and Ecosystems*, ed. C. G. Jones & J. H. Lawton, pp. 224–39. New York: Chapman and Hall.

Fryer, G. I. & Johnson, E. A. (1988). Reconstructing fire behaviour and effects in a subalpine forest. *Journal of Applied Ecology*, **25**, 1063–72.

Gao, Q. & Zhang, X. (1997). A simulation study of responses of the Northeast China Transect to elevated CO_2 and climate change. *Ecological Applications*, **7**, 470–83.

Gardner, R. H., Hargrove, W. W., Turner, M. G. & Romme, W. H. (1996). Climate change, disturbances and landscape dynamics. In *Global Change and Terrestrial Ecosystems*, ed. B.H. Walker & W.L. Steffen, pp. 149–72, Cambridge: Cambridge University Press. IGBP Book Series No. 2.

Gardner, R. H., Milne, B. T., Turner, M. G. & O'Neil. R. V.(1987). Neutral models for the analysis of broad-scale landscape pattern. *Landscape Ecology*, **1**, 19–28.

Gardner, R. H., O'Neil, R. V. & Turner, M. G. (1993). Ecological implications of landscape fragmentation. In *Humans as Components of Ecosystems: the Ecology of Subtle Human Effects and Populated Areas*, ed. S. T. A. Pickett & M. J. McDonnell, pp. 208–26 New York: Springer-Verlag.

Gardner, R. H., Turner, M. G., O'Neil, R. V. & Lavorel. S.(1992). Simulation of the scale-dependent effects of landscape boundaries on species persistence and dispersal. In *The Role of Landscape Boundaries in the Management and Restoration of Changing Environments*, ed. M. M. Holland, P. G. Risser & R. J. Naiman, pp. 76–89. New York: Chapman and Hall.

Gate, P. (1995). *Ecophysiologie du Blé*. 429 pp. Paris: TEC and DOC Lavoisier.

GCTE. (1996). *Global Change and Terrestrial Ecosystems Report No. 5*. GCTE Core Research: 1994/95 Report. Canberra, Australia: GCTE Core Project Office, pp. 51–2.

Gebauer, G. & Schulze, E.-D. (1991). Carbon and nitrogen isotope ratios in different compartments of a healthy and a declining *Picea abies* forest in the Fichtelgebirge, NE Bavaria. *Oecologia*, **87**, 198–207.

Gehrke, C., Johanson, U., Callaghan, T. V., Chadwick, D. & Robinson, C. H. (1995). The impact of enhanced ultraviolet-B radiation on litter quality and decomposition processes in Vaccinium leaves from the Subarctic. *Oikos*, **72**, 213–22.

Giesler, R. Högberg, M. & Högberg, P. (1998). Soil chemistry and plants in Fennoscandian boreal forest as exemplified by a local gradient. *Ecology*, **79**, 119–37.

Gifford, R. M. (1992). Interaction of carbon

dioxide with growth-limiting environmental factors in vegetation productivity: implications for the global carbon cycle. *Advances in Bioclimatology*, 1, 24–58.

Gifford, R. M. (1994*a*). A comparison of potential photosynthesis, productivity and yield of plant species with differing photosynthetic metabolism. *Australian Journal of Plant Physiology*, 1, 107–17.

Gifford, R. M. (1994*b*). The global carbon cycle: a viewpoint on the missing sink. *Australian Journal of Plant Physiology*, 21, 1–15.

Gifford, R. M. (1995). Whole plant respiration and photosynthesis of wheat under increased CO_2 concentration and temperature: long-term vs. short-term distinctions for modelling. *Global Change Biology*, 1, 385–96.

Gifford, R. M., Campbell, B. D. & Howden, S. M. (1996). Options for adapting agriculture to climate change. In *Greenhouse: Coping With Climate Change*, ed. W. J. Bouma, G. I. Pearman & M. R. Manning, pp. 399–416. Melbourne: CSIRO.

Gignoux J., Menaut J-C., Noble I. R. & Davies I. D. (1997) A spatial model of savanna function and dynamics: model description and preliminary results. In *Population and Community Dynamics in the Tropics*, ed. D. M. Newbery, H. H. T. Prins & N. D. Brown. (British Ecological Society Annual Symposium), pp 361–83. Cambridge: Blackwell.

Gignoux, J., Noble, I. R. & Menaut, J.-C. (1995). Modelling tree community dynamics in savannas: effects of competition with grasses and impact of disturbance. In *Functioning and Dynamics of Natural and Perturbed Ecosystems*, Ed. D. Bellan, G. Bonin & C. Emig, pp. 219–30. Paris, France: Lavoisier Publishing.

Gitay, H. & Noble, I. R. (1997). What are functional types and how should we seek them? In *Plant Functional Types: Their Relevance to Ecosystem Properties and Global Change*, ed. T. M. Smith, H. H. Shugart & F. I. Woodward, pp. 3–19. Cambridge: Cambridge University Press. IGBP Book Series Vol. 1.

Gleason, J. F., Bhartia, P. K., Herman, J. R., McPeters, R., Newman, P., Stolarsky, R. S., Flynn, L., Labow, G., Larko, D., Seftor, C., Wellemeyer, C., Komhyr, W. D., Miller, A. J. & Plantet, W. (1993). Record low global ozone in 1992. *Science*, 260, 523–6.

Golley, F. B. (1972). Energy flux in ecosystems. In *Ecosystem Structure and Function*, ed. J. A. Wiens, pp. 69–90. Corvallis, OR: Oregon University Press.

Golluscio, R. A. & Sala, O. E. (1993). Plant functional types and ecological strategies in Patagonian forbs. *Journal of Vegetation Science*, 4(6), 839–46.

Gorham, E. (1991). Northerm peatlands: role in the carbon cycle and probable responses to climatic warming. *Ecological Applications*, 1, 182–195.

Gorissen, A., van Ginkel, J. H., Keurentjies, J. J. B. & van Veen, J. A. (1995). Grass root decomposition is retarded when grass has been grown under elevated CO_2. *Soil Biology and Biochemistry*, 27, 117–21.

Goudriaan, J. (1995). Global carbon cycle and carbon sequestration. In *Carbon Sequestration in the Biosphere*, ed. M. Beran, pp. 1–18, NATO ASI Series 133. Berlin: Springer.

Goudriaan, J., van de Geijn, S. C. & Ingram, J. S. I. (1994). *GCTE Focus 3 Wheat Modelling and Experimental Data Comparison Workshop Report*. 16 pp. Oxford: GCTE Focus 3 Office.

Goudriaan, J. & Zadoks, J. C. (1995). Global climate change: modelling the potential responses of agro-ecosystems with special reference to crop protection. *Environmental Pollution*, 87(2), 215–24.

Goulden, M. J., Munger, J. W., Fan, S.-M., Daube, B. C. & Wofsy, S. C. (1996). Exchange of CO_2 by a deciduous forest: response to interannual climate variability. *Science*, 271, 1576–8.

Grace, J., Malhi, Y., Lloyd, J., McIntyre, J., Miranda, A. C., Meir, P. & Miranda, H. S. (1996). The use of eddy covariance to infer the net carbon dioxide uptake of Brazilian rain forest. *Global Change Biology*, 2(3), 209–17.

Grant, R. F., Garcia R. L., Pinter, P. J., Hunsaker, D., Wall, G. W., Kimball, B. A. & LaMorte, R. L. (1995*a*). Interaction between atmospheric CO_2 concentration and water deficit on gas exchange and crop growth: testing of ecosystems with data from the Free Air CO_2 Enrichment (FACE) experiment. *Global Change Biology*, **1**, 443–54.

Grant, R. F., Kimball, B. A., Pinter Jr, P. J., Wall, G. W., Garcia, R. L., LaMorte, R. L. & Hunsaker, D. J., (1995b). Energy exchange between the wheat ecosystem and the atmosphere under ambient vs. elevated atmospheric CO_2 concentrations: testing of the model *ecosys* with data from the free air CO_2 enrichment (FACE) experiment. *Agronomy Journal*, **87**, 446–57.

Grasshof, C., Dijkstra, P., Nonhebel, S., Schapendonk, H. C. M. & Van de Geijn, C. (1995). Effects of climate change on productivity of cereals and legumes: model evaluation of observed year-to-year variability of CO_2 response. *Global Change Biology*, **1**, 417–28.

Green, D. G. (1989). Simulated effects of fire, dispersal and spatial pattern on competition within forest mosaics. *Vegetatio*, **82(2)**, 139–53.

Gregory, P. J. & Ingram, J. S. I. (1996). The contribution of the Soil Organic Matter Network (SOMNET) to GCTE. In *Evaluation of Soil Organic Matter Models*, ed. D. S. Powlson, P. Smith & J. U. Smith, pp. 13–26. NATO ASI Series, Vol. I 38, Heidelberg: Springer-Verlag.

Gregory, S. V., Swanson, F. J., McKee, W. A. & Cummins, K. W. (1991). An ecosystem perspective of riparian zones. *BioScience*, **41**, 540–50.

Grime, J. P. (1977). Evidence for the existence of three primary strategies in plants and its relevance to ecological and evolutionary theory. *American Naturalist*, **111**, 1169–94.

Grime, J. P. (1979). *Plant Strategies and Vegetation Processes*. Chichester: Wiley.

Grime, J. P., Hodgson, J. G., Hunt, R., Thompson, K., Hendry, G. A. F., Campbell, B. D., Jalili, A., Hillier, S. H., Diaz, S. & Burke, M. J. W. (1997). Functional types: testing the concept in Northern England. In *Plant Functional Types: Their Relevance to Ecosystem Properties and Global Change*, ed. T. M. Smith,

H. H. Shugart & F. I. Woodward, pp. 122–50. Cambridge: Cambridge University Press. IGBP Book Series Vol. 1.

Grootes, P. M., Stuiver, M., White, W. C., Johnsen, S. & Jouzel, J. (1993). Comparison of oxygen isotope records from the GISP2 and GRIP Greenland ice cores. *Nature*, **366**, 552–4.

Grulke, N. E., Riechers, G. H., Oechel, W. C., Hjelm, U. & Jaeger, C. (1990). Carbon balance in tussock tundra under ambient and elevated atmospheric CO_2. *Oecologia*, **83**, 485–94.

Gunderson, C. A. & Wullschleger, S. C. (1994). Photosynthetic acclimation in trees to rising atmospheric CO_2: A broader perspective. *Photosynthesis Research*, **39**, 369–88.

Gustafson, E. J. & Gardner, R. H. (1996). The effect of landscape heterogeneity on the probability of patch colonization. *Ecology*, **77**, 94–107.

Gutierrez, A. P. (1996). Applied population ecology: a supply-demand approach. New York: John Wiley & Sons.

Gwynn-Jones, D., Bjorn, L. O., Callaghan, T. V., Gehrke, C., Johanson, U., Lee, J. A. & Sonesson, M. (1996). Effects of enhanced UV-B radiation and elevated concentrations of CO_2 on a subarctic heathland. In *Carbon Dioxide, Populations, and Communities*, ed. Ch. Körner & F. A. Bazzaz, pp. 197–207. San Diego: Academic Press.

Hadley, P., Batts, G. R., Ellis, R. H., Morison, J. I. L., Pearson, S., Wheeler, T. R. (1995). Temperature gradient chambers for research on global environment change: II. A twin-wall tunnel system for low-stature, field-grown crops using a split heat pump. *Plant, Cell and Environment*, **18**, 1055–63.

Hagen, L. J. (1991). A wind erosion prediction system to meet user needs. *Journal of Soil and Water Conservation*, **46**, 106–11.

Hairston, N. G., Smith, F. E. & Slobodkin L. B. (1960). Community structure, population control and competition. *American Naturalist*, **94**, 421–25.

Hall, F. G., Sellers, P. J. & Williams, D. L. (1996). Initial results from the boreal ecosystem–atmosphere experiment, BOREAS. *Silvia Fennica*, **30**, 109–21.

Halliwell, D.H., Apps, M.J. & Price, D.T. (1995). A survey of the forest site characteristics in a transect through the central Canadian boreal forest. *Water, Air and Soil Pollution*, **82**, 257–70.

Hänninen, H. (1995). Assessing ecological implications of climatic change: can we rely on our simulation models? *Climatic Change*, **31**(1), 1–4.

Hanski, I., Clobert, J. & Reid. W. (1995*a*). Ecology of extinctions. In *Global Biodiversity Assessment*, ed. UNEP, pp. 232–45 Cambridge: Cambridge University Press.

Hanski, I., Pakkala, T., Kuussaari, M. & Lei, G. (1995*b*). Metapopulation persistence of an endangered butterfly in a fragmented landscape. *Oikos*, **72**, 21–8.

Hanson, J. D., Parton, W. J. & Innis, G. S. (1985) Plant growth and production of garssland ecosystems: a comparison of modelling approaches. *Ecological Modelling*, **29**, 131–44.

Happel, J. D. & Chanton, J. P. (1993). Carbon remineralisation in a north Florida swamp forest: effects of water level on the pathways and rates of soil organic matter decomposition. *Global Biogeochemical Cycles*, **7**, 475–90.

Harrington, R. & Stork, N. E. (1995). *Insects in a Changing Environment*. 535 pp. London: Academic Press.

Harrison, K. & Broekner, W. (1993). A strategy for estimating the impact of CO_2 fertilisation on soil carbon storage. *Global Biogeochemical Cycles*, **7**, 69–80.

Harte, J. & Shaw, R. (1995). Shifting dominance within a montane vegetation community: results of a climate-warming experiment. *Science*, **267**, 876–80.

Harte, J., Torn, M. S., Chang, F.-R., Feifarek, B., Kinzig, A. P., Shaw, R. & Shen, K. (1995). Global warming and soil microclimate: results from a meadow-warming experiment. *Ecological Applications*, **5**, 132–50.

Hastings, D. (1996). The global land 1-km base elevation digital elevation model: a progress report. *IGBP Newsletter 1996*, (27), 11–12.

Hättenschwiler, S. & Körner, C. (1996). System-level adjustments to elevated CO_2 in model spruce ecosystems. *Global Change Biology*, **2**, 377–87.

Hättenschwiler, S., Miglietta, F., Raschi, A. & Körner, Ch. (1997). Thirty years of in situ tree growth under elevated CO_2 a model for future forest responses. *Global Change Biology*, **3**, 463–71.

Haumaier, L. & Zech, W. (1995) Black carbon – possible source of highly aromatic components of soil humic acids. *Organic Geochemistry*, **23**, 191–96.

Havström, M., Callaghan, T. V. & Jonasson, S. (1993). Differential growth responses of *Cassiope tetragona*, an arctic dwarf-shrub, to environmental perturbations among three contrasting high-and sub-arctic sites. *Oikos*, **66**, 389–402.

Haxeltine, A. & Prentice, I. C. (1996*a*). A general model for the light use efficiency of primary production. *Functional Ecology*, **10**, 551–61.

Haxeltine, A. & Prentice, I. C. (1996*b*). BIOME3: an equilibrium biosphere model based on ecophysiological constraints, resource availability and competition among plant functional types. *Global Biogeochemical Cycles*, **10**, 693–709.

Haxeltine, A., Prentice, I. C. & Creswell, I. D. (1996). A coupled carbon and water flux model to predict vegetation structure. *Journal of Vegetation Science*, **7**(5), 651–66.

Heagle, A. S., Body, D. E. & Neeley, G. E. (1974). Injury and yield responses of soybean to chronic doses of ozone and sulfur dioxide in the field. *Phytopathology*, **64**, 132–6.

Heal, O. W., Struwe, S. & Kjller, A. (1996). Diversity of soil biota and ecosystem function. In *Global Change and Terrestrial Ecosystems*, ed. B.H. Walker & W.L. Steffen, pp.385–402. Cambridge: Cambridge University Press. IGBP Book Series No. 2.

Heck, W. W., Taylor, O. C. & Tingey, D. T. (1988). *Assessment of Crop Loss from Air Pollutants*. London: Elsevier Applied Science.

Henderson-Sellers, A. (1993). Continental vegetation as a dynamic component of a global climate model – a preliminary assessment. *Climatic Change,* **23**(4), 337–77.

Henderson-Sellers, A. R., Dickinson, R. E., Dirbidge, T. B., Kennedy, P. J., McGuffie, K. & Pitman, A. J. (1993). Tropical deforestation: Modeling local- to regional-scale climate change. *Journal of Geophysical Research,* **98**, 7289–315.

Henderson-Sellers, A. & McGuffie, K. (1995). Global climate models and 'dynamic' vegetation changes. *Global Change Biology,* **1**(1), 63–75.

Henderson-Sellers, A., McGuffie, K. & Gross, C. (1995). The sensitivity of global climate model simulations to increased stomatal resistance and CO_2 increases. *Journal of Climate,* **8**, 1738–56.

Hengeveld, R. & Van den Bosch, F. (1997). Invading into an ecologically non-uniform area. In *Past and Future Rapid Environmental Changes: The Spatial and Evolutionary Responses of Terrestrial Biota,* ed. B. Huntley, W. Cramer, A. V. Morgan, H. C. Prentice & J. R. M. Allen, Berlin: Springer-Verlag. NATO ASI Series Vol. 47.

Heong, K. L., Song, Y. H., Pimsamarn, S., Zhang, R. & Bae, S. D. (1995). Global warming and rice arthropod communities. In *Climate Change and Rice,* ed. S. Peng, K. T. Ingram, H. U. Neue & L. H. Ziska. Berlin: Springer-Verlag.

Heywood, V. H. & Watson, R. T. (1995). *Global Biodiversity Assessment. United Nations Environment Programme.* 1140 pp. Cambridge: Cambridge University Press.

Hibbard, J. M., Whitbread, R. & Farrar, J. F. (1994). Elevated atmospheric CO_2 concentrations and powdery mildew of barley. *Abstract of Papers Presented at the Proceedings of BSPP Climate Change Conference,* Newport, UK.

Hildebrand, E. (1994). Deposition as site factor for forest stands in Southwest Germany. *Nova Acta Leopoldina,* **288**, 418–23.

Hillier, S. H., Sutton, F. and Grime, J. P. (1994). A new technique for the experimental manipulation of temperature in plant communities. *Functional Ecology,* **8**, 755–62.

Hirschel, G., Körner, Ch. & Arnone, J. A. III. (1997). Will rising atmospheric CO_2 affect litter quality and in situ decomposition rates in native plant communities? *Oecologia,* **110**, 387–92.

Hobbie, S. E. (1992). Effects of plant species on nutrient cycling. *Trends in Ecology and Evolution,* **7**, 336–9.

Hobbie, S. E. (1995). Direct and indirect effects of plant species on biogeochemical processes in arctic ecosystems. In *Arctic and Alpine Biodiversity: Patterns, Causes and Ecosystems Consequences,* ed. F. S. Chapin, III & C. Körner, pp. 213–224. Berlin: Springer-Verlag.

Hobbs, R. J. (1989). The nature and effects of disturbance relative to invasions in. *Biological Invasions: A Global Perspective,* ed. J. A. Drake, H. A. Mooney, F. di Castri, R. H. Groves, F. J. Kruger, M. Rejmanek & M. Williamson, pp. 389–401. Chichester, UK: John Wiley & Sons.

Hobbs, R. J. & Mooney, H. A. (1991). Effects of rainfall variability and gopher disturbance on serpentine annual grassland dynamics. *Ecology,* **72**, 59–68.

Hochberg, M. E., Menaut, J.-C. & Gignoux, J. (1994). The influences of tree biology and fire in the spatial structure of the West African savannah. *Journal of Ecology,* **82**(2), 217–226.

Hogg, E. D., Lieffers V. J. & Wein, R. W. (1992). Potential carbon losses from peat profiles: effects of temperature, drought cycles and fire. *Ecological Applications,* **2**, 298–306.

Holden, J., Peacock, J. & Williams, T. (1993). Genes, crops and the environment. Cambridge: Cambridge University Press.

Holdridge, L. R. (1947). *Life Zone Ecology.* San Josè, Costa Rica: Tropical Science Center.

Holling, C. S., Peterson, G., Marples, P., Sendzimir, J., Redford, K., Gunderson, L. & Lambert, D. (1996). Self-organization in ecosystems: lumpy geometries, periodicities and morphologies. In *Global Change and Terrestrial Ecosystems,* ed. B. Walker & W. Steffen, pp.346–84.Cambridge: Cambridge University Press. IGBP Book Series Vol. 2.

Hollinger, D. Y., Kelliher, F. M., Schulze,

E.-D., Bauer, G., Arneth, A., Byers, J. N., Hunt, J. E., McSeveny, T. M. Kobak, K. I., Milyokova, I., Sogachev, A., Tatarinov, F., Verlagin, A., Ziegler, W. & Vygodskaya, N. N. (1997). Forest–atmosphere carbon dioxide exchange in eastern Siberia, *Agriculture and Forest Meteorology* (in press).

Hollinger, D. Y., Kelliher, F. M., Schulze, E.-D., Vygodskaya, N. N., Varlagin, A., Milyukova, I., Byers, J. N., Sogachov, A., Hunt, J.E., McSeveny, T. M., Kobak, K.I., Bauer, G. & Arneth, A. (1995). Initial assessment of multi-scale measures of CO_2 and H_2O flux in the Siberian taiga. *Journal of Biogeography*, **22**, 425–31.

Horie, T. (1993). Predicting the effects of climatic variation and effect of CO_2 on rice yield in Japan. *Journal of Agricultural Meteorology (Tokyo)*, **48**, 567–74.

Horie, T., Nakagawa, H., Ohnishi, M. & Nakno, J. (1995). Rice production in Japan under current and future climates. In *Modeling the Impact of Climate Change on Rice Production in Asia*, ed. R. B. Matthews, M. J. Kropff, D. Bachelet & H. H. van Laar, pp. 143–64. CAB International/IRRI.

Horn, H. S. (1975). Markovian properties of forest succession. In *Ecology and Evolution in Communities*, ed. M. L. Cody & J. M. Diamond, pp. 196–211. Cambridge: Harvard University Press.

Horn, R. (1993). The effect of aggregation of soils on water, gas, and heat transport. In *Flux Control in Biological Systems*, ed. E.-D. Schulze, pp. 335–64. San Diego: Academic Press.

Hornung, M. & Reynolds, B. (1995). The effects of natural and anthropogenic environmental changes on ecosystem processes at the catchment scale. *Trends in Ecological Evolution*, **10**, 443–49.

Hornung, M. (1992). The European network of catchments organized for research on ecosystems (ENCORE). In *Responses of Forest Ecosystems to Environmental Changes*, ed. A.Teller, P. Mathy, J. N. R. Jeffers, pp. 315–24. London: Elsevier Applied Science.

Hossain, M. (1997). Rice supply and demand in Asia: a socioeconomic and biophysical analysis. In *Applications of Systems Approaches at the Farm and Regional Levels, Vol. 1*, ed, P. S. Teng, M. J. Kropff, H. F. M. ten Berge, J. B. Dent, F. P. Lansigan & H. H. van Laar, pp. 263–79. Dordrecht, The Netherlands: Kluwer Academic Publishers.

Houghton, J. T., Filho, L. G. M., Callander, B. A., Harris, N., Kattenberg, A. & Maskell, K. eds. (1996). *Climate Change 1995. The Science of Climate Change*, 572 pp. Cambridge: Cambridge University Press.

Houghton, R. A. (1993). Is carbon accumulating in the North Temperate zone? *Global Biogeochemical Cycles*, **7**, 611–7.

Houghton, R. A. (1994). The worldwide extent of land-use change. *BioScience*, **44**, 305–13.

Houghton, R. A. (1995). Land-use change and the global carbon cycle. *Global Change Biology*, **1**, 275–87.

Hudson, R. J. M., Gherini, S. A. & Goldenstein, R. A. (1994). Modelling the global carbon cycle. Nitrogen fertilisation of the terrestrial biosphere and the 'missing' CO_2 sink. *Global Biogeochemical Cycles*, **8**, 307–33.

Hulme, M. (ed.)(1996). Climate change and southern Africa: an exploration of some potential impacts and implications for the SADC region, 104 pp. Norwich, UK: CRU/WWF.

Hummel, J. R. & Reck, R. A. (1979). A global surface albedo model. *Journal of Applied Meteorology*, **18**, 239–53.

Humphries, S. E., Groves, R. H. & Mitchell, D. S. (1991). *Plant Invasions of Australian Ecosystems*. Australian National Parks and Wildlife Service, Endangered Species Program Project. Report: 58.

Hungate, B. A., Jackson, R. B., Field, C. B. & Chapin, F. S. III. (1996). Field CO_2 enrichment experiments lack statistical power to detect changes in soil carbon. *Plant and Soil*, **187**, 135–45.

Hungate, B., Chapin, F. S., III, Zhong, H., Holland, E. A. & Field, C. B. (1997). Stimulation of grassland nitrogen cycling under elevated CO_2. *Oecologia*, (in press).

Hunt Jr., E. R., Piper, S. C., Nemani, R.,

Keeling, C. D., Otto, R. D. & Running, S. W. (1996). Global net carbon exchange and intra-annual atmospheric CO_2 concentrations predicted by an ecosystem process model and three-dimensional atmospheric transport model. *Global Biogeochemical Cycles*, **10**(3), 431–56.

Hunt, H. W., Trlica, M. J., Redente, E. F. (1991). Simulation model for the effects of climate change on temperate grassland ecosystems. *Ecological Modelling*, **53**, 205–46

Hunt, L. A. (1994a). Users Manual for the Cropsim Wheat Simulation Model. Crop Science Publication No. LAH-02-94, 170 pp. Guelph, Ontario, Canada: University of Guelph.

Hunt, L. A. (1994b). Data requirements for crop modelling. In *Crop Modelling and Related Environmental Data, A Focus on Application for Arid and Semiarid Regions in Developing Countries*, ed. P. F. Uhlir & G. C. Carter, pp. 15–26. Paris: Codata.

Hunt, L. A. & Boote, K. J. (1997). Data for model operation, calibration and validation. In *The IBSNAT Decade*. Dordrecht: Kluwer.

Hunt, L. A., Jones, J. W., Ritchie, J. T. & Teng, P. S. (1990). Genetic coefficients for the IBSNAT Crop Models. In *Proceedings of the IBSNAT Symposium: Decision Support System for Agrotechnology Transfer*, pp. 15–29. Dept. of Agronomy and Soil Science, College of Tropical Agriculture and Human Resources, Univ. of Hawaii, Honolulu, Hawaii.

Hunt L. A., Jones J. W., Tsuji, G. Y. & Uehara, G. (1994). A minimum data set for field experiments. In *Application in Modelling in the Semi-Arid Tropics*, pp. 27–33. Paris: CoData, ICSU.

Hunt, L. A. & Pararajasingham, S. (1995). Cropsim-Wheat: A model describing the growth and development of wheat. *Canadian Journal of Plant Science*, **75**, 619–32.

Hunt, L. A., Pararajasingham, S. & Wiersma, J. V. (1996). Effects of planting date on the development and yield of spring wheat: simulation of field data. *Canadian Journal of Plant Science*, **76**, 51–8.

Huntley, B. & Birks, H. J. B. (1983). *An Atlas of Past and Present Pollen Maps for Europe 0–13 000 Years Ago*. Cambridge: Cambridge University Press.

Huntley, B., Cramer, W., Morgan, A. V., Prentice, H. C. & Allen, J. R. M. (1997). Past and future rapid environmental changes: the spatial and evolutionary responses of terrestrial biota – Introduction. In *Past and Future Rapid Environmental Changes: The Spatial and Evolutionary Responses of Terrestrial Biota*. ed. B. Huntley, W. Cramer, A. V. Morgan, H. C. Prentice & J. R. M. Allen. Berlin: Springer-Verlag. NATO ASI Series Vol. 47.

Huston, M. (1997). Hidden treatments in ecosystem experiments: re-evaluating the ecosystem function of biodiversity. *Oecologia*, **110**, 449–60.

Huston, M. & Smith, T. (1987). Plant succession: life history and competition. *American Naturalist*, **130**, 168–98.

Hutchinson, M. F. (1995). Stochastic space-time weather models from ground-based data. *Agricultural and Forest Meteorology*, **73**(3–4), 237–64.

Hutchinson, M. F. & Bischof, R. J. (1983). A new method for estimating the spatial distribution of mean seasonal and annual rainfall applied to the Hunter Valley, New South Wales. *Australian Meteorological Magazine*, **31**, 179–84.

Hutchinson, M. F. & Gessler, P. E. (1994). Splines – more than just a smooth interpolator. *Geoderma*, **62**, 45–67.

Idso, K. E. & Idso, S. B. (1994). Plant responses to atmospheric CO_2 enrichment in the face of environmental constraints: a review of the past 10 years' research. *Agricultural and Forest Meteorology*, **69**, 153–203.

Idso, S. B. & Kimball, B. A. (1992). Effects of atmospheric carbon dioxide enrichment on photosynthesis, respiration, and growth of sour orange trees. *Plant Physiology*, **99**, 341–43.

Idso, S. B. & Kimball, B. A. (1993). Tree growth in carbon dioxide enriched air and its implications for global carbon cycling and maximum levels of CO_2. *Global Biogeochemical Cycles*, **7**, 537–55.

Ingestad, T. & Ågren, G. I. (1988). Nutrient

uptake and allocation at steady-state nutrition. *Physiologia Plantarum*, **72**, 450–59.

Ingram, J. S. I. (1995). *Report of the GCTE 'Rice Ecosystems' Workshop*, IRRI, Los Banos, 21–22 March 1994. 29 pp.

Ingram, J. (1996). *GCTE Focus 3 Wheat Network*, Report No 2. 257 pp. Canberra, Australia: GCTE Core Project Office.

Ingram, J. S. I. (1997). Food security in the face of global change: the GCTE Rice Network as a framework for international collaborative research. *Journal of Agricultural Meteorology of Japan*, **52**, 759–67.

Ingram, J., Lee, J. & Valentin, C. (1996). The GCTE Soil Erosion Network. A multi-participatory research programme. *Journal of Soil and Water Conservation*, **51**, 377–80.

Inouye, R. S. & Tilman, D. (1988). Convergence and divergence of old-field plant communities along experimental nitrogen gradients. *Ecology*, **69**, 995–1004.

INPE (Instituto Nacional de Pesquisas Espaciais) (1992). *Deforestation in Brazilian Amazonia*. 3 pp. Brazil: INPE Publication.

International Rice Research Institute (1995). *SARP Research Reports*.

IPCC (1994). *Climate Change 1994: Radiative Forcing of Climate Change and an Evaluation of the IPCC IS92 Emission Scenarios*, ed. J. T. Houghton, L. G. Meira Filho, J. Bruce, Hoesung Lee, B. A. Callander, E. Haites, N. Harris & K. Maskell. 339 pp. Cambridge: Cambridge University Press.

IPCC (1995) *IPCC Guidelines for National Greenhouse Gas Inventories*. Intergovernmental Panel on Climate Change. Bracknell: Meteorological Office.

IPCC (1996a). *Climate Change 1995 Working Group II*, ed. R. T. Watson, M. C. Zinyowera & R. H. Moss. 879 pp.

IPCC (1996b). *Climate Change 1995: The Science of Climate Change*. Contribution of Working Group I to the Second Assessment Report of the Intergovernmental Panel on Climate Change, ed. J. J. Houghton, L. G. Meiro Filho, B. A. Callendaer, N. Harris, A. Kattenberg & K. Maskell, 584 pp. Cambridge: Cambridge University Press.

Jackson, R. B., Canadell, J., Ehleringer, J. R., Mooney, H. A. Sala, O. E. & Schulze, E.-D. (1996a). A global analysis of root distributions for terrestrial biomes. *Oecologia*, **108**, 389–411.

Jackson, R. B., Canadell, J., Ehleringer, J. R., Mooney, H. A., Sala, O. E., Schulze, E.-D. (1996b). Maximum rooting depth of vegetation types at the global scale. *Oecologia*, **108**, 583–95.

Jackson, R. B., Mooney, H. A. & Schulze, E.-D. (1997). A global budget for fine root biomass, surface area, and nutrient contents. *Proceedings of the National Academy of Science, USA*, **94**, 7362–66.

Jackson, R. B., Sala, O. E., Field, C. B. & Mooney, H. A. (1994). CO_2 alters water use, carbon gain, and yield for the dominant species in a natural grassland. *Oecologia*, **98**, 257–62.

Jacoby, G. C. & Darrigo, R. D. (1995). Three-ring width and density evidence of climate and potential forest change in Alaska. *Global Biogeochemical Cycles*, **9**, 227–34.

James, R. L., Cobb, F. W., Jr, Miller, P. R. & Parmeter, J. R., Jr (1980). Effects of oxidant air pollution on susceptibility of pine roots to *Fomes annosus*. *Phytopathology*, **70**, 560–3.

Jamieson, P. D., Brooking, I. R., Porter, J. R. & Wilson, D. R. (1995). Prediction of leaf appearance in wheat: a question of temperature. *Field Crops Research*, **41**, 35–44.

Jamieson, P. D., Porter, J. R., Goudriaan, J., Ritchie, J. T., van Keulen, H. & Stol, W. (1998a). A comparison of the models AFRCWHEAT2, CERES-Wheat, Sirius, SUCROS2 and SWHEAT with measurements from wheat grown under drought. *Field Crops Research*, **55**, 23–44.

Jamieson, P. D., Semenov, M. A., Brooking, I. R. & Francis, G. S. (1998b). Sirius: a mechanistic model of wheat response to environmental variation. *European Journal of Agronomy* (in press).

Jarvis, P. G. (1987). Water and carbon fluxes in ecosystems. *Ecological Studies*, **61**, 50–67.

Jarvis, P. G. (1995a). The role of temperate trees and forests in CO_2 fixation. *Vegetatio*, **121**, 157–74.

Jarvis, P. G. (1995*b*). The likely impact of rising CO₂ and temperature on European forests. ECOCRAFT – Final Report. Edinburgh: University of Edinburgh.

Jarvis, P. G. & McNaughton, K. G. (1986). Stomatal control of transpiration: scaling up from leaf to region. *Advances in Ecological Research*, **15**,1–49.

Jefferies, R. L. & Maron, J. L. (1997). The embarrassment of riches: atmospheric deposition of nitrogen and community and ecosystem processes. *Trends in Ecology and Evolution*, **12**, 74–8.

Jeffers, J. N. R. (1993). Problems of statistics in ecology. In *Design and Execution of Experiments on CO₂ Enrichment*. Ecosystem Research Report 6, ed. E.-D Schulze & H. A. Mooney, pp. 117–126. Brussels: Commission of the European Communities.

Jenkins, A. & Wright, R. F. (1995). The CLIMEX project: performance of the experimental facility during the first year of treatment. In *Ecosystem Manipulation Experiments: Scientific Approaches, Experimental Design, and Relevant Results*, ed. A. Jenkins, R. C. Ferrier & C. Kirby. Ecosystem Research Report 20, pp. 323–27, Brussels: Commission of the European Communities.

Jenkins, P. T. (1996). Free trade and exotic species introductions. *Conservation Biology*, **10**, 300–2.

Jenny, H. (1980). *The Soil Resources: Origin and Behavior*. New York: Springer-Verlag.

Johnson, W.C. & Adkisson, C. S. (1985). Dispersal of beech nuts by blue jays in fragmeted landscapes. *American Midland Naturalist*, **113**, 319–24.

Jones, H. G. (1992). *Plants and Microclimate*. Cambridge: Cambridge University Press.

Jones, J. W. & Kropff, M. J. (1996). *Developing Concepts and Methods for ICASA Systems Tools and Applications*. ICASA.

Jones, M. B., Jongen, M. & Doyle, T. (1996). Effects of elevated carbon dioxide concentrations on agricultural grassland production. *Agricultural and Forest Meteorology*, **79**, 243–52.

Jones, P. D., Raper, S. C. B., Cherry, B. S.

G., Goodess, C. M. & Wigley, T. M. (1986). *A Grid Point Surface Air Temperature Data Set for the Southern Hemisphere, 1851–1984*. 73 pp. U.S. Department of Energy, Carbon Dioxide Research Division. TR027.

Jones, P. D., Raper, S. C. B., Santer, B. D., Cherry, B. S. G., Goodess, C. M., Kelly, P. M., Wigley, T. M., Bradley, R. S. & Diaz, H. F. (1985). *A Grid Point Surface Air Temperature Data Set for the Northern Hemisphere*, 251 pp. U.S. Department of Energy, Carbon Dioxide Research Division. TR022.

Jordan, T. E. & Weller, D. E. (1996). Human contributions to terrestrial nitrogen flux. *BioScience*, **4**, 655–64.

Kane, D. L., Hinzman, L. D., Woo, M. & Everett, K. R. (1992). Arctic hydrology and climate change. In *Arctic Ecosystems in a Changing Climate: An Ecophysiological Perspective*, ed. F. S. Chapin III, R. L. Jefferies, J. F. Reynolds, G. R. Shaver & J. Svoboda, pp. 35–57. San Diego: Academic Press.

Kartschall, Th., Grossman, S., Wechsung, F., Poschenrieder, W. & Graefe, J. (1996). *Investigation of the Climate Change Impacts on Agricultural Ecosystems*, Potsdam, Germany: PIK.

Kasischke, E. S., Christensen N. L. & Stocks., B. J. (1995). Fire, global warming, and the carbon balance of boreal forests. *Ecological Applications*, **5**, 437–51.

Kasischke, E.S., Melack, J.M. & Dobson, M.C. (1997). The use of imaging radars for ecological applications – a review. *Remote Sensing of Environment*, **59**, 141–56.

Kattenberg, A., Giorgi, F., Grassl, H., Meehl, G. A., Mitchell, J. F. B., Stouffer, R. J., Tokioka, T., Weaver, A. J. & Wigley, T. M. L. (1996). Climate models – projections of future climate. In *Climate Change 1995 – The Science of Climate Change*, ed. J. T. Houghton, L. G. Meira Filho, B. A. Callander, N. Harris, A. Kattenberg & K. Maskell, pp. 285–357.Cambridge: Cambridge University Press.

Kauppi, P. E., Miellikäinen, K. & Kuusela, K. (1992). Biomass and carbon budget of

European forests, 1971 to 1990. *Science*, **256**, 70–4.

Keeling, C. D., Chin, J. F. S. & Whorf, T. P. (1996*a*). Increased activity of northern vegetation inferred from atmospheric CO_2 measurements. *Nature*, **382**, 146–9.

Keeling, R. F., Piper, S. C. & Heimann, M. (1996*b*). Global and hemispheric CO_2 sinks deduced from changes in atmospheric O_2 concentration. *Nature*, **381**, 218–21.

Keeling, R. F. & Shertz, S. R. (1992). Seasonal and interannual variations in atmospheric oxygen and implications for the global carbon cycle. *Nature*, **358**, 723–27.

Kelliher, F. M, Hollinger, D. Y., Schulze, E.-D, Vygodskaya, N. N., Byers, J. N., Hunt, J. E., McSeveny, T.M., Milukova, I., Sorgachov, A., Varlagin, A., Ziegler, W., Arneth, A. & Bauer, G. (1997). Evaporation from eastern Siberian larch Forest. *Agricultural and Forest Meteorology*, **85**, 135–47.

Kelliher, F. M., Leuning R., Raupach, M. R. & Schulze, E. -D. (1995). Maximum conductances for evaporation from global vegetation types. *Agricultural and Forest Meteorology*, **73**, 1–16.

Kelliher, F. M., Leuning, R. & Schulze, E.-D. (1993). Evaporation and canopy characteristics of coniferous forests and grasslands. *Oecologia*, **95**, 153–63.

Kemp, P. R., Waldecker, D. G., Owensby, C. E., Reynolds, J. F. & Virginia, R. A. (1994). Effects of elevated CO_2 and nitrogen fertilization pretreatments on decomposition on tallgrass prairie leaf litter. *Plant and Soil*, **165**, 115–27.

Kennedy, A. D. (1994). Simulated climate change: a field manipulation study of polar microarthropod community response to global warming. *Ecography*, **17**, 131–40.

Kennedy, A. D. (1995). Simulated climate change: are passive greenhouses a valid microcosm for testing the biological effects of environmental perturbations? *Global Change Biology*, **1**, 29–42.

Kerr, Y.H. & Njoku, E.G. (1993). On the use of passive microwaves at 37 GHz in remote sensing of vegetation. *International Journal of Remote Sensing*, **14**(10),

1931–43.

Kheshgi, H. S., Jain, A. K. & Weubbles, D. J. (1996). Accounting for the missing carbon sink with the CO_2 fertilisation effect. *Climatic Change*, **33**, 31–62.

Kielland, K. (1990). Processes controlling nitrogen release and turnover in arctic tundra. Ph.D. thesis, University of Alaska, Fairbanks.

Kienast, F. (1991). Simulated effects of increasing atmospheric CO_2 and changing climate on the successional characteristics of alpine forest ecosystems. *Landscape Ecology*, **5**(4), 225–38.

Kimball, B. A., Pinter, P. J., Jr, Garcia, R. L., LaMorte, R. L., Wall, G. W., Hunsaker, D. J., Wechsung, G., Wechsung, F. & Karschall, T. (1995). Productivity and water use of wheat under free-air CO_2 enrichment. *Global Change Biology*, **1**, 429–42.

Kindermann, J., Lüdeke, M. K. B., Badeck, F.-W., Otto, R. D., Klaudius, A., Häger, C., Würth, G., Lang, T., Dönges, S., Habermehl, S. & Kohlmaier, G. H. (1993). Structure of a global and seasonal carbon exchange model for the terrestrial biosphere – The Frankfurt Biosphere Model (FBM). *Water, Air and Soil Pollution*, **70**(1–4), 675–684.

King, G. A. & Herstrom, A. A. (1997). Holocene tree migration rates objectively determined from fossil pollen data. In *Past and Future Rapid Environmental Changes: The Spatial and Evolutionary responses of Terrestrial Biota*, ed. B. Huntley, W. Cramer, A. V. Morgan, H. C. Prentice & J. R. M. Allen. Berlin: Springer-Verlag. NATO ASI Series Vol. 47.

King, G. and Schnell, S. (1994). Effect of increasing atmospheric methane concentration on ammonium inhibition of soil methane consumption. *Nature*, **370**, 282–4.

Kingsolver, J. G. (1989). Weather and the population dynamics of insects: integrating physiological and population ecology. *Physiological Zoology*, **62**, 314–34.

Kirschbaum, M. U. F. (1996). Ecophysiological, ecological, and soil processes in terrestrial ecosystems: a primer on general concepts and relationships. In *Climate Change 1995. Impacts, Adaptations and Mitigation of Climate Change: Scientific-Technical Analyses*, ed. R. T. Watson, M. C. Zinyowera, R. H. Moss & D. J. Dokken, pp. 57–74. Cambridge: Cambridge University Press.

Kirschbaum, M. U. F., Bullock, P., Evans, J. R., Goulding, K., Jarvis, P. G., Noble, I. R., Rounsevell, M. & Sharkey, T. D. (1996a). Ecophysiological, ecological and soil processes in terrestrial ecosystems: a primer on general concepts and relationships. In *Climate Change 1996: Impacts, Adaptations and Mitigation of Climate Change: Scientific-Technical Analysis*. ed. R. T. Watson, M. C., Zinyowera, R. H. Moss & D. J. Dokken, pp. 57–74. Cambridge: Cambridge University Press.

Kirschbaum, M. U. F., Fischlin, A., Cannell, M. G. R., Cruz, R. V. O., Cramer, W., Alvarez, A., Austin, M. P., Bugmann, H. K. M., Booth, T. H., Chipompha, N. W. S., Cisela, W. M., Eamus, D., Goldammer, J. G., Henderson-Sellers, A., Huntley, B., Innes, J. L., Kaufmann, M. R., Kräuchi, N., Kile, G. A., Kokorin, A. O., Körner, C., Landsberg, J., Linder, S., Leemans, R., Luxmoore, R. J., Markham, A., McMurtrie, R. E., Neilson, R. P., Norby, R. J., Odera, J. A., Prentice, I. C., Pitelka, L. F., Rastetter, E. B., Solomon, A. M., Stewart, R., van Minnen, J., Weber, M. & Xu, D. (1996b). Climate change impacts on forests. In *Climate Change 1995 – Impacts, Adaptations and Mitigation of Climate Change: Scientific-Technical Analyses*, ed. R. T. Watson, M. C. Zinyowera & R. H. Moss, pp. 95–129. Cambridge: Cambridge University Press.

Kirschbaum, M. U. F., King, D. A., Comins, H. N., McMurtrie, R. E., Raison, R. J., Pongracic, S., Murty, D., Keith, H., Medlyn, B. E., Khanna, P. K. & Sheriff, D. W. (1994). Modelling forest response to increasing CO_2 concentration under nutrient-limited conditions. *Plant, Cell and Environment*, **17**,1081–99.

Kittel, T. G. F., Rosenbloom, N. A., Painter, T. H., Schimel, D. S., Melillo, J. M., Pan, Y. D., Kicklighter, D. W.,

McGuire, A. D., Neilson, R. P., Chaney, J., Ojima, D. S., McKeown, R., Parton, W. J., Pulliam, W. M., Prentice, I. C., Haxeltine, A., Running, S. W., Pierce, L. L., Nemani, R. R., Hunt, E. R., Smith, T. M., Rizzo, B. & Woodward, F. I. (1995). The VEMAP integrated database for modelling United States ecosystem/vegetation sensitivity to climate change. *Journal of Biogeography*, **22**, 857–62.

Kleidon, A. & Heimann, M. (1998). A method of determining rooting depth from a terrestrial biosphere model and its impacts on the water and carbon cycle. *Global Change Biology*, **4**, 275–86.

Klein Goldewijk, C. G. M. & Battjes, J. J. (1995). *The IMAGE 2 Hundred Year (1890–1900) Data Base of the Global Environment (HYDE)*.

Klein Goldewijk, K. & Leemans R. (1995). Systems models of terrestrial carbon cycling. In *Carbon Sequestration in the Biosphere*, ed. M Beran, pp. 129–139. NATO ASI Series 133. Berlin: Springer.

Klein Goldewijk, K., Van Minnen, J. G., Kreileman, G. J. J., Vloedbeld, M. & Leemans, R. (1994). Simulating the carbon flux between the terrestrial environment and the atmosphere. *Water, Air and Soil Pollution*, **76**(1–2), 199–230.

Koch, G. (1993). The use of natural situations of CO_2 enrichment in studies of vegetation responses to increasing atmospheric CO_2. In *Design and Execution of Experiments on CO_2 Enrichment*. Ecosystems Research Report 6, ed. E.-D. Schulze & H. A. Mooney, pp. 381–92, Brussels: Commission of the European Communities.

Koch, G. W. & Mooney, H. A. (1996). *Carbon Dioxide and Terrestrial Ecosystems*. San Diego: Academic Press.

Koch, G. W., Scholes, R. J., Steffen, W. L., Vitousek, P. M. & Walker, B. H. (1995b). *The IGBP Terrestrial Transects: Science Plan*. IGBP Report 36. Stockholm: International Geosphere-Biosphere Programme.

Koch, G. W., Vitousek, P. M., Steffen, W. L. & Walker, B. H. (1995a). Terrestrial transects for global change research. *Vegetatio*, **121**, 53–65.

Kolchugina, T. P. & Vinson, T. S. (1993). Carbon sources and sinks in forest biomes of former Soviet Union. *Global Biogeochemical Cycles*, **7**, 291–304.

Körner, C. (1993). CO_2 fertilization: The great uncertainty in future vegetation development. In *Vegetation Dynamics and Global Change*, ed. A. M. Solomon & H. H. Shugart, pp.53–70. New York: Chapman & Hall.

Körner, C. (1994). Leaf diffusive conductances in the major vegetation types of the globe. In *Ecophysiology of Photosynthesis*, ed. E.-D Schulze & M. M. Caldwell, Ecological Studies Vol. 100. pp. 463–90. Berlin, Heidelberg, New York: Springer.

Körner, Ch. (1996). The responses of complex multispecies systems to elevated CO_2. In *Global Change and Terrestrial Ecosystems*, ed. B. Walker & W. Steffen, pp. 20–42. Cambridge: Cambridge University Press.

Körner, Ch. & Arnone, J. A., III. (1992). Responses to elevated carbon dioxide in artificial tropical ecosystems. *Science*, **257**, 1672–5.

Körner, Ch. & Bazzaz, F. A. (1996). *Carbon Dioxide, Populations, and Communities*. 465 pp. San Diego: Academic Press.

Körner, Ch. & Würth, M. (1996). A simple method for testing leaf responses of tall tropical forest trees to elevated CO_2. *Oecologia*, **107**, 421–5.

Körner, Ch., Diemer, M., Schäppi, B., Niklaus, P. & Arnone, J. (1997). The response of alpine grassland to four seasons of CO_2 enrichment: a synthesis. *Acta Oecologia – International Journal of Ecology*, **18**(3), 165–75.

Köstner, B. M. M., Schulze, E.-D., Kelliher, F. M., Hollinger, D. Y., Byers, J. N., Hunt, J. E., McSeveny, T. M., Meserth, R. & Weir, P. L. (1992). Transpiration and canopy conductance in a pristine broad-leaved forest of Nothofagus: an analysis of xylem sap flow and eddy correlation measurements. *Oecologia*, **91**, 350–9.

Kot, M., Lewis, M. A. & Van den Driessche, P. (1996). Dispersal data and the spread of invading organisms. *Ecology*, **77**, 2027–42.

Kotanen, P. M. (1995). Responses of vegetation to a changing regime of disturbance: effects of feral pigs on a California coastal prairie. *Ecography*, **18**, 190–9.

Kozlowski, T. T. & Pallardy, S. G. (1997). *Physiology of Woody Plants*. San Diego: Academic Press.

Kramer, P. J. & Boyer, J. S. (1995). *Water Relations of Plants and Soils*. San Diego: Academic Press.

Kropff, M. J. & van Laar, H. H. eds. (1993). *Modelling Crop–Weed Interactions*. 274 pp. Wallingford, England: CAB International.

Küchler, A. W. (1949). A physiognomic classification of vegetation. *Annals of the Association of American Geographers*, **39**, 201–10.

Kuhlbusch, T. A. J., Andreae, M. O., Cachier, H., Goldammer, J. G., Lacaux, J.-P., Shea, R. & Crutzen, P. J. (1996). Black carbon formation by savanna fires: measurements and implications for the global carbon cycle. *Journal of Geophysical Research*, **101**, 23651–65.

Kullman, L. (1996). Norway spruce present in the Scandes Mountains, Sweden at 8000 BP: new light on Holocene tree spread. *Global Ecology and Biogeography Letters*, **5**, 94–101.

Kurz, W. A., Apps, M. J., Beukema, S. J. & Lekstrum, T. (1995). 20th century carbon budget of Canadian forests. *Tellus*, **47B**, 170–7.

Kutschera, L (1960). *Wurzelatlas mitteleuropäischer Ackerunkräuter und Kulturpflanzen*. Frankfurt DLG: Verlag.

Kutzbach, J. E. & Guetter, P. J. (1986). The influence of changing orbital parameters and surface boundary conditions on climatic simulations for the past 18 000 years. *Journal of the Atmospheric Sciences*, **43**, 1726–59.

Kutzbach, J. E., Bonan, G., Foley, J. A. & Harrison, S. P. (1996). Vegetation and soil feedbacks on the response of the African monsoon to orbital forcing in the early to middle Holocene. *Nature*, **384**, 623–6.

LaBau, V. J. (1996) A preliminary test of the U.S. national forest health monitoring procedures in the boreal forest of Alaska. Paper presented at the International

Boreal Forest Research Assocation (IBFRA) Conference, Saint Petersburg, Russia, 19–23 August 1996.

Lachenbruch, A. H. & Marshall, B. V. (1986). Climate change: geothermal evidence from permafrost in the Alaskan arctic. *Science*, **34**, 689–96.

Lambers, H. & Poorter, H.(1992). Inherent variation in growth rate between higher plants: a search for physiological causes and ecological consequences. *Advances in Ecological Research*, **23**, 187–261.

Lambers, H., Stulen, I. & Van der Werf, A. (1996). Carbon use in root respiration as affected by elevated atmospheric CO_2. *Plant and Soil*, **187**, 251–63.

Landsberg, J. J. & Hingston, F. J. (1996). Evaluating a simple radiation/dry matter conversion model using data from *Eucalyptus globulus* plantations in Western Australia. *Tree Physiology*, **16**, 801–08.

Landsberg, J. J. & Waring, R. H. (1997). A generalised model of forest productivity using simplified concepts of radiation-use efficiency, carbon balance and partitioning. *Forest Ecology and Management*, **95**(3), 209–28.

Lane, L. J. & Nearing, M. A. (1989). *USDA-Water Erosion Prediction Project: Profile Model Documentation*. NSERL. 2.

Langan, S. J., Ewers, F. W. & Davis, S. D. (1997). Xylem disfunction caused by water stress and freezing in two species of co-occurring chaparral shrubs. *Plant, Cell and Environment*, **20**(4), 425–37.

Lange, O. L., Weikert, R. M., Wedler, M., Gebel, J. & Heber, U. (1989). Photosynthese und nährstoffversorgung von fichten aus einem waldschadensgebiet auf basenarmem untergrund. *Allg. Forst Zeitschrift*, **3**, 54–63.

Larigauderie, A. & Körner, Ch. (1995). Acclimation of leaf dark respiration to temperature in alpine and lowland plant species. *Annals of Botany*, **76**, 245–52.

Last, F. T. & Watling, R. (1991). *Acid Deposition. Its Nature and Impacts*, 340 pp. Edinburgh: The Royal Society of Edinburgh.

Lauenroth, W. K. (1979). Grassland primary production: North American grasslands in perspective. In *Perspectives in Grassland Ecology*, ed. N. R. French. New York: Springer-Verlag. Ecological Studies Vol. 32.

Lauenroth, W. K. (1996). Application of patch models to examine regional sensitivity to climate change. *Climatic Change*, **34**(2), 155–60.

Lauenroth, W. K. & Sala, O. E. (1992). Long term forage production of North American shortgrass steppe. *Ecological Applications*, **2**(4), 397–403.

Lauenroth, W. K., Coffin, D. P., Burke, I. C. & Virginia, R. A. (1997). Interactions between demographic and ecosystem processes in a semi-arid and an arid grassland: a challenge for plant functional types. In *Plant Functional Types: Their Relevance to Ecosystem Properties and Global Change*, ed. T. M. Smith, H. H. Shugart & F. I. Woodward, pp.234–54. Cambridge: Cambridge University Press. IGBP Book Series Vol. 1.

Lauenroth, W. K., Dodd, J. L. & Sims, P. L. (1978). The effects of water- and nitrogen-induced stresses on plant community structure in a semiarid grassland. *Oecologia*, **36**, 211–22.

Lavorel, S., Gardner, R. H. & O'Neill, R. V. (1993). Analysis of patterns in hierarchically structured landscapes. *Oikos*, **67**(3), 521–8.

Lavorel, S., Gardner, R. H. & O'Neil, R. V. (1995). Dispersal of annual plants in hierarchically structured landscapes. *Landscape Ecology*, **10**, 277–89.

Lavorel, S., McIntyre, S., Landsberg, J. & Forbes, T. D. A. (1997). Functional attributes underlying species traits. *Trends in Ecology and Evolution*, **12**, 474–8.

Lawton, J. H. & Brown, V. K. (1993). Redundancy in ecosystems. In *Biodiversity and Ecosystem Function*, ed. E.- D. Schulze & H. A. Mooney, pp. 255–70 Berlin, Heidelberg, New York: Springer-Verlag.

Lawton, J. H. & Jones. C. G. (1995). Linking species and ecosystems: organisms as ecosystem engineers. In *Linking Species and Ecosystems*, ed. C. G. Jones & J. H. Lawton, pp. 141–50. New York: Chapman and Hall.

LBA Science Planning Group (1996). *The Large Scale Biosphere–Atmosphere Ex-*

periment in Amazonia (LBA). Concise Experimental Plan. LBA Coordinating Office, Centro de Previsão de Tempo e Estudos Climáticos, Instituto Nacional de Pespquisas Espaciais, Cachoeira Paulista, SP, Brazil.

Le Houerou, H. N. (1989). *The Grazing Land Ecosystems of the African Sahel.* 282 pp. Berlin, New York: Springer-Verlag.

Le Houerou, H. N. (1993). Changements climatiques et desertisation. *Secheresse*, **4**, 95–111.

Leadley, P. W. & Körner Ch. (1996). Effects of elevated CO_2 on plant species dominance in a highly diverse calcareous grassland. In *Carbon Dioxide, Populations, and Communities*, ed. Ch. Körner & F. A. Bazzaz, pp. 159–75. San Diego: Academic Press.

Leakey, R. & Lewin, R. (1995). *The Sixth Extinction: Patterns of Life and the Future of Human Kind.* 271 pp. New York: Doubleday.

Lean, J. & Warrilow, D. A. (1989) Climatic impact of Amazon deforestation. *Nature*, **342**, 311–13.

Lebel, L. & Steffen, W. (1998). *Human Driving Forces of Environmental Change in Southeast Asia and the Implications for Sustainable Development: An Integrated SARCS Study.* Science Plan. Southeast Asian Regional Committee for START (SARCS) (in press).

Leemans, R. (1997). The use of plant functional type classifications to model global land cover and simulate the interactions between the terrestrial biosphere and the atmosphere. In *Plant Functional Types: Their Relevance to Ecosystem Properties and Global Change*, ed. T. M. Smith, H. H. Shugart & F. I. Woodward, pp. 289–316. Cambridge: Cambridge University Press. IGBP Book Series Vol. 1.

Leemans, R. & Cramer, W. (1991). The IIASA database for mean monthly values of temperature, precipitation and cloudiness of a global terrestrial grid. International Institute for Applied Systems Analysis (IIASA), RR–91–18.

Leemans, R., Cramer, W. & Van Minnen, J. G. (1996). Prediction of global biome distribution using bioclimatic equilibrium models. In *Global Change: Effects on Coniferous Forests and Grasslands*, ed. A. I. Breymeyer, D. O. Hall, J. M. Melillo & G. I. Ågren, pp. 413–50, Chichester, New York, Brisbane, Toronto, Singapore: SCOPE, John Wiley & Sons. Vol. 56.

Leemans, R. & Prentice, I. C. (1989). *FORSKA, A General Forest Succession Model*. Department of Plant Ecology, Uppsala University. Meddelanden från Växtbiologiska Institutionen, Uppsala Universitet: 89/2.

Leemans, R. & Solomon, A. M. (1993). The potential response and redistribution of crops under a doubled CO_2 climate. *Climate Research*, **3**(1–2), 79–96.

Leemans, R. & Van den Born, G. J. (1994). Determining the potential distribution of vegetation, crops and agricultural productivity. *Water, Air and Soil Pollution*, **76**(1–2), 133–61.

Legates, D. R. & Willmott, C. J. (1990). Mean seasonal and spatial variability in global surface air temperature. *Theoretical and Applied Climatology*, **41**, 11–21.

Leishman, M. R., Hughes, L., French, K., Armstrong, D. & Westoby, M. (1992). Seed and seedling biology in relation to modelling vegetation dynamics under global climate change. *Australian Journal of Botany*, **40**, 599–613.

Lekkerkerk, L. J. A., van de Geijn, S. C. & Van Veen, J. A. (1990). Effects of elevated atmospheric CO_2-levels on the carbon economy of a soil planted with wheat. In *Soils and the Greenhouse Effect*, ed. A. F. Bouwman, pp. 423–29. Chichester: John Wiley & Sons.

Leuning, R. (1995). A critical appraisal of a combined stomatal-photosynthesis model for C-3 plants. *Plant, Cell and Environment*, **18**(4), 339–55.

Leuning, R., Kelliher, F. M., DePury, D. G. G. & Schulze E.-D. (1995) Leaf nitrogen, photosynthesis, conductance and transpiration: scaling from leaves to canopies. *Plant, Cell and Environment*, **18**, 1183–200.

Lewin, K. F., Hendrey, G. R. & Kolber, Z. (1992) Brookhaven National Laboratory free-air carbon dioxide enrichment facility. *Critical Reviews in Plant Sciences*,

11, 135–41.

Li, J. & Wang, G. (1993). *Collection of Research Results in Yaojinzi Natural Preserve of* Aneurolepidium chinese *Grassland*. The Natural Preserve Information Compiling Committee, Jilin, China (in Chinese).

Li, W. H., Deng, K. M. & Li, F. (1981). A study on the biomass production of the main ecosystems in Changbai Mountain. In *Forest Ecosystem Research*, Vol. 2, ed. Z. Wang, pp. 34–50. Changbai Forest Ecosystem Research Station of the Chinese Academy of Sciences, Shenyang (in Chinese).

Li, X. (1995). Modelling the responses of vegetation in North East China Transect to global change. *Journal of Biogeography*, 22, 515–22.

Lieth, H. (1964). Versuch einer kartographischen Darstellung der Produktivität der Pflanzendecke der Erde. In *Geographisches Taschenbuch*. pp.72–80. Wiesbaden: Franz Steiner Verlag.

Lieth, H. (1973). Primary production: terrestrial ecosystems. *Human Ecology*, 1, 303–32.

Lightfoot, C. J. & Norval, R. A. (1981). Tick problems in wildlife in Zimbabwe. 1. The effects of tick parasitism on wild ungulates. *South African Journal of Wildlife Research*, 11, 41–5.

Likens, G. E., Discoll, C. T. & Buso D. C. (1996). Long-term effects of acid rain: response and recovery of a forest ecosystem. *Science*, 272, 244–6.

Lin, G., Phillips, S. L. & Ehleringer, J. R. (1996). Monsoonal precipitation responses of shrubs in a cold desert community on the Colorado Plateau. *Oecologia*, 106, 8–17.

Linder, S. (1987). Responses to water and nutrition in coniferous ecosystems. In *Potentials and Limitations of Ecosystem Analysis. – Ecological Studies*, 61, ed. E.-D. Schulze & H. Zwölfer, pp. 180–202. Berlin, Heidelberg, New York: Springer-Verlag.

Linder, S. (1995). Foliar analysis for detecting and correcting nutrient imbalances in Norway spruce. *Ecological Bulletins (Copenhagen)*, 44, 178–90.

Linder, S., Benson, M. L., Myers, B. J. & Raison, R. J. (1987). Canopy dynamics and growth of *Pinus radiata*. I. Effects of irrigation and fertilisation during a drought. *Canadian Journal of Forest Research*, 10, 1157–65.

Linder, S., McMurtrie, R. E. & Landsberg, J. J. (1996). Global change impacts on managed forests. In *Global Change and Terrestrial Ecosystems*, ed. B. H. Walker & W. L Steffen, pp. 275–90. Cambridge: Cambridge University Press. IGBP Book Series No. 2.

Lindroth, R. L. (1996). Consequences of elevated atmospheric CO_2 for forest insects. In *Carbon Dioxide, Populations, and Communities*, ed. Ch. Körner & F. A. Bazzaz, pp. 347–61. San Diego: Academic Press.

Lloyd, J. & Farquhar, G. D. (1994). [13]C discrimination by the terrestrial biosphere. *Oecologia*, 99, 201–15.

Lloyd, J. & Farquhar, G. D. (1996). The CO_2 dependence of photosynthesis, plant growth responses to elevated atmospheric CO_2 concentrations and their interaction with plant nutrient status. *Functional Ecology*, 10, 4–32.

Lloyd, J., Kruit, B., Hollinger, D. Y., Grace, J., Francey, R. J., Wong, S.-C., Kelliher, F. M., Miranda, A. C., Gash, J. H. C, Vygodskaya, N. N., Wright, I. R., Miranda, H. S., Farquhar, G. D. & Schulze, E.-D. (1996). Vegetation effects on the isotopic composition of atmospheric CO_2 at local and regional scales: theoretical aspects and a comparison between rain forest in Amazonia and a boreal forest in Siberia. *Australian Journal of Plant Physiology*, 23, 371–99.

Lo Seen, D., Mougin, E. S. R., Gaston, A. & Hiernaux, P. (1995). A regional Sahelian grassland model to be coupled with multispectral satellite data. II: Toward the control of its simulations by remotely sensed indices. *Remote Sensing of Environment*, 52, 194–206.

Loehle, C. (1995). Anomalous responses of plants to CO_2 enrichment. *Oikos*, 73, 181–7.

Loehle, C. & Leblanc, D. (1996). Model-based assessments of climate change effects on forests – a critical review. *Ecological Modelling*, **90**(1), 1–31.

Long, S. P., Jones, M. B. & Roberts, M. J. (eds) (1992). *Primary Productivity of Grass Ecosystems of the Tropics and Subtropics*. London: Chapman and Hall.

Lonsdale, D. & Gibbs, J. N. (1994). Effects of climate change on fungal diseases of trees. In *Fungi and Environmental Change*, ed. J. C. Frankland, N. Magan & G. M. Gadd, pp. 1–19. Cambridge: Cambridge University Press.

Lonsdale, W. M., Miller, I. L. & Forno, I. W. (1989). The biology of Australian weeds. 20. *Mimosa pigra* L. *Plant Protein Quarterly*. **4**, 119–31.

Lonsdale, W. M., Farrell, G. & Wilson, C. G. (1995). Biological control of a tropical weed: a population model and experiment for *Sida acuta*. *Journal of Applied Ecology*, **32**, 391–99.

Loveland, T. R., Merchant, J. W., Ohlen, J. F. & Brown, J. F. (1991). Development of a land-cover characteristics database for the conterminous US. *Photogrammetric Engineering and Remote Sensing*, **57**, 1453–63.

Lüdeke, M. K. B., Badeck, F.-W., Otto, R. D., Häger, C., Dönges, S., Kindermann, J., Würth, G., Lang, T., Jäkel, U., Klaudius, A., Ramge, P., Habermehl, S. & Kohlmaier, G. H. (1994). The Frankfurt Biosphere Model. A global process oriented model for the seasonal and long-term CO_2 exchange between terrestrial ecosystems and the atmosphere. I. Model description and illustrative results for cold deciduous and boreal forests. *Climate Research*, **4**(2), 143–66.

Lüdeke, M. K. B., Ramge, P. H. & Kohlmaier, G. H. (1996). The use of satellite NDVI data for the validation of global vegetation phenology models: application to the Frankfurt Biosphere Model. *Ecological Modelling*, **91**, 255–70.

Lükewille, A. & Wright, R. F. (1997). Experimentally increased soil temperture causes release of nitrogen at a boreal forest catchment in southern Norway. *Global Change Biology*, **3**, 13–21.

Luo, Y. Q. (1996). Sensitivity of leaf photosynthesis to CO_2 concentration is an invariant function for C3 plants: a test with experimental data and global applications. *Global Biogeochemical Cycles*, **10**, 209–22.

Luo, Y. & Mooney, H. A. (1998). *Carbon Dioxide and Environmental Stress*. San Diego: Academic Press (in press).

Luo, Y., TeeBest, D. O. & Fabellar, N. G. (1995). Simulation studies on risk analysis of rice leaf blast epidemics associated with global climatic change in several Asian countries. *Journal of Biogeography*, **22**, 673–8.

Lutz, W. (ed.) (1994). *Population–Development–Environment*. 400 pp. Berlin: Springer-Verlag.

Lutz, W. (Ed.) (1996). *The Future Projections of the World: What Can We Assume Today?* Earthscan.

MacDonald, I. A. W. (1992). Global change and alien invasions: implications for biodiversity and protected area management. In *Biodiversity and Global Change* ed. O. T. Solbrig, H. M. Van Emden, & P. G. W. J. Van Oort, IUBS Monographs Vol. No. 8, pp.197–207. Paris: The International Union of Biological Sciences (IUBS).

Mack, R. N. (1986). Alien plant invasion into the Intermountain West: A case history. In *Ecology of Biological Invasions of North America and Hawaii*, ed. H. A. Mooney, & J. A. Drake, pp. 191–212. New York: Springer-Verlag.

Madronich, S. McKenzie, R. L., Caldwell, M. M. & Bjorn, L. O. (1995). Changes in ultraviolet radiation reaching the earth's surface. *Ambio*, **24**, 143–52.

Maisongrande, P., Ruimy, A., Dedieu, G. & Saugier, B. (1995). Monitoring seasonal and interannual variations of gross primary productivity, net primary productivity and net ecosystem productivity using a diagnostic model and remotely-sensed data. Tellus Series B – *Chemical and Physical Meteorology* 47B(1–2), 178–90.

Malanson, G. P. (1984). Intensity as a third factor of disturbance regime and its effect on species diversity. *Oikos*, **43**, 411–13.

Malanson, G. P. & Armstrong, M. P. (1996). Dispersal probability and forest diversity in a fragmented landscape. *Ecological Modelling*, **87**(1–3), 91–102.

Manning, W. J. & Tiedemann, A. V. (1995). Climate change: potential effects of increased atmospheric carbon dioxide (CO_2), ozone (O_3), and ultraviolet-B (UV-B) radiation on plant diseases. *Journal of Environmental Pollution*, **88**, 219–45.

Marks, G. C. & Smith, I. W. (1991). The cinnamon fungus in Victorian forests. *Victorian Lands & Forests*, Bulletin 31, 33 pp.

Marschner, H. (1995). *Mineral Nutrition of Higher Plants*. London: Academic Press.

Marshall, P. R., McCall, D. G. & Johns, K. L. (1991). Stockpol: A decision support model for livestock farms. *Proceedings of the New Zealand Grassland Association*, **53**, 137–40.

Martin, P. (1992). EXE – a climatically sensitive model to study climate change and CO_2 enhancement effects on forests. *Australian Journal of Botany*, **40**(4–5), 717–35.

Martin, P. H. & Guenther, A. B. (1995). Insights into the dynamics of forest succession and non-methane hydrocarbon trace emissions. *Journal of Biogeography*, **22**, 493–500.

Matthews, E. (1983). Global vegetation and land use: New high-resolution data bases for climate studies. *Journal of Climate and Applied Meteorology*, **22**, 474–87.

Matthews, R. B., Horie, T., Kropff, M. J., Bachelet, D., Centeno, H. G., Shin, J. C., Mohandass, S., Singh, S., Defeng, Z. & Moon Hee Lee. (1995). A regional evaluation of the effect of future climate change on rice production in Asia. In *Modeling the Impact of Climate Change on Rice Production in Asia*, ed. R. B. Matthews, M. J. Kropff, D. Bachelet & H. H. van Laar, pp. 95–139. CAB International/IRRI.

Mauser, W., Schädlich, S. & Wege, C. (1997). Modeling the spatial distribution of evapotranspiration using remote sensing data and PROMET. *Journal of Hydrology* (in press).

Maxwell, B. (1992). Arctic climate: potential for change under global warming. In *Arctic Ecosystems in a Changing Climate: an*

Ecophysiological Perspective, ed. F. S. Chapin, R. L. Jefferies, J. F. Reynolds, G. R. Shaver & J. Svoboda, pp. 11–34. San Diego: Academic Press.

May, R. M. (1992). Past efforts and future prospects towards understanding how many species there are. In *Biodiversity and Global Change*, ed. O. T. Solbrig, H. M van Emden, & P. G. W. J. van Oordt, Wallingford, UK: CAB International.

Mayeux, H. S., Johnson, H. B., Polley, H. W., Dumesnil, M. J. & Spanel, G. A. (1993). A controlled environment chamber for growing plants across a subambient carbon dioxide. *Functional Ecology*, **7**, 125–33.

McBride, J. R., Semion, V. P. & Miller, P. R. (1975). Impact of air pollution on the growth of Ponderosa pine. *California Agriculture*, **29**, 8–9.

McConnaughay, K. D. M., Bassow, S. L., Berntson, G. M. & Bazzaz, F. A. (1996). Leaf senescence and decline of end-of-season gas exchange in five temperate deciduous tree species grown in elevated CO_2 concentrations. *Global Change Biology*, **2**, 25–33.

McCown, R. L., Hammer, G. L., Hargreaves, J. N. G., Holzworth, D. P. & Freebairn, D. M. (1996). APSIM: a novel software system for model development, model testing and simulation in agricultural systems research. *Agricultural Systems*, **50**, 255–71.

McGuire, A. D., Melillo, J. M. & Joice, L. A. (1995a). The role of nitrogen in the response of forest net primary production to elevated atmospheric carbon dioxide. *Annual Review of Ecology and Systematics*, **26**, 473–503.

McGuire, A. D., Melillo, J. M., Kicklighter, D. W. & Joyce, L. A. (1995b). Equilibrium responses of soil carbon to climate change – empirical and process-based estimates. *Journal of Biogeography*, **22**(4–5), 785–96.

McKane, Rastetter, E. B. Melillo, J. M., Shaver, G. R., Hopkinson, C. S., Fernandes, D. N., Skole, D. L. & Chomentowski, W. H. (1993). Effects of global change on carbon storage in tropical forests of South America. *Global*

Biogeochemical Cycles **9**, 329–50.

McKeon, G. M., Howden, S. M., Abel, N. O. J. & King, J. M. (1993). Climate change: adapting tropical and sub-tropical grasslands. *Proceedings of XVIIth International Grasslands Congress,* Palmerston North, New Zealand, pp. 1181–90.

McLeod, A. R. & Skeffington, R. A. (1995). The Liphook Forest fumigation project: an overview. *Plant, Cell and Environment,* **18**, 327–35.

McMurtrie, R. E. (1993). Modelling of canopy carbon and water balance. In *Photosynthesis and Production in a Changing Environment: a Field and Laboratory Manual,* ed. D. O. Hall, J. M. O. Scurlock, H. R Bothar-Nordenkampf, R. C. Leegood and S. P. Long, pp. 220–31. London: Chapman & Hall.

McMurtrie, R. E., Benson, M. L., Linder, S., Running, S. W., Talsma, T., Crane, W. J. B. & Myers, B. J. (1990a). Water–nutrient interactions affecting the productivity of stands of *Pinus radiata*. *Forest Ecology and Management,* **30**, 415–23.

McMurtrie, R. E. & Comins, H. N. (1996). The temporal response of forest ecosystems to doubled atmospheric CO_2 concentration. *Global Change Biology,* **2**, 49–57.

McMurtrie, R. E., Comins, H. N., Kirschbaum, M. U. F. & Wang, Y.-P. (1992a). Modifiying existing forest growth models to take account of direct effects of elevated CO_2. *Australian Journal of Botany,* **40**, 657–77.

McMurtrie, R. E., Gholz, H. L., Linder, S. & Gower, S. T. (1994). Climatic factors controlling the productivity of pine stands: a model-based analysis. *Ecological Bulletins (Copenhagen),* **43**, 173–88.

McMurtrie, R. E., Leuning, R., Thompson, W.A. & Wheeler, A.M. (1992b). A model of canopy photosynthesis and water use incorporating a mechanistic formulation of leaf CO_2 exchange. *Forest Ecology and Management,* **52**, 261–78.

McMurtrie, R. E., Rook, D. A. & Kelliher, F. M. (1990b). Modelling the yield of *Pinus radiata* on a site limited by water and nitrogen. *Forest Ecology and Management,* **30**, 381–413.

McNaughton, S. J. (1977). Diversity and stability of ecological communities: a comment on the role of empiricism in ecology. *The American Naturalist,* **111**, 515–25.

McNaughton, S. J. (1985). Ecology of a grazing ecosystem: the Serengeti. *Ecological Monographs,* **53**, 259–94.

McNaughton, S. J. (1988). Mineral nutrition and spatial concentrations of African ungulates. *Nature,* **334**, 343–5.

McNaughton, S. J. (1993). Biodiversity and function in grazing ecosystems. *Ecological Studies,* **99**, 361–84.

Meeson, B. W., Corpew, F. E., McManus, J. M. P., Myers, D. M., Closs, J. W., Sun, K.-J., Sunday, D. J. & Sellers, P. J. (1995). ISLSCP Initiative I–Global Data Sets for Land-atmosphere Models, 1987–1988. CD-ROM. NASA.

Meinzer, F. C. (1993). Stomatal control of transpiration. *Trends in Ecology and Evolution,* **8**, 289–94.

Melillo, J. M., Aber, J. D. & Muratore, J. F. (1982). Nitrogen and lignin control of hardwood leaf litter decomposition dynamics. *Ecology,* **63**, 621–6.

Melillo, J. M., McGuire, A. D., Kicklighter D. W., Moore, B., III, Vörösmarty, C. J. & Schloss, A. L. (1993). Global climate change and terrestrial net primary production. *Nature,* **363**, 234–40.

Melillo, J. M., Newkirk, K. M., Catricala, C. E., Steudler, P. A., Aber, J. B., Nadelhofer, K. J. & Boone, R. D. (1996a). The soil warming experiment at Harvard Forest 1991–1996. *Bulletin of The Ecological Society of America,* **77**, 300.

Melillo, J. M., Prentice, I. C., Farquar, G. D., Shulze, E.-D. & Sala, O. E. (1996b). Terrestrial biotic responses to environmental change and feedbacks to climate. In *Climate Change 1995. The Science of Climate Change,* ed. J. T. Houghton L. G. M. Filho, B. A. Callander, N. Harris, A. Kattenberg, & K. Maskell, pp. 445–81. Cambridge: Cambridge University Press.

Menaut, J. C., Flouzat, G. & Valentin, C. (1995). Modélisation, cartographie et télédétection. Programme SALT (CNRS, ORSTOM, CNES). In *Ecosystèmes & Changements Globaux. Les dossiers de l'Environnement de l'INRA,* No. 8, pp.

261–79.

Menaut, J.-C., Gignoux, J., Prado, C. & Clobert, J. (1990). Tree community dynamics in a humid savanna of the Côte-d'Ivoire: modelling the effects of fire and competition with grass and neighbours. *Journal of Biogeography*, **17**, 471–81.

Miglietta, F., Raschi, A., Bettarini, I. & van Gardingen, P. (1993). Carbon dioxide springs and their use for experimentation. In *Design and Execution of Experiments on CO₂ Enrichment*. Ecosystems Research Report, ed. E.-D. Schulze & H. A. Mooney, pp. 393–406. Brussels: Commission of the European Communities.

Miller, P. R. & Elderman, M. J. (1977). *Photochemical Oxidant Air Pollutant Effects on a Mixed Conifer Forest Ecosystem: A Progress Report, 1976*. Environmental Research Laboratory, EPA report no. EPA-600/3-77-104. Corvallis: U.S. Environmental Protection Agency.

Miller, P. R., Taylor, O. C. & Wilhour, R. G. (1982). *Oxidant Air Pollution Effects on a Western Coniferous Forest Ecosystem*. Environmental Research Laboratory; EPA report no. EPA-600/D-82-276. Corvallis: U. S. Environmental Protection Agency.

Mitchell, J. F. B., Johns, T. C., Gregory, J. M. & Tett, S. F. B. (1995). Climate response to increasing levels of greenhouse gases and sulphate aerosols. *Nature*, **376**, 501–04.

Mitchell, P. L. (1996). Comparison of five models of rice yield showing the effects of change in temperature and in carbon dioxide concentration. In *Report of the GCTE Rice Network Experimentation Planning Workshop*, ed. J. S. I. Ingram. GCTE Working Document, 19.

Mohandass, S., Kareem, A. A., Ranganathan, T. B. & Jeyaraman, S. (1995). Rice production in India under current and future climates. In *Modeling the Impact of Climate Change on Rice Production in Asia*, ed. R. B. Matthews, M. J. Kropff, D. Bachelet & H. H. van Laar, pp. 165–81. CAB International/IRRI.

Mohr, H. & Muentz, K. (1994). The terrestrial nitrogen cycle as influenced by man. *Nova Acta Leopoldina*, **70**, 418–23.

Mohren, G. M. J., Bartelink, H. H., Jorritsma, I. T. M. & Kramer, K. (1993). A process-based growth model (FORGRO) for analysis of forest dynamics in relation to environmental factors. In *Forest Reserves Workshop*, ed. M. Broekmeijer, W. Vos, & H. G. J. M. Koop, pp. 273–80. Wageningen, The Netherlands, May 6–8, 1992: PUDOC, Wageningen.

Molina, M. J. & Rowland, F. S. (1974). Stratospheric sink for chlorofluoromethanes: chlorine atom-catalysed destruction of ozone. *Nature*, **249**, 810–12.

Monsi, M. (1960). Dry-matter reproduction in plants. I. Schemata of dry-matter reproduction. *Botanical Magazine*, **73**, 81–90.

Monteith, J. L & Ingram, J. S. I. (1998). Climate variability and crop growth. In *Proceedings of the International Crop Science Congress*, New Delhi, November 1996. Delhi: Oxfore & IBH Publishing Co.

Monteith, J.L. & Unsworth, M.H. (1990). *Principles of Environmental Physics*. London: Edward Arnold.

Mooney, H. A. (1996). Biological invasions and global change. In *Proceedings, Norway/UN Conference on Alien Species*, ed. O. T. Sandlund, P. J. Schei & A. Viken, pp. 123–5. Trondheim, Norway: Directorate for Nature Management.

Mooney, H. A., Lubchenko. J., Dirzo. R. & Sala O. E. (1995). Biodiversity and ecosystem functioning: ecosystem analyses. In *Global Biodiversity Assessment*. ed. V. H. Heywood & R. T. Watson, pp. 275–326. Cambridge: Cambridge University Press.

Mooney, H. A., Medina, E., Schindler, D. W, Schulze, E.-D. & Walker B. H. (1991). *Ecosystem Experiments*. Vol. 45, 268 pp. Chichester: J. Wiley and Sons.

Morgan, R. P. C., Quinton, J. N., Smith, R. E., Govers, G., Poesen, J. W. A., Chisci, G. & Torri, D. (1998). The EUROSEM model. In *Modelling Soil Erosion by Water*, ed. J. Boardman & D. T. Favis-Mortlock. Berlin: Springer-Verlag, NATO-ASI Series I–55 (in press).

Morison, J. I. L. (1993). Responses of plants to CO₂ under water limited conditions.

Vegetatio, **104/105**, 193–209.

Morison, J. I. L. & Gifford, R. (1984). Plant growth and water use with limited water supply in high CO_2 concentrations. I. Leaf area, water use and transpiration. *Australian Journal of Plant Physiology*, **11**, 361–74.

Morris, D. W. (1995). Earth's peeling veneer of life. *Nature*, **373**(6509), 25.

Mott, J. J. (1985). Australian savanna ecosystems. In *Ecology and Management of the World's Savannas*. Proceedings of the International Savanna Symposium, Brisbane, May, 1984, ed. J. C. Tothill & J. J. Mott, pp. 56–164.

Mott, J. J. (1987). Patch grazing and degradation in native pastures of the tropical savannas in northern Australia. In *Grazing Lands Research at the Plant–Animal Interface*. Proceedings of the 15th International Grassland Congress (Kyoto, Japan: August 30, 1985), ed. F. P. Horn, J. Hodgson & R. W. Brougham, pp. 153–61.

Moulin, S., Kergoat, L., Viovy, N. & Dedieu, G. (1997). Global scale assessment of vegetation phenology using NOAA/AVHRR satellite measurements. *Journal of Climate*, **10**(6), 1154–70.

Mousseau, M. & Enoch, H. Z. (1989). Carbon dioxide enrichment reduces shoot growth in sweet chestnut seedlings (*Castanea sativa* Mill.). *Plant, Cell and Environment*, **12**, 927–32.

Moyle, P. B. (1996). Effects of invading species on freshwater and estuarine ecosystems. Conference on Alien Species, ed. O. T. Sandland, P. J. Schei & A. Viken. Trondheim, Norway, Directorate for Nature Management and Norwegian Institute for Nature Research, *Nature Research*, 86–92.

Mund, M. (1996). Wachstum und oberirdische biomasse von *fichtenbestaenden* (*Picea abies* (L.) Karts.) in einer periode anthropogenerstickstoffeinträge. Diplomarbeit, 106 pp. Germany: University Bayreuth.

Murray, D. F. (1980). Balsam poplar in arctic Alaska. *Canadian Journal of Anthropology*, **1**, 29–32.

Myneni, R. B., Keeling, C. D., Tucker, C. J.

Asrar, G. & Nemani, R. R. (1997). Increased plant growth in the northern high latitudes from 1981 to 1991. *Nature*, **386**, 698–702.

Nadelhoffer, K. J., Bouwman, A. F., Delaney, M., Melillo, J. M., Schafer, W., Scholes, M. C., Scholes, R. J., Sonntag, Ch., Sunda, W. G., Veldkamp, E. & Welch, E. B. (1995). Effects of climate change and human perturbations on interactions between nonliving organic matter and nutrients. In *The Role of Non-Living Organic Matter in the Earth's Carbon Cycle*, ed. R. Zepp & Ch. Sonntag, pp. 227–56. Chichester: John Wiley.

Nadelhoffer, K. J., Giblin, A. E., Shaver, G. R. & Laundre, J. A. (1991). Effects of temperature and substrate quality on element mineralization in six arctic soils. *Ecology*, **72**, 242–53.

Naeem, S., Thompson, L. J., Lawler, S. P., Lawton, J. H. & Woodfin, R. M. (1994). Declining biodiversity can alter the performance of ecosystems. *Nature*, **368**, 734–7.

Naeem, S., Thompson, L. F., Lawler, S. P., Lawton, J. H. & Wooddfin R. M. (1995). Empirical evidence that declining species diversity may alter the performance of terrestrial ecosystems. *Philosophical Transactions of the Royal Society of London, Series B*, **347**, 249–62.

NAS [National Academy of Sciences](1992). *Rethinking the Ozone Problem in Urban and Regional Air Pollution*. Washington: National Academy Press.

Navas, M. -L., Guillerm, J. -L., Fabreguettes, J. & Roy, J. (1995). The influence of elevated CO_2 on community structure, biomass and carbon balance of mediterranean old-field microcosms. *Global Change Biology*, **1**, 325–35.

Neilson, R. P. (1993). Vegetation redistribution: a possible biosphere source of CO_2 during climatic change. *Water, Air and Soil Pollution*, **70**(1–4), 659–73.

Neilson, R. P. (1995). A model for predicting continental-scale vegetation distribution and water balance. *Ecological Applications*, **5**(2), 362–85.

Neilson, R. P., King, G. A. & Koerper, G. (1992). Toward a rule-based biome model. *Landscape Ecology*, **7**(1), 27–43.

Neilson, R. P. & Running, S. W. (1996). Global dynamic vegetation modelling: coupling biogeochemistry and biogeography models. In *Global Change and Terrestrial Ecosystems*, ed. B. H. Walker & W. L. Steffen, pp. 451–65. Cambridge: Cambridge University Press. IGBP Book Series.

Nemani, R.R. & Running, S.W. (1989). Testing a theoretical climate–soil–leaf area hydrologic equilibrium of forests using satellite data and ecosystem simulation. *Agricultural and Forest Meteorology*, **44**, 245–60.

Nemani, R. & Running, S. W. (1996). Implementation of a hierarchical global vegetation classification in ecosystem function models. *Journal of Vegetation Science*, **7**(3), 337–46.

Nepstad, D. C., de Carvalho, C. R. & Davidson, E. A. (1994). The role of deep roots in the hydrological and carbon cycles of Amazonian forests and pastures. *Nature*, **372**, 666–9.

Nepstad, D. C., Uhl, C. & Serrão, E. A. S. (1991). Recuperation of a degraded Amazonian landscape: forest recovery and agricultural restoration. *Ambio*, **20**, 248–55.

Neue, H.-U. and Sass, R. L. (1994). Trace gas emissions from rice fields. In *Global Atmospheric–Biospheric Chemistry*, ed. R.G. Prinn. New York: Plenum.

Nevison, C. D., Esser, G. & Holland, E. A. (1996). A global model of changing N_2O emissions from natural and perturbed soils. *Climatic Change*, **32**, 327–78.

Newton, P. C. D., Clark, H., Bell, C. C., Glasgow, E. M., Tate, K. R., Ross, D. J., Yeates, G. W. & Saggar, S. (1995). Plant growth and soil processes in temperate grassland communities at elevated CO_2. *Journal of Biogeography*, **22**, 235–40.

Nicks, A. D. & Lane, L. J. (1989). Weather generator. In *USDA Water Erosion Prediction Project: Profile Model Documentation*, ed. L. J. Lane & M. A. Nearing, pp.1–19. West Lafayette: NSERL.

Nix, H. A. (1984). Minimum data sets for agrotechnology transfer. In *International Crops Research Institute for the Semi-Arid Tropics (ICRISAT)*, pp. 181–8. Patancheru.

Noble, I. R. & Gitay, H. (1996). A functional classification for predicting the dynamics of landscapes. *Journal of Vegetation Science*, **7**(3), 329–36.

Noble, I. R. & Slatyer, R. O. (1980). The use of vital attributes to predict successional changes in plant communities subject to recurrent disturbances. *Vegetatio*, **43**, 5–21.

Nobre, C. A., Sellers, P. J. & Shukla, J. (1991) Amazonian deforestation and regional climate change. *Journal of Climate*, **4**, 957–88.

Norby, R. J. (1994). Issues and perspectives for investigating root responses to elevated atmospheric carbon dioxide. *Plant and Soil*, **165**, 9–20.

Norby, R. J., Gunderson, C. A., Wullschleger, S. D., O'Neill, E. G. & McCracken, M. K. (1992). Productivity and compensatory responses of yellow-poplar trees in elevated CO_2. *Nature*, **357**, 322–4.

Norby, R. J., Wullschleger, S. D. & Gunderson, C. A. (1996). Tree responses to elevated CO_2 and implications for forests. In *Carbon Dioxide and Terrestrial Ecosystems*, ed. G. W. Koch & H. A. Mooney, pp. 1–21. San Diego: Academic Press.

Noss, R. F. (1983). A regional landscape approach to maintain diversity. *BioScience*, **33**, 700–6.

Noss, R. F. (1996). Conservation of biodiversity at the landscape scale. In *Biodiversity in Managed Landscapes: Theory and Practice*, ed. R. C. Szaro & D. W. Johnston, pp. 574–89. New York: Oxford University Press.

NRC (1994). *The Role of Terrestrial Ecosystems in Global Change: a Plan for Action*. Washington: National Academy Press.

NSF (National Science Foundation) (1992). *Soil Warming Experiments in Global Change Research*. The report of a workshop held in Woods Hole, Massachusetts, 27–28 September 1991. National Science Foundation, Ecosystem Studies Program, Washington, D.C.

Oberbauer, S. F. & Oechel, W. C. (1989). Maximum CO_2-assimilation rates of vascular plants on an Alaskan arctic tundra slope. *Holarctic Ecology*, **12**, 312–16.

Odada, E., Totolo, O., Stafford Smith, M. & Ingram, J. S. I. (eds.) (1996). Global change and subsistence rangelands in Southern Africa: the impacts of climatic variability and resource access on rural livelihoods. *GCTE Working Document No.20*, 99 pp. Canberra, Australia: GCTE Core Project Office.

Oechel, W. C., Cowles, S., Grulke, N., Hastings, S .J., Lawrence, B., Prudhomme, T., Riechers, G; Strain, B., Tissue, D. & Vourlitis, G. (1994). Transient nature of CO_2 fertilization in Arctic tundra. *Nature*, **371**, 500–3.

Oechel, W. C., Hastings, S. J., Vourlitis, G., Jenkins, M., Riechers, G. & Grulke, N. (1993). Recent change of Arctic tundra ecosystems from a net carbon dioxide sink to a source. *Nature*, **361**, 520–3.

Oechel, W. C. & Vourlitis, G. L. (1996). Direct effects of elevated CO_2 on Arctic plant and ecosystem function. In *Carbon Dioxide and Terrestrial Ecosystems*, ed. G. W. Koch & H. A. Mooney, pp. 163–76. San Diego: Academic Press.

Oerke, E.-C., Dehne, H.-W., Schonbeck, F. & Weber, A. (1994). *Crop Production and Crop Protection*. Amsterdam: Elsevier.

Ojima, D. S., Parton, W. J., Coughenour, M. B., Scurlock, J. M. O., Kirchner, T. B., Kittel, T. G. F., Hall, D. O., Schimel, D. S., Garcia Moya, E., Gilmanov, T. G., Seastedt, T. R., Kamnalrut, A., Kinyamario, J. I., Long, S. P., Menaut, J.-C., Sala, O. E., Scholes, R. J. & Van Veen, J. A. (1996). Impact of climate and atmospheric carbon dioxide on grasslands of the world. In *Global Change: Effects on Coniferous Forests and Grasslands*, ed. A. I. Breymeyer, D. O. Hall, J. M. Melillo & G. I. Ågren, pp. 271–311. Chichester, New York, Brisbane, Toronto, Singapore: SCOPE, John Wiley and Sons. Vol. 56.

Oldeman, L. R. (1993). Global extent of soil degradation. In *Soil Resilience and Sustainable Land Use*, ed D. J. Greenland & I. Szabolcs, pp. 99–118, Wallingford, UK: CAB International.

Oldeman, L. R., Hakkeling, R. T. A. & Sombroek, W. G. (1991). *World Map of the Status of Human-Induced Soil Degra-dation: An Explanatory Note*. 34 pp. ISRIC, Wageningen, UNEP, Nairobi.

Olson, D. & Prince, S. (1996). Global primary production data initiative update. *IGBP Newsletter 1996*, **27**, 13.

Olson, J., Watts, J. A. & Allison, L. J. (1983). *Carbon in Live Vegetation of Major World Ecosystems*. Oak Ridge National Laboratory, USA.

Olson, J., Watts, J. A. & Allison, L. J. (1985). *Major World Ecosystem Complexes Ranked by Carbon in Live Vegetation: A Database*. Carbon Dioxide Information Center, Oak Ridge National Laboratory, USA.

Olson, R. J., Scurlock, J. M. O., Cramer, W., Prince, S. D. & Parton, W. J. (1997). Global Primary Production Data Initiative Workshop: from sparse field observations to a consistent global dataset on net primary production. *IGBP-DIS Working Paper*, **16**, 23 pp.

O'Neill, E. G. (1994). Response of soil biota to elevated atmospheric carbon dioxide. *Plant and Soil*, **165**, 55–65.

O'Neill, E. G., Luxmoore, R. J. & Norby, R. J. (1987). Elevated atmospheric CO_2 effects on seedling growth, nutrient uptake, and rhizosphere bacterial populations of *Liriodendron tulipifera* L. *Plant and Soil*, **104**, 3–11.

O'Neill, R. V. (1988). Hierarchy theory and global change. In *Scales and Global Change*, ed. T. Rosswall, R. G. Woodmansee & P. G. Risser, pp. 29–45. Tennessee: John Wiley & Sons Ltd.

O'Neill, R. V., Gardner, R. H., Turner, M. G. & Romme, W. H. (1992). Epidemiology theory and disturbance spread on landscapes. *Landscape Ecology*, **7**, 19–26.

Oren, R., Schulze, E. -D., Werk, K. S. & Meyer, J. (1988). Performance of two *Picea abies* (L.) Karst. stands at different stages of decline: VII. Nutrient relations and growth. *Oecologia*, **77**, 163–74.

Oren, R., Werk, K. S., Meyer, J. & Schulze, E.-D. (1989). Potentials and limitations of forest studies on forest decline with anthropogenic pollution. In *Ecological Studies*, **77**, ed. E.-D. Schulze, O. L. Lange & R. Oren, pp. 23–36. Berlin, Heidelberg: Springer.

Oreskes, N., Shrader-Frechette, K. & Belitz,

K. (1994). Verification, validation, and confirmation of numerical models in the Earth sciences. *Science*, **263**, 641–6.

Orth, A. B., Teramura, A. H. & Sisler, H. D. (1990). Effects of ultraviolet-radiation on fungal diseases development in *Cucumis sativus*. *American Journal of Botany*, **77**, 1188–92.

Osawa, A., Maximov, T. C. & Ivanov, B. I. (1994). Forest fire history and tree growth patterns in East Siberia. In *Proceedings of the Second Symposium on the Joint Siberian Permafrost Studies between Japan and Russia in 1993*, ed. G. Inoue, pp. 159–63. National Institute for Environmental Studies, Forest and Forest Products Research Institute, and Hokkaido University, Japan.

Osterkamp, T. E. & Romanovsky, V. E. (1996). Effects of global change on permafrost. Paper presented at IGBP Northern Eurasia Study: Far East Siberian Transect Workshop, Yakutsk, Sakha Republic, Russia, 9–12 October 1996.

Overpeck, J. T., Rind, D. & Goldberg, R. (1990). Climate-induced changes in forest disturbance and vegetation. *Nature*, **343**(6253), 51–3.

Owensby, C. E. (1993). Climate change and grasslands: ecosystem-level responses to elevated carbon dioxide. *Proceedings of the* XVII *International Grassland Congress*, pp. 1119–24.

Owensby, C. E., Auen, L. M. & Coyne, P. I. (1994). Biomass production in a nitrogen-fertilised tallgrass prairie ecosystem exposed to ambient and elevated levels of CO_2. *Plant and Soil*, **165**, 105–13.

Owensby, C. E., Cochran, R. C. & Auen, L. M. (1996b). Effects of elevated carbon dioxide on forage quality for ruminants. In *Carbon Dioxide, Populations, and Communities*, ed. Ch. Körner & F. A. Bazzaz, pp. 363–71. San Diego: Academic Press.

Owensby, C. E., Coyne, P. I., Ham, J. M., Auen, L. M. & Knapp, A. K. (1993). Biomass production in a tallgrass prairie ecosystem exposed to ambient and elevated CO_2. *Ecological Applications*, **3**, 644–53.

Owensby, C. E., Ham, J. M., Knapp, A., Rice, C. W., Coyne, P. I. & Auen, L. M.

(1996a). Ecosystem-level responses of tallgrass prairie to elevated CO_2. In *Carbon Dioxide and Terrestrial Ecosystems*, ed. G. W. Koch & H. A. Mooney, pp. 147–62. San Diego: Academic Press.

Owen-Smith, R. N. (1988). Megaherbivores: the influence of very large body size on ecology. Cambridge: Cambridge University Press.

Pacala, S.W., Canham, C.D. & Silander Jr, J.A. (1993). Forest models defined by field measurements: I. The design of a northeastern forest simulator. *Canadian Journal of Forest Research*, **23**, 1980–8.

Palm, C. A. & Sanchez, P. A. (1990). Decomposition and nutrient release patterns of the leaves of three tropical legumes. *Biotropica*, **22**, 330–8.

Pan, Y., Melillo, J. M., McGuire, A. D., Kicklighter, D. W., Pitelka, L. F., Hibbard, K., Pierce, L.L., Running, S.W., Ojima, D.S., Parton, W.J., Schimel, D.S. & other VEMAP Members (1997). Response of terrestrial ecosystems to elevated atmospheric CO_2: A comparison of simulation studies among biogeochemistry models. *Oecologia*, in press.

Pararajaysingham, S. & Hunt, L. A. (1996). Effects of photoperiod on leaf appearance rate and leaf dimensions in winter and spring wheats. *Canadian Journal of Plant Science*, **76**, 43–50.

Parker, C. (1977). Prediction of new weed problems, especially in the developing world. In *Origins of Pest, Parasite, Disease, and Weed Problems*, ed. J. M. Cherrett & G. R. Sagar, pp. 249–64. Oxford: Blackwell Scientific Publications.

Parsons, A. J., Thornley, J. H. M., Newman, J. & Penning, P. D. (1994). A mechanistic model of some physical determinants of intake rate and diet selection in a two-species temperate grassland sward. *Functional Ecology*, **8**, 187–204.

Parton, W. J., Cole, C. V., Stewart, W. B., Ojima, D. S. & Schimel, D.S. (1989). Simulating regional patterns of soil C, N, and P dynamics in the U.S. central grasslands region. In *Ecology of Arable Land*, ed. M. Clarholm & L. Bergström, pp. 99–108. Kluwer Academic Publishers.

Parton, W. J., Mosier, A. R., Ojima, D.

S.,Valentine, D. W., Schimel, D. S., Weier, K. & Kulmala, A. E. (1996). Generalised model for N₂ and N₂O production from nitrification and denitrification. *Global Biogeochemical Cycles*, **10**, 401–12.

Parton, W. J., Ojima, D. S. and Cole, C. V. (1994*a*). A general model for soil organic matter dyanamics: sensitivity to litter chemistry, texture and management. *Soil Science Society of America Special Publication No. 39*, 147–67.

Parton, W. J., Ojima, D. S. & Schimel, D. S. (1994*b*). Environmental change in grasslands: assessment using models. *Climatic Change*, **28**(1–2), 111–41.

Parton, W. J., Ojima, D. S., Schimel, D. S. & Kittel, T. G. F. (1993*a*). Development of simplified ecosystems models for applications in Earth system studies: the Century experience. In *Modeling the Earth System*, ed. D. S. Ojima, pp. 281–302. Boulder, CO: UCAR/Office for Interdisciplinary Earth Studies. Global Change Institute Vol. 3.

Parton, W. J., Schimel, D. S., Cole, C. V. & Ojima, D. S. (1987). Analysis of factors controlling soil organic matter levels in Great Plains grasslands. *Soil Science Society of America Journal*, **51**, 1173–9.

Parton, W. J., Scurlock, J. M. O., Ojima, D. S., Gilmanov, T. G., Scholes, R. J., Schimel, D. S., Kirchner, T., Menaut, J. -C., Seastedt, T., Moya, E. G., Kamnalrut, A. & Kinyamario, J. I. (1993*b*). Observations and modeling of biomass and soil organic matter dynamics for the grassland biome worldwide. *Global Biogeochemical Cycles*, **7**, 785–809.

Parton, W. J., Scurlock, J. M. O., Ojima, D. S., Schimel, D. S., Hall, D. O., Coughenour, M. B., Garcia, E. M., Gilmanov, T. G., Kamnalrut, A., Kinyamario, J. I., Kirchner, T., Long, S. P., Menaut, J. -C., Sala, O. E., Scholes, R. J. & Van Veen, J. A. (1995). Impact of climate change on grassland production and soil carbon worldwide. *Global Change Biology*, **1**, 13–22.

Parton, W. J., Stewart, J. W. B. & Cole, C. V. (1988). Dynamics of C, N, P and S in grassland soils: a model. *Biogeochemistry*, **5**, 109–31.

Pastor, J. & Post, W. M. (1988). Response of northern forests to CO_2-induced climate change. *Nature*, **334**, 55–8.

Pastor, J. & Post, W. M. (1993). Linear regressions do not predict the transient responses of eastern North American forests to CO_2 induced climate change. *Climate Change*, **23**, 111–9.

Patterson, D. T. & Flint, E. P. (1990). Implications of increasing carbon dioxide and climate change for plant communities and competition in natural and managed ecosystems. In *Impact of Carbon Dioxide, Trace Gases, and Climate Change on Global Agriculture*. American Society of Agronomy Special Publication No. 53.

Pearcy, R. W, Ehleringer, J. R., Mooney, H. A. & Rundel, P. W. (1989). Plant physiological ecology, field methods and instrumentation. London, New York: Chapman and Hall.

Peng, S., Ingram, K. T., Neue, H. -U. & Ziska, L. H. (eds) (1995). *Climate Change and Rice*. Los Baños, The Philippines: IRRI.

Penman, H. L. (1948). Natural evaporation from open water, bare soil and grass. *Proceedings of the Royal Society of London, A*, **194**; 120–45.

Pereira, J. S., Linder, S., Araújo, M. C., Pereira, H., Ericsson, T., Borallho, N. & Leal, L. (1989). Optimization of biomass production in *Eucalyptus globulus* – a case study. In *Biomass Production by Fast-Growing Trees*, ed. J. S. Pereira & J. J. Landsberg, pp. 101–21. Kluwer Academic Publishers.

Pereira, J. S., Madeira, M. V., Linder, S., Ericsson, T., Tomé, M. & Araújo, M. C. (1994). Biomass production with optimized nutrition in *Eucalyptus globulus* plantations. In *Eucalyptus for Biomass Production*, ed. J. S. Periera & H. Pereira, pp. 13–30. Published by Commission of the European Communities, Brussels.

Perry, D. A. (1995). Self-organizing systems across scales. *Trends in Ecology and Evolution*, **10**(6), 241–4.

Perry, G. D., Duffy, P. B. & Miller, N. L. (1996). An extended data set of river discharges for validation of general circulation models. *Journal of Geophysical*

Research, **101**(D16), 21339–49.

Peterjohn, W. T. & Correll, D. L.(1984). Nutrient dynamics in an agricultural watershed: observations on the role of a riparian forest. *Ecology*, **65**, 1466–75.

Peterjohn, W. T., Melillo, J. M., Bowles, F. P. & Steudler, P. A. (1993). Soil warming and trace gas fluxes: experimental design and preliminary flux results. *Oecologia*, **93**, 18–24.

Peters, R. L. (1991). Consequences of global warming for biological diversity. In *Global Climate Change and Life on Earth*, ed. R. L. Wyman, pp. 99–118. New York: Routledge, Chapman and Hall.

Pettersson, R. & McDonald, A. J. S. (1992). Effects of elevated carbon dioxide concentration on photosynthesis and growth of small birch (*Betula pendula*) at optimal nutrition. *Plant, Cell and Environment*, **15**, 911–19.

Phillips, O. L. & Gentry, A. H. (1994). Increasing turnover through time in tropical forests. *Science*, **263**, 954–8.

Pickett, S. T. A. & Cadenasso, M. L. (1995). Landscape ecology: spatial heterogeneity in ecological systems. *Science*, **269**, 331–4.

Pickett, S. T. A., Kolasa, J., Armesto, J. J. & Collins. S. L. (1989). The ecological concept of disturbance and its expression at various hierarchical levels. *Oikos*, **54**, 129–36.

Pielke, R. A. & Avissar, R. (1990). Influence of landscape structure on local and regional climate. *Landscape Ecology*, **4**, 133–5.

Pimm, S. L. (1993). Biodiversity and the balance of nature. *Ecological Studies*, **99**, 347–60.

Pimm, S. L., Russell, G. J., Gittleman, J. L. & Brooks, T. M. (1995). The future of biodiversity. *Science*, **269**, 347–50.

Pinter, P. J. Jr., Kimball, B. A., Garcia, R. L., Wall, G. W., Hunsaker, D. J. & LaMorte, R. L. (1996). Free-air CO_2 enrichment: responses of cotton and wheat crops. In *Carbon Dioxide and Terrestrial Ecosystems*, ed. G. W. Koch & H. A. Mooney, pp. 215–49. San Diego: Academic Press.

Piper, S. C. & Stewart, E. F. (1996). A gridded global data set of daily temperature and precipitation for terrestrial biospheric modeling. *Global Biogeochemical Cycles*, **10**(4), 757–82.

Pitelka, L. F. & Plant Migration Workshop Group. (1997). Plant migration and climate change. *American Scientist*, **85**, 464–73.

Plentinger, M. C. & Penning de Vries, F. W. T. (1996). *CAMASE Register of Agroecosystem models*, Wageningen, The Netherlands: AB-DLO.

Plotnick, R. E. & Gardner, R. H. (1993). Lattices and landscapes. In *Lectures on Mathematics in the Life Sciences: Predicting Spatial Effects in Ecological Systems*, ed. R. H. Gardner, pp. 129–57. Providence, NY: American Mathematical Society.

Pockman, W. T. & Sperry, J. S. (1997). Freezing-induced xylem cavitation and the northern limit of *Larreatridentata*. *Oecologia*, **109**, 19–27.

Pollard, D. & Thompson, S. L. (1995). Use of a land surface transfer scheme (LSX) in a global climate model – the response to doubling stomatal resistance. *Global Planetary Change*, **10**, 129–61.

Polley, H. W., Johnson, H. B. & Mayeux, H. S. (1994). Increasing CO_2: comparative responses of the C_4 grass *Schizachyrium* and grassland invader *Prosopis*. *Ecology*, **75**, 976–88.

Poorter, H. (1993). Interspecific variation in the growth response of plants to an elevated ambient carbon dioxide. *Vegetatio*, **104/105**, 77–97.

Porter, G. R. (1993). AFRCWHEAT2: A model of the growth and development in wheat incorporating responses to water and nitrogen. *European Journal of Agronomy*, **2**, 69–82.

Porter, J. R., Leigh, R. A., Semenov, M. A. & Miglietta, F. (1995). Modelling the effect of climatic change and genetic modification on nitrogen use by wheat. *European Journal of Agronomy*, **4**(4), 419–29.

Post, W. M. & Pastor, J. (1996). LINKAGES – An individual-based forest ecosystem model. *Climatic Change*, **34**, 253–61.

Post, W. M., Emanuel, W. R. & King, A. W. (1994). Soil organic matter dynamics and the global carbon cycle. In *World Inventory of Soil Emission Potentials*, ed. N.H.

Batjes & E.M. Bridges, pp. 107–19. Wageningen: ISRIC.

Poth, M., Anderson, I. C., Miranda, H. S., Miranda, A. C. & Riggan, P. J. (1995). The magnitude and persistance of soil NO, N$_2$O, CH$_4$ and CO$_2$ fluxes from burned tropical savanna in Brazil. *Global Biogeochemical Cycles*, **9**, 503–13.

Potter, C. S., Randerson, J. T., Field, C. B., Matson, P. A., Vitousek, P. M., Mooney, H. A. & Klooster, S. A. (1993). Terrestrial ecosystem production – a process model based on global satellite and surface data. *Global Biogeochemical Cycles*, **7**(4), 811–41.

Potvin, C. & Vasseur, L. (1997). Long-term CO$_2$ enrichment of a pasture community: species diversity and dominance pattern. *Ecology*, **78**, 666–7.

Power, M. E. (1992). Hydrologic and trophic controls of seasonal algal blooms in northern California rivers. *Archiev für Hydrobiologie*, **125**, 385–410.

Powlson, D. S., Smith, P. & Smith, J. U. (eds.) (1996). Evaluation of soil organic matter models using long-term datasets. *NATO ASI Series, Series I: Global Environmental Change, No. 38*. Springer.

Prather, M., Derwent, R., Ehhalt, D., Fraser, P. Sanhueza. E. & Zhou, X. (1995). Other trace gases and atmospheric chemistry. In *Climate Change 1994: Radiative Forcing of Climate Change*, ed. J. T. Houghton, L. G. Meira Filho, J. Bruce, Lee Hoesung, B.A. Callander, E. Haites, N. Harris & K. Maskell, pp. 73–125. Cambridge: IPCC/Cambridge University Press.

Pregitzer, K. S., Zakn D. R., Curtis, P. S., Kubiske, M. E., Teeri, J. A. & Vogel, C. S. (1995). Atmospheric CO$_2$, soil nitrogen and turnover of fine roots. *New Phytologist*, **129**, 579–85.

Prentice, I. C. & Leemans, R. (1990). Pattern and process and the dynamics of forest structure. *Journal of Ecology*, **78**, 340–55.

Prentice, I. C., Cramer, W., Harrison, S. P., Leemans, R., Monserud, R. A. & Solomon, A. M. (1992). A global biome model based on plant physiology and dominance, soil properties and climate. *Journal of Biogeography*, **19**, 117–34.

Prentice, I. C., Sykes, M. T. & Cramer, W. (1993). A simulation model for the transient effects of climate change on forest landscapes. *Ecological Modelling*, **65**(1–2), 51–70.

Prentice, I. C., Webb, R. S., Ter-Mikhaelian, M. T., Solomon, A. M., Smith, T. M., Pitovranov, S. E., Nikolov, N. T., Minin, A. A., Leemans, R., Lavorel, S., Korzuhin, M. D., Helmisaari, H. O., Hrabovszky, J. P., Harrison, S. P., Emanuel, W. R. & Bonan, G. B. (1989). Developing a Global Vegetation Dynamics Model: Results of an IIASA Summer Workshop. Laxenburg, Austria: International Institute for Applied Systems Analysis.

Prentice, K. C. & Fung, I. Y. (1990). The sensitivity of terrestrial carbon storage to climate change. *Nature*, **346**, 48–51.

Price, D. T. & Apps, M. J. (1995). The Boreal Forest Transect Case Study: global change effects on ecosystem processes and carbon dynamics in boreal Canada. *Water, Air and Soil Pollution*, **82**, 203–14.

Priestley, C. H. B. & Taylor, R. J. (1972). On the assessment of surface heat flux and evaporation using large scale parameters. *Monthly Weather Review*, **100**(2), 81–92.

Prince, S. D., Justice, C. O. & Moore, B., III (1994). Monitoring and Modelling of Terrestrial Net and Gross Primary Production. 56 pp. *IGBP-DIS Working Paper*, **8**.

Prospero, J. M. & Nees, R. T. (1977). Dust concentration in the atmosphere of the equatorial North Atlantic: possible relationship to the Sahelian drought. *Science*, **196**, 1196–8.

Quay, P. D., King, S. L., Stutsman, J., Wilbur, D. O., Steele, L. P., Fung, I., Gammon, R. H., Brown, T. A., Farewell, G. W., Grootes, P. M. & Schmidt, F. H. (1991). Carbon isotopic composition of atmospheric methane: fossil and biomass burning strength. *Global Biogeochemical Cycles*, **5**, 25–47.

Rabbinge, R. (1993). The ecological background of food production. In *Crop Protection and Sustainable Agriculture*, ed. D. J. Chadwick & J. Marsh, pp. 2–29. Chichester, UK: John Wiley & Sons Ltd.

Ciba Foundation Symposia Vol. 177.

Rabbinge, R. & van Ittersum, M. K. (1994). Tension between aggregation levels. In *The Future of the Land: Mobilising and Integrating Knowledge From Land Use Options*, ed. L. O. Fresco, L. Stroosnijder, J. Bouma & H. van Keulen. Chichester: John Wiley and Sons Ltd.

Raison, R. J., Myers, B. J. & Benson, M. L. (1992). Dynamics of *Pinus radiata* foliage in relation to water and nitrogen stress. 1: Needle production and properties. *Forest Ecology and Management*, **52**, 139–58.

Ramakrishnan, P. S. (1992). Shifting agriculture and sustainable development: an interdisciplinary study from north-eastern India. Park Ridge, NJ: Parthenon Publishing Group.

Ramakrishnan, P. S., Campbell, J., Demierre, L., Gyi, A., Malhotra, K. C., Mehndiratta, S., Rai, S. N. & Sashidharan, E. M. (1994). *Ecosystem Rehabilitation of the Rural Landscape in South and Central Asia: An Analysis of Issues*. Special Publication, ed. M. Hadley, New Delhi: UNESCO (ROSTCA).

Rasmussen, D. I. (1941). Biotic communities of Kaibab Plateau, Arizona. *Ecological Monographs*, **3**, 229–75.

Raunkiær, C. (1907). *Planterigets Livsformer*. Copenhagen/Kristiania: Gyldendalske Boghandel & Nordisk Forlag.

Raunkier, C. (1934). *The Life-Forms of Plants and Statistical Plant Geography*. Oxford University Press, Oxford.

Raupach, M. R., Denmead O. T. & Dunin, F. X. (1992). Challenges to linking atmospheric CO_2 concentrations to fluxes at local and regional scales. *Australian Journal of Botany*, **40**, 697–716.

Raupach, M. R. & Finnigan, J. J. (1988). Single-layer models of evaporation from plant canopies are incorrect but useful, whereas multilayer models are correct but useless: discuss. *Australian Journal of Plant Physiology*, **15**, 705–16.

Raynaud, D., Jouzel, J., Barnola, J. M., Chappellaz, J., Delmas, R. J. & Lorius, C. (1993). The ice core record of greenhouse gases. *Science*, **259**, 926–34.

Read, D. J. (1993). Plant–microbe mutualism and community structure. *Ecological Studies*, **99**, 181–210.

Read, J. J. & Morgan, J. A. (1996). Growth and partitioning in *Pascopyrum smithii* (C_3) and *Bouteloua gracilis* (C_4) as influenced by carbon dioxide and temperature. *Annals of Botany*, **77**, 487–96.

Reed, B. C., Brown, J. F., Vanderzee, D., Loveland, T. R., Merchant, J. W. & Ohlen, D. O. (1994). Measuring phenological variability from satellite imagery. *Journal of Vegetation Science*, **5**(5), 703–14.

Reekie, E. G. & Bazzaz, F. A. (1989). Competition and patterns of resource use among seedlings of five tropical trees grown at ambient and elevated carbon dioxide. *Oecologia*, **79**, 212–22.

Reich, P. B. & Amundson, R. G. (1985). Ambient levels of ozone reduce net photosynthesis in tree and crop species. *Science*, **230**, 566–70.

Reid, W. V. (1992). How many species will there be? In *Tropical Deforestation and Species Extinction*, ed. T. C. Whitmore & J. A. Sayer, pp. 55–74. London: Chapman & Hall.

Reid, W.V. & Miller, K.R. (1989) *Keeping Options Alive. The Scientific Basis for Conserving Biodiversity*. 128 pp. Washington, DC: World Resources Institute.

Reynolds, J. F., Virginia, R. A. & Schlesinger, W. H. (1997). Defining functional types for models of desertification. In *Plant Functional Types: Their Relevance to Ecosystem Properties and Global Change*, ed. T. M. Smith, H. H. Shugart & F. I. Woodward, pp. 195–216. Cambridge: Cambridge University Press. IGBP Book Series Vol. 1.

Richards, L. A. (1931). Capillary conduction of liquids through porous mediums. *Physics*, **1**, 318–33.

Richardson, C. W. (1981). Stochastic simulation of daily precipitation, temperature, and solar radiation. *Water Resources Research*, **17**(1), 182–90.

Ricklefs, R. E. (1995). The distribution of biodiversity. In *Global Biodiversity Assessment*, ed. V. H. Heywood & R. T. Watson, pp. 139–173. Cambridge: Cambridge University Press.

Ritchie, J. C. & MacDonald, G. M. (1986). The pattern of post-glacial spread of white spruce. *Journal of Biogeography*, 13, 527–40.

Ritchie, J. T. & Otter S. (1985). Description and performance of CERES-Wheat: a user-oriented wheat yield model. USDA-ARS, ARS-38, pp. 159–75.

Robles, M. & Chapin, F. S., III. (1995). Comparison of the influence of two exotic species on ecosystem processes in the Berkeley hills. *Madrono*, 42, 349–57.

Rodda, G. H. & T. H. Fritts. (1992). The impact of the introduction of the Colubrid Snake *Bioga irregularis* on Guam's lizards. *Journal of Herpetology*, 26, 166–74.

Rodin, L. E. & N. I. Bazilevich. (1967). *Production and Mineral Cycling in Terrestrial Vegetation*. Edinburgh, UK: Oliver and Boyd.

Rogers, H. H., Prior, S. A., Runion, G. B. & Mitchell, R. J. (1996). Plant response to atmospheric CO_2 enrichment: allocation patterns in crops. *Plant and Soil*, 187, 229–48.

Romme, W. H. (1982). Fire and landscape diversity in subalpine forests of Yellowstone National Park. *Ecological Monographs*, 52, 199–221.

Romme, W. H. & Turner, M. G. (1991) Implications of global climate change for biogeographic patterns in the greater Yellowstone ecosystem. *Conservation Biology*, 5, 373–86.

Rosenzweig, C. & Parry, M. L. (1994). Potential impact of climate change on world food supply. *Nature*, 367, 133–8.

Rossing, W. A., van Oijen, H. M., van der Werf, W., Bastiaans, L. & Rabbinge, R. (1992). Modelling the effects of foliar pests and pathogens on light interception, photosynthesis, growth rate and yield of field crops. In *Pests and Pathogens, Plant Responses to Foliar Attack*, ed. P. G. Ayres, pp. 161–80. BIOS Scientific Publishers.

Roth, S. K. & Lindroth, R. L. (1995). Elevated atmospheric CO_2: effects on phytochemistry, insect performance and insect–parasitoid interactions. *Global Change Biology*, 1, 73–182.

Rotmans, J. & den Elzen, K. (1993). Modelling feedback mechanisms in the carbon cycle – balancing the carbon budget. *Tel-*

lus Series B, 45, 301–20.

Rundel, P. W., Ehleringer, J. R. & Nagy, K. A. (1989). Stable isotopes in ecological research. *Ecological Studies*, 68, 525 pp. Heidelberg: Springer-Verlag.

Running, S. W. & Coughlan, J. C. (1988). A general model of forest ecosystem processes for regional applications: I. Hydrologic balance, canopy gas exchange and primary production processes. *Ecological Modelling*, 42, 125–54.

Running, S. W. & Gower, S. T. (1991). FOREST-BGC, A general model of forest ecosystem processes for regional applications. II. Dynamic carbon allocation and nitrogen budgets. *Tree Physiology*, 9, 147–60.

Running, S. W. & Hunt Jr., E. R. (1993). Generalization of a forest ecosystem process model for other biomes, BIOME-BGC, and an application for global scale model. In *Scaling Physiological Process Leaf to Globe*, ed. J. R. Ehleringer & C. B. Field, pp. 141–58. San Diego: Academic Press.

Running, S. W., Loveland, T. R., Pierce, L. L., Nemani, R. & Hunt, E. R. (1995). A remote sensing based vegetation classification logic for global land cover analysis. *Remote Sensing of Environment*, 51(1), 39–48.

Running, S. W., Justice, C. O., Salomonson, V. S., Hall, D., Barker, J., Kaufmann, Y. J., Strahler, A. H., Huete, A. R., Muller, J.-P., Vanderbilt, V., Wan, Z. M. & Teillet, P. (1994). Terrestrial remote sensing science and algorithms planned for EOS/MODIS. *International Journal of Remote Sensing*, 15(17), 3587–620.

Russell-Smith, J. & Bowman, D. M. J. S. (1992). Conservation of monsoon rainforest isolates in the Northern Territory, Australia. *Biological Conservation*, 59, 51–64.

Rutherford, M. C. (1980). Annual plant production–precipitation relations in arid and semiarid regions. *Southern African Journal of Science*, 76, 53–6.

Ryan, M. G. (1991). A simple method for estimating gross carbon budgets for vegetation in forest ecosystems. *Tree Physiology*, 9, 255–66.

Ryan, M. G., Hunt, E. R., McMurtrie, R. E., Ågren, G. I., Aber, J. D., Friend, A. D., Rastetter, E. B., Pulliam, W. M., Raison, R. J. & Linder, S. (1996a). Comparing models of ecosystem function for temperate conifer forests. I. Model description and validation. In *Global Change: Effects on Coniferous Forests and Grasslands*, ed. A. I. Breymeyer, D. O. Hall, J. M. Melillo & G. I. Ågren, pp. 313–62. SCOPE56. Chichester: John Wiley and Sons Ltd.

Ryan, M. G., McMurtrie, R. E., Ågren, G. I., Hunt, E. R., Aber, J. D., Friend, A. D., Rastetter, E. B. & Pulliam, W. M. (1996b). Comparing models of ecosystem function for temperate conifer forests. II. Simulations of the effect of climate change. In *Global Change: Effects on Coniferous Forests and Grasslands*, ed. A. I. Breymeyer, D. O. Hall, J. M. Melillo & G. I. Ågren, pp. 363–87. SCOPE56. Chichester: John Wiley and Sons Ltd.

Rygiewicz, P. T. & Andersen, C. P. (1994). Mycorrhizae alter quality and quantity of carbon allocation belowground. *Nature*, **369**, 58–60.

Sage, R. F. (1994). Acclimation of photosynthesis to increasing atmospheric CO_2: the gas exchange perspective. *Photosynthesis Research*, **39**, 351–68.

Sala, O. E., Golluscio, R. A., Lauenroth, W. K. & Soriano, A. (1989). Resource partitioning between shrubs and grasses in the Patagonian Steppe. *Oecologia*, **81**, 501–5.

Sala, O. E., Lauenroth, W. K. & Golluscio, R. A. (1997). Plant functional types in temperate semi-arid regions. In *Plant Functional Types: Their Relevance to Ecosystem Properties and Global Change*, ed. T. M. Smith, H. H. Shugart & F. I. Woodward, pp. 217–33. Cambridge: Cambridge University Press. IGBP Book Series Vol. 1.

Sala, O. E., Lauenroth, W. K., McNaughton, S. J., Rusch, G. & Zhang, X. (1995). Temperate grasslands. In *Global Biodiversity Assessment*, ed. UNEP, pp. 361–6, Cambridge: Cambridge University Press.

Sala, O., Parton, W. J., Joyce, L. A. &

Lauenroth, W. K. (1988). Primary production of the central grassland region of the United States: spatial pattern and major controls. *Ecology*, **69**, 40–5.

Samuel, W. M. & Welch, D. A. (1991). Winter ticks on moose and other ungulates: factors influencing their population size. *Alces*, **27**, 169–82.

Sanchez, P. A. (1976). *Properties and Management of Soils in the Tropics*. New York: Wiley Interscience. 618 pp.

Sandermann, H., Jr, Wellburn, A. R. & Heath, R. L. (1997). Forest decline and ozone: a comparison of controlled chamber and field experiments. *Ecological Studies*, **127**. Heidelberg: Springer-Verlag.

Sanders, I. R. (1996). Plant-fungal interactions in a CO_2-rich world. In *Carbon Dioxide, Populations, and Communities*, ed. Ch. Körner & F. Bazzaz, pp. 265–72. San Diego: Academic Press.

Sarmiento, J. L., Orr, J. C. & Siegenthaler, U. (1992). A perturbation simulation of CO_2 uptake in an ocean general circulation model. *Journal of Geophysical Research*, **97**, 3621–45.

Sasek, T. W. & Strain, B. R. (1990). Implications of atmospheric CO_2 enrichment and climatic change for the geographical distribution of two introduced vines in the USA. *Climatic Change*, **16**, 31–51.

Savidge, J. A. (1987). Extinction of an island forest avifauna by an introduced snake. *Ecology*, **68**, 660–8.

Sayer, J. A. & Collins, N. M. (1991) A future for tropical forests. In *The Conservation Atlas of Tropical Forests. Asia and the Pacific*, ed. N. M. Collins, J.A. Sayer & T.C. Whitmore, pp. 77–81. London: MacMillan.

Schäppi, B. & Körner, Ch. (1996). Growth responses of an alpine grassland to elevated CO_2. *Oecologia*, **105**, 43–52.

Schimel, D. S. (1995). Terrestrial ecosystems and the carbon cycle. *Global Change Biology*, **1**, 77–91.

Schimel, D. S., Alves, D., Enting, I., Heimann, M., Joos, F., Raynaud, D., Wigley, T., Prather, M., Derwent, R., Ehhalt, D., Fraser, P., Sanhaueza, E., Zhou, X., Jonas, P., Charlson, R., Rodhe, H., Sadasivan, S., Shine, K. P.,

Fouquart, Y., Ramaswamy, V., Solomon, S., Srinivasan, J., Albritton, D., Isaksen, I., Lal, M. and Wuebbels, D. (1996). Radiative forcing of climate change. In *Climate Change 1995*, ed. J. T. Houghton, L. G. Meira Filho, B. A. Callender, N. Harris, A. Kattenberg & K. Maskell, pp. 69–131. Cambridge: Cambridge University Press.

Schimel, D. S., Braswell, B. H., Holland, E. A., McKeown, R., Ojima, D. S., Painter, T. H., Parton, W. J. & Townsend, A. R. (1994). Climatic, edaphic and biotic controls over storage and turnover of carbon in soils. *Global Biogeochemical Cycles*, 8, 279–93.

Schimel, J. (1995). Ecosystem consequences of microbial diversity and community structure. In *Arctic and Alpine Biodiversity: Patterns, Causes and Ecosystem Consequences*, ed. F. S. Chapin, III & C. Körner, pp. 239–54. Berlin: Springer-Verlag.

Schindler, D. W. (1991). Whole lake experiments at the experimental lakes area. In *Ecosystem Experiments*, ed. H. A. Mooney, E. Medina, D. W. Schindler, E.-D. Schulze, B. H. Walker. SCOPE Vol 45, pp.121–40. Chichester: John Wiley and Sons.

Schindler, D. W. & Bayley, S. E. (1993). The biosphere as an increasing sink for atmospheric carbon: estimates from increased nitrogen deposition. *Global Biogeochemcal Cycles*, 7, 717–33.

Schlesinger, W. H. (1991). *Biogeochemistry: an Analysis of Global Change*. San Diego: Academic.

Schlesinger, W. H., Reynolds, J. F., Cunningham, G. L., Huenneke, L. F., Jarrell, W. M., Virginia, R. A. & Whitford. W. G. (1990). Biological feedbacks in global desertification. *Science*, **247**, 1043–8.

Schmitt, W. R. (1965). The planetary food potential. *Annals of the New York Academy of Sciences*, 118, 645–718.

Scholes, M. C., Martin, R., Scholes, R. J., Parsons, D. & Winstead, E. (1997). NO and N$_2$O emissions from savanna soils following the first rains of the season. *Nutrient Cycling in Agroecosystems*, 7, 17–24.

Scholes, R. J. (1993). Nutrient cycling in semi-arid grasslands and savannas: its influence on pattern, productivity and stability. *Proceedings of the* XVII *International Grasslands Congress*, pp. 1331–4. Palmerston North, New Zealand: International Grasslands Society.

Scholes, R. J & Archer, S. R. (1997). Tree–grass interactions in savannas. *Annual Review of Ecology and Systematics*, 28, 517–44.

Scholes, R. J. & Hall, D. O. (1996). The carbon budget of tropical savannas, woodlands and grasslands. In *Global Change: Effects on Coniferous Forests and Grasslands*, ed. A. I. Breymeyer, D. O. Hall, J. M. Melillo & G. I. Agren, pp. 70–100. Chichester: John Wiley, SCOPE 56.

Scholes, R. J. & Scholes, M. C. (1994). The effect of land use on non-living organic matter in the soil. In *The Role of Non-living Organic Matter in the Earth's Carbon Cycle*, ed. R. Zepp & Ch. Sonntag, pp. 209–25. Chichester: John Wiley.

Scholes, R. J. & Walker, B. H. (1993). *An African Savanna: Synthesis of the Nylsvley Study*, 306 pp. Cambridge: Cambridge University Press.

Scholes, R. J., Justice, C. O. & Ward, D. (1996a). Trace gas emissions from biomass burning in southern-hemisphere Africa. *Journal of Geophysical Research*, 101, 23677–82.

Scholes, R. J., Kendall, J. & Justice, C. O. (1996b). The quantity of biomass burned in southern Africa. *Journal of Geophysical Research*, 101, 23667–76.

Scholes, R. J., Skole, D. & Ingram, J. S. I. (1995). A Global Database of Soil Properties: Proposal for Implementation. *IGBP DIS Working Paper No. 10*.

Scholes, R. J. & Parsons, D. B. (ed.) (1997). *The Kalahari Transect: Research on Global Change and Sustainable Development in Southern Africa*. IGBP Report No. 42, 61 pp. Stockholm: International Geosphere–Biosphere Programme.

Schulze, E.-D. (1982). Plant life forms as related to plant carbon, water and nutrient relations. In *Water Relations and Photosynthetic Productivity*, ed. O. L. Lange, P. S. Nobel, C. B. Osmond & H.

Ziegler, pp. 615–76. Berlin, Heidelberg: Springer-Verlag. Encyclopedia of Plant Physiology, Physiological Plant Ecology Vol. 12B.

Schulze, E.-D. (1994a). Die Wirkung von Immissionen auf den Wald. *Leopoldina* (Reihe 3), **39**, 227–43.

Schulze, E.-D. (1994b). *Flux Controlling Biological Systems*. San Diego: Academic Press.

Schulze, E.-D. (1994c). The impact of increasing nitrogen deposition on forests and aquatic ecosystems. *Nova Acta Leopoldina NF*, **70**, 515–36.

Schulze, E.-D. (1994d). The regulation of plant transpiration: interactions of feed-forward, feedback, and futile cycles. In *Flux Control in Biological Systems. From Enzymes to Populations and Ecosystems*, ed. E.-D Schulze, pp. 203–236. San Diego: Academic Press.

Schulze, E.-D. (1995a). Flux control at the ecosystem level. *Trends in Ecology and Evolution*, **10**, 40–3.

Schulze, E.-D. (1995b). Herkunft, wirkung und verbleib des stickstoffs in waldökosystemen. Bayerisches Landesministerium für Ernahrung, Landwirtschaft und Forsten, pp. 39–47. *Proceedings from Forest Decline Symposium*, Munch, May 3, 1995.

Schulze, E.-D., Bauer, G., Buchmann, N., Sala, O., Canadell, J., Ehleringer, J. R., Jackson, R. B., Jobbagy, E., Loreti, J., Mooney, H. A. & Oesterheld, M. (1997a). Water availability, rooting depth, and vegetation zones along an aridity gradient in Patagonia. *Œcologia*, **108**, 503–11.

Schulze, E.-D., Beck, E., Steudle, E., Stitt, M. & Zwölfer, H. (1994a). Flux control in biological systems: a comparative view. In *Flux Control in Biological Systems: From Enzymes to Populations and Ecosystems*, ed. E.-D. Schulze, pp. 471–85. Berlin: Springer.

Schulze, E.-D. & Chapin, F. S. III. (1987). Plant specialization to environments of different resource availability. *Ecological Studies*, **61**, 116–47.

Schulze, E.-D., Chapin, F. S., III & Gebauer, G. (1994c). Coexisting species tab isotopically distinct nitrogen pools in N- and P-limited Alaskan spruce forests. *Oecologia*, **100**, 406–12.

Schulze, E.-D, Ellis, R., Schulze, W., Trimborn, P. & Ziegler, H. (1996a). Diversity, metabolic types and d¹³C carbon isotope ratios in the grass flora of Namibia in relation to growth form, precipitation and habitat conditions. *Oecologia*, **106**, 352–69.

Schulze, E.-D. & Freer-Smith, P. H. (1991). An evaluation of forest decline based on field observations focussed on Norway spruce, *Picea abies*. In *Acid Deposition – Its Nature and Impacts*, ed. F. T. Last & R. Watling, pp. 155–68. The Royal Society of Edinburgh.

Schulze, E.-D. & Hall, A. E. (1982). Stomatal control of water loss. In *Water Relations and Photosynthetic Productivity*, Ed. O. L. Lange, P. S. Nobel, C. B. Osmond & H. Ziegler, pp. 181–230. Heidelberg: Springer Verlag. Encyclopedia of Plant Physiology, Physiological Plant Ecology, Vol. 12B.

Schulze, E.-D. & Heimann, M. (1998). Carbon and water exchange of terrestrial ecosystems. In *Asian Change in the Context of Global Change*, ed. J. N. Galloway & J. M. Melillo. Cambridge: Cambridge University Press.

Schulze, E.-D., Kelliher, F. M., Körner, C., Lloyd, J. & Leuning, R. (1994b). Relationships among maximum stomatal conductance, ecosystem surface conductance, carbon assimilation rate and plant nitrogen nutrition: a global ecology scaling exercise. *Annual Review of Ecology and Systematics*, **25**, 629–60.

Schulze, E.-D. & Mooney, H. A. (1993). *Design and Execution of Experiments on CO₂ Enrichment*. Ecosystem Research Report 6. Commission of the European Communities, Brussels, 420 pp.

Schulze, E.-D., Mooney, H. A., Sala, O. E., Jobbagy, E., Buchmann, N., Bauer, G., Canadell, J., Jackson, R., Loreti, B. J., Oesterheld, M. & Ehleringer, J. R. (1996b). Rooting depth, water availability, and vegetation cover along an aridity gradient in Patagonia. *Oecologia*, **108**, 503–11.

Schulze, E.-D., Schulze, W., Kelliher, F. M., Vygodskaya, N. N., Ziegler, W., Kobak, K. I., Koch, H., Arneth, A., Kusnetsova, W. A., Sogachev, A., Issajev, A., Bauer, G. & Hollinger, D. Y. (1995). Above-ground biomass and nitrogen nutrition in a chrono sequence of pristine Dahurian larix stands in Eastern Siberia. *Canadian Journal of Forest Research*, **25**, 943–60.

Schulze, E.-D. & Ulrich, B. (1991) Acid rain – a large scale, unwanted experiment in forest ecosystems. In *Ecosystem Experiments*, ed. H. A. Mooney, E. Medina, D. W. Schindler, E.-D. Schulze & B. H. Walker, pp. 89–106. SCOPE 45. Chichester: Wiley.

Schulze, E.-D, Vries, W. de, Hauhs, M., Rosén, K., Rasmussen, L., Tamm, C. O. & Nilsson, J. (1989). Critical loads for nitrogen deposition on forest ecosystems. *Water, Air, and Soil Pollution*, **48**, 451–6.

Schulze, E.-D., Williams, R., Farquhar, G. D., Schulze, W., Langridge, J., Miller, J. & Walker, B. (1997a). Carbon and nitrogen isotope discrimination and nitrogen nutrition along the Northern Australian Tropical Transect (NATT). *Australian Journal of Plant Physiology* (in press).

Schweingruber, F. H., Briffa, K. R. & Nogler, P. (1993). A tree-ring densitometric transect from Alaska to Labrador. Comparison of ring-width and maximum-latewood-density chronologies in the conifer belt of northern North America. *International Journal of Biometeorology*, **37**,151–69.

Sedjo, R. A. (1992). Temperate forest ecosystems in the global forest cycle. *Ambio*, **21**, 274–77.

Sellers, P. J., Bounoua, L., Collatz, G. J., Randall, D. A., Dazlich, D. A., Los, S. O., Berry, J. A., Fung, I., Tucker, C. J., Field, C. B. & Jensen, T. G. (1996). Comparison of radiative and physiological effects of doubled atmospheric CO_2 on climate. *Science*, **271**, 1402–6.

Sellers, P. J., Dickinson, R. E., Randall, D. A., Betts, A. K., Hall, F. G., Berry, J. A., Collatz, G. J., Denning, A. S., Mooney, H. A., Nobre, C. A., Sato, N., Field, C. B. & Henderson, A. (1997). Modeling the exchanges of energy, water, and carbon between continents and the atmosphere. *Science*, **275**, 502–9.

Sellers, P., Hall, F., Margolis, H., Kelly, B., Baldocchi, D., den Hartog, G., Cihlar, J., Ryan, M. G., Goodison, B., Crill, P., Ranson, K. J., Lettenmaier, D. & Wickland, D. E. (1995b). The Boreal Ecosystem–Atmosphere Study (BOREAS): an overview and early results from the 1994 field year. *Bulletin of the American Meteorological Society*, **76**, 1549–77.

Sellers, P. J., Meeson, B. W., Hall, F. G., Asrar, G., Murphy, R. E., Schiffer, R. A., Bretherton, F. P., Dickinson, R. E., Ellingson, R. G., Field, C. B., Huemmrich, K. F., Justice, C. O., Melack, J. M., Roulet, N. T., Schimel, D. S. & Try, P. D. (1995a). Remote sensing of the land surface for studies of global change: models – algorithms – experiments. *Remote Sensing of Environment*, **51**(1), 3–26.

Sellers, P. J., Mintz, Y., Sud, Y. C. & Dalcher, A. (1986). A simple biosphere model (SiB) for use with general circulation models. *Journal of the Atmospheric Sciences*, **43**(6), 505–31.

Sellers, P. J., Tucker, C. J., Collatz, G. J., Los, S. O., Justice, C. O., Dazlich, D. A. & Randall, D. A. (1994). A global 1 degrees-by-1 degrees NDVI data set for climate studies.2. The generation of global fields of terrestrial biophysical parameters from the NDVI. *International Journal of Remote Sensing*, **15**(17), 3519–45.

Shantz, H. L. (1954). The place of grasslands in the Earth's cover of vegetation. *Ecology*, **35**, 143–5.

Shaver, G. R., Chapin III, F. S. & Gartner, B. L. (1986). Factors limiting seasonal growth and peak biomass accumulation in *Eriophorum vaginatum* in Alaskan tussock tundra. *Journal of Ecology*, **74**, 257–78.

Shaver, G. R., Giblin, A. E., Nadelhoffer, K. J. & Rastetter, E. B. (1997). Plant functional types and ecosystem change in arctic tundras. In *Plant Functional Types: Their Relevance to Ecosystem Properties and Global Change*, ed. T. M. Smith, H. H. Shugart & F. I. Woodward. pp. 153–173. Cambridge: Cambridge University Press. IGBP Book Series Vol. 1.

Shugart, H. H. (1984). *A Theory of Forest Dynamics: The Ecological Implications of Forest Succession Models.* 278 pp. New York: Springer-Verlag.

Shugart, H. H. (1997). *Terrestrial Ecosystems in Changing Environments.* Cambridge: Cambridge University Press.

Shugart, H. H. & Emanuel, W. R. (1985). Carbon dioxide increase: the implications at the ecosystem level. *Plant, Cell and Environment*, 8, 381–6.

Shukla, J., Nobre, C. & Sellers, P. (1990). Amazon deforestation and climate change. *Science*, 247, 1322–5.

Siegenthaler, U. & Oeschger, H. (1987). Biospheric CO_2 emissions during the past 200 years reconstructed by deconvolution of ice core data. *Tellus*, 39B, 140–54.

Siegenthaler, U. and Sarmiento J. L. (1993). Atmospheric carbon dioxide and the ocean. *Nature*, 365, 119–25.

Skidmore, E. L. (1965). Assessing wind erosion forces: directions and relative magnitudes. *Soil Science Society of America Journal*, 29, 587–90.

Skidmore, E. L. & Williams, J. R. (1991). Modified EPIC wind erosion model. In *Modelling Plant and Soil Systems*, pp. 457–69. ASA–CSSA–SSSA, Agronomy Monograph, Vol. 31.

Skole, D. & Tucker, C. (1993) Tropical deforestation and habitat fragmentation in the Amazon from satellite data from 1978 to 1988. *Science*, 260, 1905–10.

Skole, D. L., Chomentowski, W. A., Salas, W. A. & Nobre, A. D. (1994). Physical and human dimensions of deforestation in Amazonia. *Bioscience*, 44, 314–22.

Smith, J. A. C. & Griffiths, H. (1993). *Water Deficits Plant Responses – From Cell To Community.* Oxford: BIOS Scientific Publishers.

Smith, P., Powlson D. S. & Glendining, M. J. (1996a). Establishing a European GCTE Soil Organic Matter Network (SOMNET). In *Evaluation of Soil Organic Matter Models Using Existing Long-Term Datasets*, ed. D. S. Powlson, P. Smith & J. U. Smith, pp. 81–97. Berlin: Springer-Verlag.

Smith, P., Powlson, D. S., Smith, J. U. &

Glendining, M. J. (1996b). The GCTE SOMNET. A global network and database of soil organic matter models and long-term experimental datasets. *Soil Use and Management*, 12, 104.

Smith, P., Smith, J. U., Powlson, D. S., Arah, J. R. M., Chertov, O. G., Coleman, K., Franko, U., Frolking, S., Jenkinson, D. S., Jensen, L. S., Kelyy, R. H., Kelin-Gunnewick, H., Komarov, A. S., Li, C., Molina, J. A. E., Mueller, T., Parton, W. S., Thornley, J. H. M. & Whitmore, A. P. (1997). A comparison of the performance of nine soil organic matter models using datasets from seven long-term experiments. In *Geoderma Special Issue Evaluation and Comparison of Soil Organic Matter Models Using Datasets From Seven Long-term Experiments*, ed. P. Smith, D. S. Powlson, J. U. Smith & E. T. Elliott. Elsevier.

Smith, T. M. (ed.) (1996). *The Application of Patch Models in Global Change Research.* 182 pp. Dordrecht, Boston, London: Kluwer Academic Publishers.

Smith, T. M., Halpin, P. N., Shugart, H. H. & Secrett, C. M. (1995). Global Forests. In *If Climate Changes: International Impacts of Climate Change*, ed. K. M. Strzepek & J. B. Smith. Cambridge: Cambridge University Press.

Smith, T. M., Leemans, R. & Shugart, H. H. (1992b). Sensitivity of terrestrial carbon storage to CO_2 induced climate change: comparison of four scenarios based on general circulation models. *Climatic Change*, 21, 367–84.

Smith, T. M. & Shugart, H. H. (1993). The transient response of terrestrial carbon storage to a perturbed climate. *Nature*, 361, 523–6.

Smith, T. M. & Shugart, H. H. (1996). The application of patch models in global change research. In *Global Change and Terrestrial Ecosystems*, ed. B. Walker & W. Steffen pp. 127–48. Cambridge: Cambridge University Press. IGBP Book Series Vo. 2.

Smith, T. M., Shugart, H. H. & Halpin, P. N. (1992). Computer models of forest dynamics and global changes in the environment. In *Responses of Forest Ecosystems to*

Environmental Changes, Ed. A. Teller, P. Mathy & J. N. R. Jeffers, pp. 91–102. London, New York: Elsevier Applied Science.

Smith, T. M., Shugart, H. H. & Woodward, F. I. (eds.) (1997). *Plant Functional Types: Their Relevance to Ecosystem Properties and Global Change.* 369 pp. Cambridge: Cambridge University Press, IGBP Book Series Vol. 1.

Smith, T. M., Shugart, H. H., Woodward, F. I. & Burton, P. J. (1993). Plant functional types. In *Vegetation Dynamics and Global Change*, A. M. Solomon & H. H. Shugart, pp. 272–92. New York: Chapman and Hall.

Smith, T. M., Weishampel, J. F., Shugart, H. H. & Bonan, G. B. (1992*a*). The response of terrestrial C storage to climate change: modelling C dynamics at varying temporal and spatial scales. *Water, Air, and Soil Pollution*, 64(1–2), 307–26.

Solomon, A. M. (1986). Transient responses of forests to CO_2-induced climate change: simulation modeling in eastern North America. *Oecologia*, 68, 567–79.

Solomon, A. M. (1997). Natural migration rates of trees: Global terrestrial carbon cycle implications. In *Past and Future Rapid Environmental Changes: The Spatial and Evolutionary Responses of Terrestrial Biota*, ed. B. Huntley, W. Cramer, A. V. Morgan, H. C. Prentice & J. R. M Allen. Berlin: Springer-Verlag. NATO ASI Series Vol. 47.

Solomon, A. M. & Tharp, M. L. (1985). Simulation experiments with late-Quaternary carbon storage in mid-latitude forest communities. In *The Carbon Cycle and Atmospheric Carbon Dioxide: Natural Variations Archaen to Present*, ed. E. T. Sundquist & W. S. Broecker, pp. 235–50. Washington, DC: American Geophysical Union. Geophysical Monographs Series Vol. 32.

Solomon, A. M., Prentice, I. C., Leemans, R. & Cramer, W. (1993). The interaction of climate and land use in future terrestrial carbon storage and release. *Water, Air and Soil Pollution*, 70(1–4), 595–614.

Solomon, A. M., West, D. C. & Solomon, J. A. (1981). Simulating the role of climate change and species immigration in forest succession. In *Forest Succession. Concepts and Application*, ed. D. C. West, H. H. Shugart & D. B. Botkin, pp.154–77. New York: Springer-Verlag.

Sousa, W. P. (1984). The role of disturbance in natural communities. *Annual Review of Ecology and Systematics*, 15, 353–91.

Soussana, J. F. & Hartwig, U. (1996). The effects of elevated CO_2 on symbiotic N_2 fixation: a link between the carbon and nitrogen cycles. *Plant and Soil*, 182, 321–32.

Sperry, J. S. (1997). Hydraulic constraints on plant gas exchange. *Agricultural and Forest Meteorology* (in press).

Spieker, H., Miellikänen, K., Köhl, M. & Skovsgaard, J. P. (1996). *Growth Trends in European Forests.* Heidelberg: Springer-Verlag.

Spitters, C. J. T. & Schapendonk, A. H. C. M. (1990). Evaluation of breeding strategies for drought tolerance in potato by means of crop growth simulation. *Plant and Soil*, 123, 193–203.

Stafford Smith, M. (1996). Management of rangelands: paradigms at their limits. In *The Ecology and Management of Grazing Systems*, ed. J. Hodgson & A. Illius, pp. 325–57. Wallingford, UK: CAB International.

Stafford Smith, M., Archer, S. & Campbell, B. D. (1995). Understanding the effects of global change on rangelands and improved pastures: an implementation plan for international research. *Proceedings Fifth International Rangelands Congress*, pp. 553–34.

Stafford Smith, M., Campbell, B., Steffen, W. & Archer, S. (1994). State-of-the-Science Assessment of the Likely Impacts of Global Change on the Australian Rangelands. *GCTE Working Document No.14*, 72 pp. Canberra, GCTE Core Project Office.

Starfield, A. M. & Chapin, F. S., III. (1996). Model of transient changes in arctic and boreal vegetation in response to climate and land use change. *Ecological Applications*, 6, 842–64.

Steffen W. L., Walker, B. H., Ingram, J. S. I. & Koch. G. W. (eds). (1992). *Global*

Change and Terrestrial Ecosystems: The Operational Plan. 95 pp. Stockholm: The International Geosphere-Biosphere Programme, Report no. 21.

Steffen, W. L., Cramer, W., Plöchl, M. & Bugmann, H. K. M. (1996). Global vegetation models: incorporating transient changes to structure and composition. *Journal of Vegetation Science*, **7**(3), 321–8.

Stocker, R., Leadley, P. W. & Körner, Ch. (1997). Carbon and water fluxes in a calcareous grassland under elevated CO_2. *Functional Ecology*, **11**(2), 222–30.

Stockle, C. O., Williams, J. R., Rosenberg, N. J. & Jones, C. A. (1992). A method for estimating the direct and climatic effects of rising atmospheric carbon-dioxide on growth and yield of crops. 1. Modification of the EPIC model for climate change analysis. *Agricultural Systems*, **38**(3), 225–38.

Stomphe, T. J. (1990). Seedling Establishment in Pearl millet (*Pennisetum glaucum* (L.) R. Br); the Influence of Genotype, Physiological Seed Quality, Soil Temperature and Soil Water. PhD Thesis, 199 pp. University of Reading, UK.

Stone, C. P. (1985). Alien animals in Hawai'i's native ecosystems: towards controlling the adverse effects of introduced vertebrates. In *Hawai'i's Terrestrial Ecosystems, Preservation and Management*, ed. C. P. Stone & J. M. Scott, pp. 251–97. Honolulu: University Of Hawai'i.

Stott, P. (1995). *Journal of Biogeography*, **22**(2–5), (four special issues).

Strain, B. R. & Cure, J. D. (1994). Direct effects of atmospheric CO_2 enrichment on plants and ecosystems. An Updated Bibliographic Data Base. Carbon Dioxide Information Analysis Center. Oak Ridge National Laboratory. ORNL/CDIAC-70. Oak Ridge, USA.

Street-Perrott, F. A., Mitchell, J. F. B., Marchand, D. S. & Brunner, J. S. (1990). Milankovitch and albedo forcing of the tropical monsoons: A comparison of geological evidence and numerical simulations for 9000 BP. *Transactions of the Royal Society of Edinburgh – Earth Sciences*, **81**, 407–27.

Strong, D. R. (1992). Are trophic cascades all wet? Differentiation and donor-control in speciose ecosystems. *Ecology*, **73**, 747–54.

Sullivan, J. H. & Teramura, A. H. (1992). The effects of ultraviolet-B radiation on loblolly pine. *Trees*, **6**, 115–20.

Sutherst, R. W. (1987). The dynamics of hybrid zones between tick (ACARI) species. *International Journal of Parasitology*, **17**(4), 921–6.

Sutherst, R. W. (1991). Pest risk analysis and the greenhouse effect. *Review of Agricultural Entomology*, **79**, 1177–87.

Sutherst, R. W. (1995). The potential advance of pests in natural ecosystems under climate change: implications for planning and management. In *Impacts of Climate Change on Ecosystems and Species: Terrestrial Ecosystems*. Gland, Switzerland: International Union for Conservation of Nature and Natural Resources. ed. J. Pernetta, R. Leemans, D. Elder & S. Humphrey, pp. 83–98.

Sutherst, R. W. (ed.) (1996) *Impacts of Climate Change on Pests, Diseases and Weeds in Australia*. GCTE Working Document 25. Report of an International Workshop, Brisbane 9–12 October 1995. Canberra: CSIRO Division of Entomology.

Sutherst, R. W., Maywald, G. F. & Skarratt, D. B. (1995). Predicting insect distributions in a changed climate. In *Insects in a Changing Environment*, ed. R. Harrington & N. E. Stork, pp. 59–91. London: Academic Press.

Sutherst, R. W., Yonow, T., Chakraborty, S., O'Donnell, C. & White, N. (1996). A generic approach to defining impacts of climate change on pests, weeds and diseases in Australasia. In *Greenhouse: Coping With Climate Change*, ed. W. J. Bouma, G. I. Pearman & M. R. Manning. Melbourne: CSIRO.

Swetnam, T. W. (1993). Fire history and climate change in Giant Sequoia groves. *Science*, **262**(5135), 885–9.

Swift, M. J., Vandermeer, J., Ramakrishnan, R. S., Ong, C. K., Anderson, J. M. & Hawkins, B. (1995). Agroecosystems. In *Global Biodiversity Assessment*, ed. V. N. Heywood & R. T. Watson pp. 443–446. Cambridge: Cambridge University Press.

Sykes, M. T., Prentice, I. C. & Cramer, W. (1996). A bioclimatic model for the potential distributions of North European tree species under present and future climates. *Journal of Biogeography*, **23**(2), 203–33.

Tamm, C. O. (1991). Nitrogen in terrestrial ecosystems. *Ecological Studies*, Vol. 81. 115 pp. Heidelberg: Springer Verlag.

Tans, P. P., Fung, I. Y. & Takahashi, T. (1990). Observational constraints on the global atmospheric CO_2 budget. *Science*, **247**, 1431–38.

Tarpley, J. D., Schneider, S. R. & Money, R. L. (1984). Global vegetation indices from the NOAA-7 meteorological satellite. *Journal of Climate and Applied Meteorology*, **23**, 491–4.

Taylor, G. E., Jr (1994). Role of genotype in governing the response of *Pinus taeda* (loblolly pine) to tropospheric ozone. *Journal of Environmental Quality*, **12**, 63–82.

Taylor, G. E., Jr, Ross-Todd, B. M., Allen, E., Conklin, P., Edmonds, B., Joranger, E., Miller, E., Ragsdale, L., Shephard, J., Silsbee, D. & Swank, W. (1992). Patterns of tropospheric ozone in forested landscapes of the Integrated Forest Study. In *Atmospheric Deposition and Nutrient Cycling: a Synthesis of the Integrated Forest Study*, ed. D. W. Johnson & S. E. Lindberg, pp. 50–71. New York: Springer-Verlag.

Taylor, G. E., Jr, Johnson, D. W. & Andersen, C. P. (1994). Air pollution and forest ecosystems: a regional to global perspective. *Ecological Applications*, **4**, 662–89.

Teng, P. S., Heong, K. L., Kropff, M. J., Nutter, F. W. & Sutherst, R. W. (1996). Linked pest-crop models under global change. In *Global Change and Terrestrial Ecosystems*, ed. B. H. Walker & W. L. Steffen, pp. 291–316. Cambridge: Cambridge University Press, IGBP Book Series Vol. 2.

Teramura, A. H. (1983). Effects of ultraviolet-B radiation on the growth and yield of crop plants. *Physiologia Plantarum*, **58**, 415–27.

Teskey, R. O. (1995). A field study of the effects of elevated CO_2 on carbon assimilation, stomatal conductance and leaf and branch growth of *Pinus taeda* trees. *Plant, Cell and Environment*, **18**, 565–73.

Thiery, J. M., D'Herbes, J.-M. & Valentin, C. (1995). A model simulating the genesis of banded vegetation patterns in Niger. *Journal of Ecology*, **83**, 497–507.

Thomas, G. & Rowntree, P. R. (1992). The boreal forests and climate. *Quarterly Journal of the Royal Meteorological Society*, **118**, 469–97.

Thompson, K., Hillier, S. H., Grime, J. P., Bossard, C. C. & Band, S. R. (1996). A functional analysis of a limestone grassland community. *Journal of Vegetation Science*, **7**(3), 371–80.

Thompson, S. L. & Pollard, D. (1995). A global climate model (Genesis) with a land-surface transfer scheme (LSX). Part 1. Present climate simulation. *Journal of Climate*, **8**, 732–61.

Thornburn, P. J., Walker, G. R. & Hatton, T. J. (1992). Are river read gums taking water from soil, groundwater, or streams. In *Catchments of Green* – The proceedings of the National Conference on vegetation and water management, pp. 37–42. Adelaide, Australia.

Thornley, J. H. M. & Cannell, M. G. R. (1996). Temperate forest responses to CO_2, temperature and nitrogen: a model analysis. *Plant, Cell and Environment*, **19**, 1331–48.

Thornthwaite, C. W. & Mather, J. R. (1957). *Instructions and Tables for Computing Potential Evapotranspiration and the Water Balance*. Drexel Institute of Technology, Laboratory of Climatology.

Tilman, D. (1987). Secondary succession and the pattern of plant dominance along experimental nitrogen gradients. *Ecological Monographs*, **57**, 189–214.

Tilman, D. (1988). *Plant Strategies and the Dynamics and Function of Plant Communities*. Princeton: Princeton University Press.

Tilman, D. (1993). Community diversity and succession: the roles of competition, dispersal, and habitat modification. *Ecological Studies*, **99**, 327–46.

Tilman, D. & Downing, J. A. (1994). Biodiversity and stability in grasslands. *Nature*,

367, 363–5.

Tilman, D., May, R., Lehman C. & Nowak, M. (1994). Habitat destruction and extinction debt. *Nature*, **371**, 361–65.

Tilman, D., Wedin, D. & Knops, J. (1996). Productivity and sustainability influenced by biodiversity in grassland ecosystems. *Nature*, **379**, 718–20.

Tingey, D. T., Johnson, M. G., Phillips, D. L. & Storm, M. J. (1995). Effects of elevated CO_2 and nitrogen on ponderosa pine fineroots and associated fungal components. *Journal of Biogeography*, **22**, 281–7.

Tinker, P. B. & Ingram, J. S. I. (1996). The Work of Focus 3. In *Global Change and Terrestrial Ecosystems*, ed. B. H. Walker & W. L. Steffen, pp. 207–28. Cambridge: Cambridge University Press.

Tinker, P. B., Gregory, P. G., Ingram, J. S. I. & Canadell, J. (1996). Plant-soil carbon below ground: the effects of elevated CO_2. *Plant and Soil*, **187**.

Tissue, D. L. & Oechel, W. C. (1987). Physiological response of *Eriophorum vaginatum* to elevated CO_2 and temperature in the Alaskan tussock tundra. *Ecology*, **68**, 401–10.

Toure, A., Major, D. J. & Lindwall, C. W. (1995). Sensitivity of four wheat simulation models to climate change. *Canadian Journal of Plant Science*, **75**, 69–74.

Townsend, A. R. & Vitousek, P. M. (1995). Soil organic matter dynamics along gradients in temperature and land use on the island of Hawaii. *Ecology*, **76**, 721–33.

Townsend, A. R., Vitousek, P. M. & Holland, E. A. (1992). Tropical soils could dominate the short-term carbon cycle feedbacks to increased global temperatures. *Climatic Change*, **22**, 293–303.

Townshend, J. R. G. (1992). Improved Global Data for Land Application: A Proposal for a New High Resolution Data Set. International Geosphere-Biosphere Programme, Stockholm. IGBP Report, No. 20, 87 pp.

Trenberth, K. E. (1995). Atmospheric circulation climate changes. *Climatic Change*, **31**, 2–4.

Trumbore, S. E. (1993). Comparison of carbon dynamics in tropical and temperate soils using radiocarbon measurements. *Global Biogeochemical Cycles*, **7**, 275–90.

Trumbore, S. E., Davidson, E. A., de Camargo, P. B., Nepstad, N. C. & Martinelli, L. A. (1995). Belowground cycling of carbon in forests and pastures of Eastern Amazonia. *Global Biogeochemical Cycles*, **9**, 515–28.

Tsuji, G. Y., Uehara, G. & Balas, S. (eds.) (1994). *DSSAT v3*. Honolulu, Hawaii: University of Hawaii.

Tucker, C. J. (1979*a*). Red and photographic infrared/red radiance ratios for estimating vegetation biomass and physiological status. *Proceedings of the 11th International Symposium of Remote Sensing of the Environment*, Vol. 1, pp. 493–94.

Tucker, C. J. (1979*b*). Red and photographic infrared linear combinations for monitoring vegetation. *Remote Sensing of the Environment*, **8**, 127–50.

Turner B. L., II, Clark, W. C., Kates, R. W., Richards, J. F., Mathews, J. T. & Meyer, W. B. (eds.). (1990). *The Earth as Transformed by Human Action*. Cambridge: Cambridge University Press.

Turner B. L., II, Skole, D. L., Sanderson, S., Fischer, G., Fresco, L. & Leemans, R. (1995). Land Use and Land-Cover Change: Science/Research Plan. International Geosphere-Biosphere Programme, Stockholm, IGBP Report No. **35**.

Turner, I. M. (1996). Species loss in fragments of tropical rainforests – A review of the evidence. *Journal of Applied Ecology*, **33**(2), 200–9.

Turner, M. G. (1989). Landscape ecology: the effect of pattern on process. *Annual Review of Ecology and Systematics*, **20**, 171–97.

Turner, M. G., Costanza, R. & Sklar, F. H. (1989*a*). Methods to evaluate the performance of spatial simulation models. *Ecological Modelling*, **49**, 1–18.

Turner, M. G., Gardner, R. H., Dale V. H. & O'Neill., R. V. (1989*b*). Predicting the spread of disturbance across heterogenous landscapes. *Oikos*, **55**, 121–9.

Turner, M. G., Romme, W. H. & Gardner, R. H. (1994). Landscape disturbance models and the long-term dynamics of natural areas. *Annual Review of Ecology and Systematics*, **20**, 171–9.

Turner, M. G., Romme, W. H., Gardner, R. H., O'Neill, R. V. & Kratz, T. K. (1993). A revised concept of landscape equilibrium – disturbance and stability on scaled landscapes. *Landscape Ecology*, **8**(3), 213–27.

Tyree, M. T. & Sperry, J. S. (1989). Vulnerability of xylem to cavitation and embolism. *Annual Review on Plant Physiology and Plant Molecular Biology*, **40**, 19–38.

Ulrich, B (1989). Effects of acidic precipitation on forest ecosystems in Europe. *Advances in Environmental Science*, **2**, 189–272.

UN (1992). *The Forest Resources of the Temperate Zones. Main findings of the UN-ECE/FAO 1990 Forest Resource Assessment*. New York: United Nations Publication.

UNEP/GEMS (1994). Report of the UNEP/FAO Expert Meeting on Harmonizing Land Cover and Land Use Classifications. 43 pp.

UNESCO (1973). *International Classification and Mapping of Vegetation*. Paris, France: UNESCO Report.

Urban, D. L., Bonan, G. B., Smith, T. M. & Shugart, H. H. (1991). Spatial applications of gap models. *Forest Ecology Management*, **42**(1,2), 95–100.

USDA Forest Service (1997). *Forest Insect and Disease Conditions in Alaska – 1996*. General Technical Report R10-TP-67. Alaska: USDA.

Valentin, C. (1996). Soil erosion under global change. In *Global Change and Terrestrial Ecosystems*, ed. B. H. Walker & W. L. Steffen, pp. 317–38. IGBP Book Series, No.2. Cambridge: Cambridge University Press.

Valentin, C., Collinet, J. & Albergel, J., (1994). Assessing erosion in West African savannas under global change: overview and research needs. *XVth International Congress of Soil Science*, Acapulco, Mexico, Volume 7a: pp. 253–74.

Valentin, C. & d'Herbes, J. M. (1997). Water harvesting along a climatic transect across tiger bush patterns in Niger: structure, slope and production. *Catena* (in press).

Van Cleve, K., Chapin, F. S. III, Dryness, C. T. & Viereck, L. A. (1991). Element cycling in taiga forest: state-factor control. *BioScience*, **41**, 78–88.

Van Cleve, K., Oechel, W. C. & Hom, J. L. (1990). Response of black spruce (*Picea mariana*) ecosystems to soil temperature modification in interior Alaska. *Canadian Journal of Forest Research*, **20**, 1536–45.

Van Diepen, C. A., Wolf, J. Van Keulen, H. & Rappoldt, C. (1989). WOFOST: a simulation model of crop production. *Soil Use and Management*, **5**(1), 16–24.

Van Genuchten, M.Th, (1980). A closed-form equation for predicting the hydraulic conductivity of unsaturated soils. *Soil Science Society of America Journal*, **44**, 892–8.

van Hylckama, T. E. A. (1974). Water use by salt cedar as measured by the water budget method. *Geological Survey Professional Paper*, 491-E.

Van Keulen, H. & Seligman, N. G. (1987). Simulation of water use, nitrogen nutrition and growth of a spring wheat crop. *Simulation Monographs*. 310 pp. Wageningen: Pudoc.

Van Laar, H. H., van Keulen, H. & Goudriaan, J. (1992). Simulation of crop growth for potential and water-limited production situations, as applied to spring wheat. Simulation Reports 27., Wageningen, The Netherlands: Agricultural University/AB-DLO.

van Noordwijk, M., Tomich, T. P., Winahyu, R., Murdiyarso, D., Suyanto, Partoharjono, S. & Fagi, A. M. (1995) *Alternatives to Slash-and-Burn in Indonesia. Summary Report of Phase I*. ASB–Indonesia Report No. 4, Bogor, Indonesia (published by International Centre for Research in Agroforestry, Nairobi, Kenya), 154 pp.

Van Oijen, M. (1991). Identification of the major characteristics of potato cultivars which affect yield loss caused by late blight. PhD thesis, 116 pp. Wageningen Agricultural University.

Van Veen, J. A., Liljeroth, E., Lekkerkerk, L. J. A. & Van De Geijn, S. C. (1991). Carbon fluxes in plant-soil systems at elevated atmospheric carbon dioxide levels. *Ecological Applications*, **1**, 175–81.

Vandermeer, J. & Schultz, B. (1990). Variability, stability, and risk in intercropping: some theoretical explorations, In *Agroecology: Researching the Ecological Basis for Sustainable Agriculture*, ed. S. R. Gleissman, pp. 205–29. New York: Springer-Verlag.

VEMAP Members, Melillo, J. M., Borchers, J., Chaney, J., Fisher, H., Fox, S., Haxeltine, A., Janetos, A., Kicklighter, D. W., Kittel, T. G. F., McGuire, A. D., McKeown, R., Neilson, R., Nemani, R., Ojima, D. S., Painter, T., Pan, Y., Parton, W. J., Pierce, L., Pitelka, L., Prentice, I. C., Rizzo, B., Rosenbloom, N. A., Running, S. W., Schimel, D. S., Sitch, S., Smith, T. & Woodward, I. (1995). Vegetation/ecosystem modelling and analysis project: comparing biogeography and biogeochemistry models in a continental-scale study of terrestrial ecosystem responses to climate change and CO_2 doubling. *Global Biogeochemical Cycles*, **9**(4), 407–37.

Verardo, D. J. & Ruddiman W. F. (1996). Late Pleistocene charcoal in tropical Atlantic deep-sea sediments: climatic and geochemical significance. *Geology*, **24**, 855–7.

Vitousek, P. M. (1990). Biological invasions and ecosystem processes: towards an integration of population biology and ecosystem studies. *Oikos*, **57**, 7–13.

Vitousek, P. M. (1994). Beyond global warming: ecology and global change. *Ecology*, **75**, 1861–76.

Vitousek, P. M., Aber, J. D., Howarth, R. W., Likens, G. E., Matson, P. A., Schindler, D. W., Schlesinger, W. H. & Tilman, D. (1997*b*). Human alteration of the global nitrogen cycle: sources and consequences. *Ecological Applications*, **7**, 737–50.

Vitousek, P. M., Ehrlich, P. R., Ehrlich, A. H. & Matson, P. A. (1986). Human appropriation of the products of photosynthesis. *BioScience*, **36**, 368–73.

Vitousek, P. M. & Hooper, D. U. (1993). Biological diversity and terrestrial ecosystem biogeochemistry. In *Biodiversity and Ecosystem Function*, ed. E.-D. Schulze & H. A. Mooney, pp. 3–14. Berlin: Springer-Verlag.

Vitousek, P. M. & Howarth, R. W. (1991). Nitrogen limitation on land and in the sea: how can it occur? *Biogeochemistry*, **13**, 87–115.

Vitousek, P. M., Loope, L. & Adsersen, H. (eds.) (1995). *Islands: Biological Diversity and Ecosystem Function*. Berlin: Springer-Verlag.

Vitousek, P. M., Mooney, H. A., Lubchenco, J. & Melillo, J. M. (1997*a*). Human domination of earth's ecosystems. *Science*, **277**, 494–9.

Vitousek, P. M., Walker, L. R., Whiteaker, L. D., Mueller-Dombois, D. & Matson, P. A. (1987). Biological invasion by *Myrica faya* alters ecosystem development in Hawaii. *Science*, **238**, 802–4.

Volin, J. C. & Reich, P. B. (1996). Interaction of elevated CO_2 and O_3 on growth, photosynthesis and respiration of three perennial species grown in low and high nitrogen. *Physiologia Plantarum*, **97**, 674–84.

Walker, B. H. (1992). Biodiversity and ecological redundancy. *Conservation Biology*, **6**, 18–23.

Walker, B. H. (1996). Predicting a future terrestrial biosphere: challenges to GCTE science. In *Global Change and Terrestrial Ecosystems*, ed. B. Walker & W. Steffen, pp. 595–607. Cambridge: Cambridge University Press. IGBP Book Series Vol. 2.

Walker, B. H. (1997). Functional types in non-equilibrium ecosystems. In *Plant Functional Types: Their Relevance to Ecosystem Properties and Global Change*, ed. T. M Smith, H. H. Shugart & F. I. Woodward, pp. 91–103. Cambridge: Cambridge University Press. IGBP Book Series Vol. 1.

Walker B. H. & Steffen, W. L. (eds.) (1996). *Global Change and Terrestrial Ecosystems*. Cambridge: Cambridge University Press. IGBP Book Series No. 2.

Wall, G. W. & Kimball, B.A. (1993). Biological databases derived from Free Air Carbon Dioxide Enrichment experiments. In *Design and Execution of Ex-*

periments on CO₂ Enrichment, ed. E-E. Schulze & H. A. Mooney. Ecosystems Research Report No. 6. Brussels: Commission of the European Communities.

Walter, H. (1971). *Ecology of Tropical and Subtropical Vegetation*. Edinburgh: Oliver and Boyd.

Wang, Y. P. & Jarvis, P. G. (1990a). Description and validation of an array model – MAESTRO. *Agricultural and Forest Meteorology*, **51**(3–4), 257–80.

Wang, Y. P. & Jarvis, P. G. (1990b). Influence of crown structural properties on PAR absorption, photosynthesis, and transpiration in Sitka spruce: application of a model (MAESTRO). *Tree Physiology*, **7**, 297–316.

Wang, Z. P., Delaune, R. D., Masscheleyn, P. H. & Patrick, W. H. Jr (1993). Redox potential and pH effects on methane production in flooded rice soil. *Soil Science Society of America Journal*, **57**, 382–5.

Watson, R. T., Zinyowera, M. C. & Moss, R. H. (eds.) (1996). *Climate Change 1995: Impacts, Adaptations, and Mitigation: Scientific-Technical Analyses*. Contribution of Working Group II to the Second Assessment Report of the Intergovernmental Panel on Climate Change. 879 pp. Cambridge: Cambridge University Press.

Watt, A. D., Whittaker, J. B., Docherty, M., Brooks, G., Lindsay, E. & Salt, D. T. (1995). The impact of elevated CO₂ on herbivores. In *Insects in a Changing Environment*, ed. R. Harrington & N. E. Stork, pp. 197–271. London: Academic Press.

Webb, R. S. & Rosenzweig, C. (1993). Specifying land surface characteristics in general circulation models: soil profile data set and derived water-holding capacities. *Global Biogeochemical Cycles*, **7**(1), 97–108.

Webb, T. I. & Bartlein, P. J. (1992). Global changes during the last three million years: climatic controls and biotic responses. *Annual Review of Ecology and Systematics*, **23**, 141–74.

Webb, T., III (1987). The appearance and disappearance of major vegetational assemblages: long-term vegetational dynamics in eastern North America. *Vegetatio*, **69**, 177–87.

Wedin, D. A. & Tilman, D. (1996). Influence of nitrogen loading and species composition on the carbon balance of grasslands. *Science*, **274**, 1720–23.

Weinstein, D. A., Shugart, H. H. & Brandt, C. C. (1983). Energy flow and the persistence of human population: A simulation analysis. *Human Ecology*, **11**(2), 201–25.

Welker, J. M., Parsons, A. N., Walker, M. D., Walker, D. A., Seastedt, T. R., Robinson, C. H. & Wookey, P. A. (1995). Comparative responses of *Dryas octopetala* to simulated changes in climate from alpine, low- and high arctic ITEX sites. *Bulletin of the Ecological Society of America*, **76**, 281.

Weller, G., Chapin, F. S., Everett, K. R., Hobbie, J. E., Kane, D., Oechel, W. C., Ping, C. L., Reeburgh, W. S., Walker, D. & Walsh, J. (1995). The Arctic Flux Study: a regional view of trace gas release. *Journal of Biogeography*, **22**, 365–74.

Whalen, S. C. & Reeburgh, W. S. (1992). Interannual variations in tundra methane emission: a 4-year time series at fixed sites. *Global Biogeochemical Cycles*, **6**, 139–59.

Wheeler, T. R., Batts, G. R., Ellis, R. H., Hadley, P. & Morison, J. R. L. (1996). Growth and yield of winter wheat (*Triticum aestivum*) crops in response to CO₂ and temperature. *Journal of Agricultural Science*, **127**, 37–48.

White, F. (1983). *UNESCO Vegetation Map of Africa*. Paris, France: UNESCO.

White, R. G. & Trudell, J. (1980). Habitat preference and forage consumption by reindeer and caribou near Atkasook, Alaska. *Arctic and Alpine Research*, **12**, 511–29.

Whitlock, C. H., Charlock, W. F., Staylor, W. F., Pinker, R. T., Laszlo, I., Di Pasquale, R. C. & Ritchey, N. A. (1993). *WCRP Surface Radiation Budget Shortwave Data Product Description-Version 1.1*. NASA Technical Memorandum.

Whittaker, R. H. (1975). *Communities and Ecosystems* (2nd edn) New York: MacMillan.

Whittaker, R. H. & Likens, G. E. (1975).

The biosphere and man. In *Primary Productivity of the Biosphere*, ed. H. Lieth & R. H. Whittaker, pp. 305–28. Berlin: Springer-Verlag. Ecological Studies Vol. 14.

Wight, J. R. & Skiles, J. W. (1987). *SPUR: Simulation of Production and Utilization of Rangelands. Documentation and User Guide*. USDA-ARS, 63.

Williams, D. W. & Leibhold, A. M. (1995). Forest defoliators and climate change: potential changes in spatial distribution of outbreaks of western spruce budworm (*Lepidoptera: Tortricidae*) and gypsy moth (*Lepidoptera: Lymantriidae*). *Environmental Entomology*, **24**, 1–9.

Williams, J. B. & Batzli, G. O. (1982). Pollination and dispersion of five species of lousewort (*Pedicularis*) near Atkasook, Alaska, *U.S.A. Arctic and Alpine Research*, **14**, 59–74.

Williams, J. R., Dyke, P. T., Fuchs, W. W., Benson, V. W., Rice, O. W. & Taylor, E. D. (1990). *EPIC- Erosion/Productivity Impact Calculator. User Manual*. USDA Technical Bulletin No. 1768. 127 pp.

Williams, W. E., Garbutt, K., Bazzaz, F. A. & Vitousek, P. M. (1986). The response of plants to elevated CO_2. IV. Two deciduous-forest tree communities. *Oecologia*, **69**, 454–9.

Wilson, J. B. & Agnew, D. Q. (1992). Positive-feedback switches in plant communities. *Advances in Ecological Research*, **23**, 263–336.

Wilson, M. F. & Henderson-Sellers, A. (1985). A global archive of land cover and soils data for use in general circulation models. *Journal of Climate*, **5**, 119–43.

Witte, F., Goldschmidt, T., Wanink, J., Van Oijen, M., Goudswaard, K., Witte-Maas, E. & Bouton, N.(1992). The destruction of an endemic species flock: quantitative data on the decline of the haplochromine cochlids of Lake Victoria. *Environmental Biology of Fishes*, **34**, 1–28.

Woods, K. D. & Davis, M. B. (1989). Paleoecology of range limits: Beech in the upper peninsula of Michigan. *Ecology*, **70**(3), 681–96.

Woodward, F. I. (1987a). *Climate and Plant Distribution*. 174 pp. Cambridge: Cambridge University Press.

Woodward, F. I. (1987b). Stomatal numbers are sensitive to increase of CO_2 from pre-industrial levels. *Nature*, **327**, 617–18.

Woodward, F. I. & Cramer, W. (eds.) (1996a). *Plant Functional Types and Climatic Change*. 125 pp. Uppsala, Sweden: Opulus Press.

Woodward, F. I. & Cramer, W. (1996b). Plant functional types and climatic changes–Introduction. *Journal of Vegetation Science*, **7**(3), 306–8.

Woodward, F. I. & Kelly, C. K. (1995). The influence of CO_2 concentration on stomatal density. *New Phytologist*, **131**, 311–27.

Woodward, F. I. & Kelly, C. K. (1997). Plant functional types: towards a definition by environmental constraints. In *Plant Functional Types: Their Relevance to Ecosystem Properties and Global Change*, T. M. Smith, H. H. Shugart & F. I. Woodward, pp. 47–65. Cambridge: Cambridge University Press. IGBP Book Series Vol. 1.

Woodward, F. I. & Smith, T. M. (1994a). Predictions and measurements of the maximum photosynthetic rate, A_{max}, at the global scale. *Ecological Studies*, **100**, 491–510.

Woodward, F.I. & Smith, T.M. (1994b). Global photosynthesis and stomatal conductance: modelling the controls by soil and climate. *Advances in Botanical Research*, **20**, 1–41.

Woodward, F. I., Smith, T. M. & Emanuel, W. R. (1995). A global land primary productivity and phytogeography model. *Global Biogeochemical Cycles*, **9**(4), 471–90.

Woodward, F. I. & Steffen, W. L. (1996). Natural disturbances and human land use in Dynamic Global Vegetation Models. *IGBP Report No. 38*, International Geosphere-Biosphere Programme, Stockholm, 49 pp.

Wookey, P. A., Parsons, A. N., Welker, J. M., Potter, J. A., Callaghan, T. V, Lee, J. A. & Press, M. C. (1993). Comparative responses of phenology and reproductive development to simulated environmental change in sub-arctic and high arctic plants. *Oikos*, **67**, 490–502.

Wookey, P. A., Robinson, C. H., Parsons, A. N., Welker, J. M., Press, M. C., Callaghan, T. V. & Lee, J. A. (1995). Environmental constraints on the growth, photosynthesis and reproductive development of *Dryas octopetala* at a high Arctic polar semi-desert, Svalbard. *Oecologia*, 102, 478–89.

World Resources Institute (1992). *World Resources 1992–1993: A Guide to the Global Environment*. New York: Oxford University Press.

Worldwatch Institute (1996). *State of the World 1996*. New York: W. W. Norton & Co.

Wright, D. H. (1987). Estimating human effects on global extinction. *International Journal of Biometeorology*, 31, 293–9.

Wullschleger, S. D., Post ,W. M. & King, A. W. (1995). On the potential for a CO_2 fertilization effect in forest: estimates of the biotic growth factor based on 58 controlled-exposure studies. In *Biotic Feedbacks in the Global Climatic System*, ed. G. Woodwell and Mackenzie, pp. 85–107. New York: Oxford University Press.

Xia, J. (1997). Biological control of cotton aphid (*Aphis gossypii* Glover) in cotton (inter) cropping systems in China; a simulation study. PhD Thesis, Wageningen Agricultural University, Wageningen, The Netherlands.

Xie, P., Rudolf, B., Schneider, U. & Arkin, P. A. (1996). Gauge-based monthly analysis of global land precipitation from 1971 to 1994. *Journal of Geophysical Research*, 101(D14), 19023–34.

Xue, Y. & Shukla, J. (1993). The influence of land surface properties on Sahel climate. Part 1. Desertification. *Journal of Climate*, 6, 2232–45.

Xue, Y., Sellers, P. J., Kinter, J. L. & Shukla, J. (1991). A simplified biosphere model for global climate studies. *Journal of Climate*, 4(3), 345–64.

Yin, X., (1996). *Quantifying the Effects of Temperature and Photoperiod on Phenological Development to Flowering in Rice*. Wageningen, The Netherlands: Wageningen Agricultural University.

Yoshida, S. (1981). *Fundamentals of Rice Crop Science*. Los Baños, The Philippines: IRRI.

Yudelman, M. (1993). *Demand and Supply of Foodstuffs up to 2050 with Special Reference to Irrigation*. Colombo: International Irrigation Management Institute.

Zadoks, J. C., Chang, T. T. & Konzak, C. F. (1974). A decimal code for the growth stages of cereals. *Eucarpia Bulletin*, 7, 42–52.

Zaller, H. & Arnone, J. A. III, (1997). Activity of surface-casting earthworms in a calcareous grassland under elevated atmospheric CO_2. *Oecologia*, 111(2), 249–54.

Zhou, G. & Zhang, X. (1995). A natural vegetation NPP model. *Acta Phytoecologica Sinica*, 19, 193–200 (in Chinese).

Zimov, S. A., Chuprynin, V. I., Oreshko, A. P., Chapin, F. S. III, Reynolds, J. F. & Chapin, M. C. (1995). Steppe–tundra transition: an herbivore-driven biome shift at the end of the Pleistocene. *American Naturalist*, 146, 765–94.

Zimov, S. A., Davidov, S. P., Voropaev, Y. V., Prosiannikov, S. F., Semiletov, I. P., Chapin, M. C. & Chapin, F. S. III, (1996). Siberian CO_2 efflux in winter as a CO_2 source and cause of seasonality in atmospheric CO_2. *Climatic Change*, 33, 111–20.

Zinke, P. J., Stangenberger, A. G., Post, W. M., Emanuel, W. R. & Olson, J. S. (1984). *Worldwide Organic Soil Carbon and Nitrogen Data*. Tennessee: Oak Ridge National Laboratory.

Zobler, L. (1986). *A World Soil File for Global Climate Modeling*. Goddard Institute for Space Studies. NASA Technical Memorandum.

Zuozhong, C. (1996). Land use patterns, issues and models in the Inner Mongolian Plateau. *Proceedings of the International Conference on Temperate Grasslands for the 21st Century*, Beijing, July 1996.

Zwölfer, H. & Arnold-Rinehart, J. (1993) The evolution of interactions and diversity in plant-insect systems: the Urophora–Eurytoma food web in galls on palearctic *Cardueae*. *Ecological Studies*, 99, 211–36.

Index

Bold type indicates material in boxes, tables and illustrations

MUSE model, and patch models, vegetation
dynamics **127**

NATT, Northern Australian Tropical
Transect, (IGBP) 357–62
net biome productivity (NBP) 152
net ecosystem productivity (NEP) 6, 36,
152, 333, 350
component definitions **280**
net primary production (NPP) 6, 36, 142,
221, 281
double CO_2 **166, 188**
and evapotranspiration **288**
VEMAP experiment **293**
networks
benefits and products 63–5
benefits and responsibilities of GCTE
research **46–7**
and consortia
list **50–5**
publications 60
data and model sharing policy **62–3**
intellectual property rights 61–3
networking tools standardization 58–61
publications 60
regional, for global change impacts 49,
57–8
requirements for success 45–6
research programmes 45–87
site networks 49–55
Soil Erosion Network **52–5**
Soil Organic Matter Network
(SOMNET) 49–50
standard file formats **61**
thematic 49, 56–7
nitrogen:lignin ratio 291
nitrogen
and methane 290–1
radioisotopes 41–3
nitrogen cycle 281–3, 290–4
human alteration 8, **178, 282**
integration with carbon cycle 290–4
linkages with other cycles **274**
nitrogen effects
availability and deposition 124, 177–82
content of leaves **285**
and NPP 97–8
content of soils **285**
double CO_2 **168, 188**
nitrogen–phosphorus interactions, and
carbon cycle 295–6
nutrients as drivers of global change,

perturbations 278–81
perturbation of carbon cycle 281–3
NOAA/AVHRR *see* Advanced Very High
Resolution radiometer (AVHRR)
Normalized Differential Vegetation Index
(NVDI) 98, 103
NorthEast China *see* China

oxygen, radioisotopes 41–3
ozone
formation 291
long-term changes **185**
and solar UV-B radiation **183**
tropospheric increase, physiological
responses to change 184–7

pastures and rangelands *see* grasslands
Patagonia, transect, vegetation variations
69
patch models
modelling ecosystem structure 122–6,
194–200
MUSE model, vegetation dynamics
127
Penman–Monteith equation, soil water 95,
119
pests and diseases 114–15, 199–200, 235,
252–9, 267–8
phosphorus
carbon–nitrogen interactions 295–6
constraints on carbon cycle 294–5
content of leaves **285**
content of soils **285**
phosphorus cycle
human perturbations **294**
linkages with other cycles **274**
photosynthesis
and solar UV-B radiation **183**
see also net primary production (NPP)
Phytophthora cinnamoni 200
Picea abies, growth curves **179**
Pinus sylvestris
BFG model, Australia 187, **188**
growth curves **179**
Pinus taeda, ozone effects **185**
plant functional types
Dynamic Global Vegetation Models
(DGVMs) **215**
NPP **221**
species diversity 212–15
plant migration and global change 205–10
PNET-CN model 161